Relativity

An Introduction to Special and General Relativity

Thoroughly revised and updated, and now also including special relativity, this book provides a pedagogical introduction to relativity. It is based on lectures given by the author in Jena over the last decades, and covers the material usually presented in a three-term course on the subject. It is self-contained, but the reader is expected to have a basic knowledge of theoretical mechanics and electrodynamics. The necessary mathematical tools (tensor calculus, Riemannian geometry) are provided. The author discusses the most important features of both special and general relativity, as well as touching on more difficult topics, such as the field of charged pole–dipole particles, the Petrov classification, groups of motions, exact solutions and the structure of infinity.

The book is written as a textbook for undergraduate and introductory graduate courses, but will also be useful as a reference for practising physicists, astrophysicists and mathematicians. Most of the mathematical derivations are given in full and exercises are included where appropriate. The bibliography gives many original papers and directs the reader to useful monographs and review papers.

HANS STEPHANI (1935–2003) gained his Diploma, Ph.D. and Habilitation at the Friedrich-Schiller Universität, Jena. He became Professor of Theoretical Physics in 1992, and retired in 2000. He began lecturing in theoretical physics in 1964 and published numerous papers and articles on relativity over the years. He is also the author of four books.

T0318310

RELATIVITY

An Introduction to Special and General Relativity

Third Edition

HANS STEPHANI

CAMBRIDGE UNIVERSITY PRESS
Cambridge, New York, Melbourne, Madrid, Cape Town, Singapore, São Paulo

Cambridge University Press
The Edinburgh Building, Cambridge CB2 8RU, UK

Published in the United States of America by Cambridge University Press, New York

www.cambridge.org
Information on this title: www.cambridge.org/9780521811859

Originally published in German as *Allgemeine Relativitätstheorie* by VEB
Deutscher Verlag der Wissenschaften, Berlin, G. D. R.
© VEB Deutscher Verlag der Wissenschaften, first German edition 1977
Second German edition 1980
Third German edition 1989

First published in English as General Relativity by
Cambridge University Press 1982
Second English edition 1990
Third English edition 2004

A catalogue record for this publication is available from the British Library

ISBN 978-0-521-81185-9 hardback
ISBN 978-0-521-01069-6 paperback

Transferred to digital printing 2007

Contents

Contents　　vii

Preface

Special Relativity originally dealt with the symmetries of the electromagnetic field and their consequences for experiments and for the interpretation of space and time measurements. It arose at the end of the nineteenth century from the difficulties in understanding the properties of light when this light was tested by observers at rest or in relative motion. Its name originated from the surprise that many of the concepts of classical non-relativistic physics refer to a frame of reference ('observer') and are true only relative to that frame.

The symmetries mentioned above show up as transformation properties with respect to Lorentz transformations. It was soon realized that these transformation properties have to be the same for all interacting fields, they have to be the same for electromagnetic, mechanic, thermodynamic, etc. systems. To achieve that, some of the 'older' parts of the respective theories had to be changed to incorporate the proper transformation properties. Because of this we can also say that Special Relativity shows how to incorporate the proper behaviour under Lorentz transformation into all branches of physics. The theory is 'special' in that only observers moving with constant velocities with respect to each other are on equal footing (and were considered in its derivation).

Although the words 'General Relativity' indicate a similar interpretation, this is not quite correct. It is true that historically the word 'general' refers to the idea that observers in a general state of motion (arbitrary acceleration) should be admitted, and therefore arbitrary transformation of coordinates should be discussed. Stated more generally, for a description of nature and its laws one should be able to use *arbitrary coordinate systems*, and in accordance with the *principle of covariance* the form of the laws of nature should not depend essentially upon the choice of the coordinate system. This requirement, in the first place purely mathematical, acquires a physical meaning through the replacement of 'arbitrary coordinate system' by 'arbitrarily moving observer'. The laws of nature should be independent of the state of motion of the observer. Here also belongs the question, raised in particular by Ernst Mach, of whether an absolute acceleration (including an absolute rotation) can really be defined meaningfully, or whether every measurable rotation means a rotation relative to the fixed stars (*Mach's principle*).

But more important for the evolution of General Relativity was the recognition that the Newtonian theory of gravitation was inconsistent with Special Relativity; in it gravitational effects propagate with an infinitely large velocity. So a really new theory of gravitation had to be developed, which correctly reflects the dynamical behaviour of the whole universe and which at the same time is valid for stellar evolution and planetary motion.

General Relativity is the theory of the gravitational field. It is based on Special Relativity in that all laws of physics (except those of the gravitational field) have to be written in the proper special-relativistic way before being translated into General Relativity. It came into being with the formulation of the fundamental equations by Albert Einstein in 1915. In spite of the success of the theory (precession of the perihelion of Mercury, deflection of light by the Sun, explanation of the cosmological redshift), it had retained for a long time the reputation of an esoteric science for specialists and outsiders, perhaps because of the mathematical difficulties, the new concepts and the paucity of applications (for example, in comparison with quantum theory, which came into existence at almost the same time). Through the development of new methods of obtaining solutions and the physical interpretation of the theory, and even more through the surprising astrophysical discoveries (pulsars, cosmic background radiation, centres of galaxies as candidates for black holes), and the improved possibilities of demonstrating general relativistic effects, in the course of the last thirty years the general theory of relativity has become a true physical science, with many associated experimental questions and observable consequences.

The early neglect of relativity by the scientific community is also reflected by the fact that many Nobel prizes have been awarded for the development of quantum theory, but none for Special or General Relativity. Only in 1993, in the laudation of the prize given to J. H. Taylor, Jr. and R. A. Hulse for their detection of the binary pulsar PSR 1913+16, was the importance for relativity (and the existence of gravitational waves) explicitly mentioned.

Modern theoretical physics uses and needs ever more complicated mathematical tools – this statement, with its often unwelcome consequences for the physicist, is true also for the theory of gravitation. The language of the general theory of relativity is differential geometry, and we must learn it, if we wish to ask and answer precisely physical questions. The part on General Relativity therefore begins with some chapters in which the essential concepts and formulae of Riemannian geometry are described. Here suffix notation will be used in order to make the book easier to read for non-mathematicians. An introduction to the modern coordinate-free notation can be found in Stephani *et al.* (2003).

This book is based on the lectures the author gave in Jena through many years (one term Special and two terms General Relativity), and thus gives a rather concise introduction to both theories. The reader should have a good knowledge of classical mechanics and of Maxwell's theory.

My thanks go to all colleagues (in particular in Jena), with whom and from whom I have learnt the theory of relativity. I am especially indebted to J. Stewart and M. Pollock for the translation of most of the parts on General Relativity for the foregoing edition, M. MacCallum for his critical remarks and suggestions, and Th. Lotze for his help in preparing the manuscript.

Notation

Minkowski space: $ds^2 = \eta_{ab}\,dx^a\,dx^b = dx^2 + dy^2 + dz^2 - c^2dt^2$

$$= d\mathbf{r}^2 - c^2dt^2 = -c^2d\tau^2.$$

Lorentz transformations: $x^{n'} = L^{n'}{}_m\,x^m, \quad L^{n'}{}_a L_{n'}{}^b = \delta_a^b.$

Special Lorentz transformation:

$$x' = \frac{x - vt}{\sqrt{1 - v^2/c^2}}, \quad ct' = \frac{ct - vx/c}{\sqrt{1 - v^2/c^2}}.$$

Addition of velocities: $v = \dfrac{v_1 + v_2}{1 + v_1 v_2/c^2}.$

Four-velocity: $u^n = dx^n/d\tau.$

Riemannian space: $ds^2 = g_{ab}\,dx^a\,dx^b = -c^2d\tau^2,$

$$g^{ab}g_{bm} = \delta_m^a = g_m^a, \quad g = |g_{ab}|.$$

ε-pseudo-tensor: $\varepsilon^{abmn}; \quad \varepsilon^{1234} = 1/\sqrt{-g},$

$$\varepsilon_{abcd}\varepsilon^{abmn} = -2(g_c^n g_d^m - g_c^m g_d^n).$$

Dualization of an antisymmetric tensor: $\tilde{F}^{ab} = \frac{1}{2}\varepsilon^{abmn}F_{mn}.$

Christoffel symbols: $\Gamma^a_{mn} = \frac{1}{2}g^{ab}(g_{bm,n} + g_{bn,m} - g_{mn,b}).$

Partial derivative: $T_{a,m} = \partial T_a/\partial x^m.$

Covariant derivative: $T^a{}_{;m} = DT^a/Dx^m = T^a{}_{,m} - \Gamma^a_{mn}T_n,$

$$T_{a;m} = DT_a/Dx^m = T_{a,m} - \Gamma^n_{am}T_n.$$

Geodesic equation: $\dfrac{D^2 x^i}{D\lambda^2} = \dfrac{d^2 x^i}{d\lambda^2} + \Gamma^i_{nm}\dfrac{dx^n}{d\lambda}\dfrac{dx^m}{d\lambda} = 0.$

Parallel transport along a curve $x^i(\lambda)$: $DT^a/D\lambda = T^a{}_{;b}\,dx^b/d\lambda = 0.$

Fermi–Walker transport: $\dfrac{DT^n}{D\tau} - \dfrac{1}{c^2}T_a\left(\dfrac{dx^n}{d\tau}\dfrac{D^2 x^a}{D\tau^2} - \dfrac{dx^a}{d\tau}\dfrac{D^2 x^n}{D\tau^2}\right) = 0.$

Lie derivative in the direction of the vector field $a^k(x^i)$:

$$\mathcal{L}_\mathbf{a}T^n = T^n{}_{,k}a^k - T^k a^n{}_{,k} = T^n{}_{;k}a^k - T^k a^n{}_{;k},$$
$$\mathcal{L}_\mathbf{a}T_n = T_{n,k}a^k + T_k a^k{}_{,n} = T_{n;k}a^k + T_k a^k{}_{;n}.$$

Killing equation: $\xi_{i;n} + \xi_{n;i} = \mathcal{L}_\xi g_{in} = 0.$

Divergence of a vector field: $a^i{}_{;i} = \dfrac{1}{\sqrt{-g}}(\sqrt{-g}a^i)_{,i}.$

Maxwell's equations: $F^{mn}{}_{;n} = (\sqrt{-g}F^{mn})_{,n}/\sqrt{-g} = j^m/c,$

$$\tilde{F}^{mn}{}_{;n} = 0.$$

Curvature tensor:

$$a_{m;s;q} - a_{m;q;s} = a_b R^b{}_{msq},$$
$$R^b{}_{msq} = \Gamma^b_{mq,s} - \Gamma^b_{ms,q} + \Gamma^b_{ns}\Gamma^n_{mq} - \Gamma^b_{nq}\Gamma^n_{ms},$$
$$R_{amsq} = \tfrac{1}{2}(g_{aq,ms} + g_{ms,aq} - g_{as,mq} - g_{mq,as}) + \text{non-linear terms}.$$

Ricci tensor: $R_{mq} = R^s{}_{msq} = -R^s{}_{mqs}; \quad R^m{}_m = R.$

Field equations: $G_{ab} = R_{ab} - \tfrac{1}{2}Rg_{ab} = \kappa T_{ab}.$

Perfect fluid: $T_{ab} = (\mu + p/c^2)u_a u_b + pg_{ab}.$

Schwarzschild metric:

$$ds^2 = \frac{dr^2}{1 - 2M/r} + r^2(d\vartheta^2 + \sin^2\vartheta\,d\varphi^2) - (1 - 2M/r)c^2 dt^2.$$

Robertson–Walker metric:

$$ds^2 = K^2(ct)\left[\frac{dr^2}{1 - \varepsilon r^2} + r^2(d\vartheta^2 + \sin^2\vartheta\,d\varphi^2)\right] - c^2 dt^2.$$

Hubble parameter: $H(ct) = \dot{K}/K.$

Acceleration parameter: $q(ct) = -K\ddot{K}/\dot{K}^2.$

$\kappa = 2.07 \times 10^{-48}\,\text{g}^{-1}\text{cm}^{-1}\text{s}^2, \quad cH = 55\,\text{km/s Mpc}.$

$2M_{\text{Earth}} = 0.8876\,\text{cm}, \quad 2M_{\text{Sun}} = 2.9533 \times 10^5\,\text{cm}.$

I. Special Relativity

1

Introduction: Inertial systems and the Galilei invariance of Classical Mechanics

1.1 Inertial systems

Special Relativity became famous because of the bewildering properties of length and time it claimed to be true: moving objects become shorter, moving clocks run slower, travelling people remain younger. All these results came out from a theoretical and experimental study of light propagation as seen by moving observers. More technically, they all are consequences of the invariance properties of Maxwell's equations.

To get an easier access to invariance properties, it is appropriate to study them first in the context of Classical Mechanics. Here they appear quite naturally when introducing the so-called 'inertial systems'. By definition, an *inertial system* is a coordinate system in which the equations of motion take the usual form

$$m\ddot{x}^{\alpha} = F^{\alpha}, \quad \alpha = 1, 2, 3 \tag{1.1}$$

(Cartesian coordinates $x^1 = x$, $x^2 = y$, $x^3 = z$, $\ddot{x}^{\alpha} = \mathrm{d}^2 x^{\alpha}/\mathrm{d}t^2$). Experimentally, an inertial system can be realized in good approximation by a system in which the stars are at rest. Inertial systems are not uniquely defined; if Σ is such a system, then all systems Σ' which originate from Σ by performing a spatial translation, a rotation about a constant (time-independent) angle, a shift of the origin of time, or a motion with constant velocity, are again inertial systems. Accelerated systems such as steadily rotating systems are not inertial systems, cp. also (15.2).

We shall now study the abovementioned transformations in more detail.

1.2 Invariance under translations

Experimental results should not depend on the choice of the origin of the Cartesian coordinate system one is using ('homogeneity of space'). So if there is a system of masses m_N, then their equations of motion

$$m_N \ddot{\mathbf{r}}_N = \mathbf{F}_N \qquad (1.2)$$

should be invariant under a translation by a constant vector $\mathbf{b}$, i.e. under the substitution

$$\mathbf{r}'_N = \mathbf{r}_N + \mathbf{b}, \ \dot{\mathbf{r}}'_N = \dot{\mathbf{r}}_N, \ \mathbf{F}'_N = \mathbf{F}_N. \qquad (1.3)$$

Substituting (1.3) into (1.2), the invariance seems to hold trivially. But a closer inspection of (1.2) shows that if we write it out as

$$m_N \ddot{\mathbf{r}}_N = \mathbf{F}_N(\mathbf{r}_M, \dot{\mathbf{r}}_M, t) \qquad (1.4)$$

(the forces may depend on the positions and velocities of all masses), then the substitution $\mathbf{r}'_N = \mathbf{r}_N + \mathbf{b}$ leads to

$$m_N \ddot{\mathbf{r}}'_N = \mathbf{F}_N(\mathbf{r}'_M - \mathbf{b}, \dot{\mathbf{r}}'_M, t). \qquad (1.5)$$

This has the form (1.4) only if the force on a mass does not depend on the positions $\mathbf{r}_M$ of the (other) masses, but only on the distances $\mathbf{r}_N - \mathbf{r}_M$, because then we have $\mathbf{F}_N = \mathbf{F}_N(\mathbf{r}_N - \mathbf{r}_M, \dot{\mathbf{r}}_M, t) \rightarrow \mathbf{F}'_N = \mathbf{F}_N(\mathbf{r}'_N - \mathbf{r}'_M, \dot{\mathbf{r}}'_M, t)$; the $\mathbf{b}$ drops out. Closed systems, for which the sources of all forces are part of the system, usually have that property.

Examples of equations of motion which are invariant against translation are $m\ddot{\mathbf{r}} = \mathbf{g}$ (motion in a homogeneous gravitational field) and the motion of a planet (at position $\mathbf{r}$) in the field of the Sun (at position $\mathbf{r}_S$)

$$m\ddot{\mathbf{r}} = f\frac{\mathbf{r} - \mathbf{r}_S}{|\mathbf{r} - \mathbf{r}_S|^3}. \qquad (1.6)$$

In a similar way, experimental results should not depend on the choice of the origin of time ('homogeneity of time'), the equations of motion should be invariant under a time translation

$$t' = t + b. \qquad (1.7)$$

An inspection of equations (1.4) shows that the invariance is only guaranteed if the forces do not *explicitly* depend on time (they are then time-dependent only via the motion of the sources of the forces); this again will hold if there are no *external* sources of the forces.

We thus can state that for closed systems the laws of nature do not permit an experimental verification, or a sensible definition, of an absolute location in space and time.

1.3 Invariance under rotations

Rotations such as the simple rotation about the z-axis

$$x' = x \cos \varphi + y \sin \varphi, \quad y' = -x \sin \varphi + y \cos \varphi, \quad z' = z, \quad (1.8)$$

are best described using matrices. To do this, we first denote the Cartesian coordinates by

$$x_1 = x^1 = x, \quad x_2 = x^2 = y, \quad x_3 = x^3 = z. \quad (1.9)$$

The convention of using x^α as well as x_α for the same set of variables looks rather strange and even clumsy; the reason for this will become clear when dealing with vectors and tensors in both Special and General Relativity. As usual in relativity, we will use the Einstein summation convention: summation over two repeated indices, of which always one is lowered and one is raised.

The general rotation (orthogonal transformation) is a linear transformation and can be written in the two equivalent forms

$$x^{\alpha'} = D^{\alpha'}{}_\beta \, x^\beta, \quad x_{\alpha'} = D_{\alpha'}{}^\beta x_\beta \quad (1.10)$$

(note the position of the indices on the Ds!). Here, and on later occasions in Special and General Relativity, we prefer a notation which distinguishes the new coordinates from the old not by a new symbol (say y^α instead of x^α), but by a prime on the index. This convention is advantageous for many calculations of a general kind, although we shall occasionally deviate from it. The transformation matrices $D^{\alpha'}{}_\beta$ mediating between the two systems thus have two kinds of indices.

Rotations leave angles and lengths fixed; so if there are two arbitrary vectors x^α and ξ^α, then their scalar product has to remain unchanged. With

$$x_{\alpha'} = D_{\alpha'}{}^\beta x_\beta, \quad \xi^{\alpha'} = D^{\alpha'}{}_\gamma \xi^\gamma \quad (1.11)$$

that gives the condition

$$x_{\alpha'} \xi^{\alpha'} = D_{\alpha'}{}^\beta D^{\alpha'}{}_\gamma \, x_\beta \xi^\gamma = x_\beta \xi^\beta. \quad (1.12)$$

For arbitrary vectors $\mathbf{x}$ and $\boldsymbol{\xi}$ this can be true only for

$$D_{\alpha'}{}^\beta D^{\alpha'}{}_\gamma = \delta^\beta_\gamma, \quad \alpha, \beta, \gamma = 1, 2, 3. \quad (1.13)$$

Equation (1.13) characterizes the general orthogonal transformation. By taking the determinants on both sides of it (note that $D_{\alpha'}{}^\beta$ and $D^{\alpha'}{}_\gamma$ are numerically identical) we get

$$\left\| D_{\alpha'}{}^\beta \right\|^2 = 1. \quad (1.14)$$

The transformations with $\left\| D_{\alpha'}{}^\beta \right\| = +1$ are rotations; an example is the rotation (1.8) with

$$D_{\alpha'}{}^\beta = \begin{pmatrix} \cos\varphi & \sin\varphi & 0 \\ -\sin\varphi & \cos\varphi & 0 \\ 0 & 0 & 1 \end{pmatrix}. \qquad (1.15)$$

Transformations with $\left\| D_{\alpha'}{}^\beta \right\| = -1$ contain reflections such as, for example, the inversion

$$\begin{aligned} x' &= -x, \\ y' &= -y, \\ z' &= -z, \end{aligned} \quad \Rightarrow \quad D_{\alpha'}{}^\beta = \begin{pmatrix} -1 & 0 & 0 \\ 0 & -1 & 0 \\ 0 & 0 & -1 \end{pmatrix}. \qquad (1.16)$$

To apply a rotation to the equations of motion, we first observe that for time-independent rotations we have

$$x^{\alpha'} = D^{\alpha'}{}_\beta \, x^\beta \quad \Rightarrow \quad \ddot{x}^{\alpha'} = D^{\alpha'}{}_\beta \, \ddot{x}^\beta . \qquad (1.17)$$

We then note that the force $\mathbf{F}$ is a vector, i.e. its components F^α transform in the same way as the components of the position vector x^α. If we now multiply both sides of equation (1.1) by $D^{\alpha'}{}_\beta$, we get

$$D^{\alpha'}{}_\beta \, \ddot{x}^\beta = m\ddot{x}^{\alpha'} = D^{\alpha'}{}_\beta \, F^\beta = F^{\alpha'}; \qquad (1.18)$$

the form of the equation remains unchanged. But if we also take into account the arguments in the components of the force,

$$m\ddot{x}^{\alpha'} = D^{\alpha'}{}_\beta \, F^\alpha(x^\beta, \dot{x}^\beta, t) = F^{\alpha'}(x^\beta, \dot{x}^\beta, t), \qquad (1.19)$$

we see that the $F^{\alpha'}$ may depend on the wrong kind of variables. This will not happen if the F^α depend only on invariants, which in practice happens in most cases.

An example of an invariant equation is given by (1.6): the $\mathbf{r} - \mathbf{r}_S$ is a vector, and the distance $|\mathbf{r} - \mathbf{r}_S|$ is rotationally invariant.

We thus can state: since the force is a vector, and for closed systems the force-components depends only on invariants, the equations of motions are rotationally invariant and do not permit the definition of an absolute direction in space.

1.4 Invariance under Galilei transformations

We consider two systems which are moving with a constant velocity $\mathbf{v}$ with respect to each other:

$$\mathbf{r}'_N = \mathbf{r}_N - \mathbf{v}t, \quad t' = t \qquad (1.20)$$

(*Galilei transformation*). Because of $\ddot{\mathbf{r}}'_N = \dot{\mathbf{r}}_N - \mathbf{v}$, $\ddot{\mathbf{r}}'_N = \ddot{\mathbf{r}}_N$, the equations of motion (1.4) transform as

$$m_N \ddot{\mathbf{r}}'_N = m_N \ddot{\mathbf{r}}_N = \mathbf{F}_N (\mathbf{r}'_M + \mathbf{v}t, \dot{\mathbf{r}}'_M + \mathbf{v}, t). \qquad (1.21)$$

Although the constant $\mathbf{v}$ drops out when calculating the acceleration, the arguments of the force may still depend on $\mathbf{v}$. The equations are invariant, however, if only relative positions $\mathbf{r}_M - \mathbf{r}_N$ (as discussed above) and relative velocities $\dot{\mathbf{r}}_N - \dot{\mathbf{r}}_M$ enter. This is usually the case if the systems are closed and the equations are properly written. Take for example the well known example of a motion in a constant gravitational field $\mathbf{g}$ under the influence of friction,

$$m\ddot{\mathbf{r}} = -a\dot{\mathbf{r}} - m\mathbf{g}. \qquad (1.22)$$

At first glance, because of the explicit $\dot{\mathbf{r}}$ occurring in it, this equation seems to be a counterexample. But what is really meant, and is the cause of the friction, is the relative velocity with respect to the air. The equation (1.22) should correctly be written as

$$m\ddot{\mathbf{r}} = -a(\dot{\mathbf{r}} - \mathbf{v}_{\text{Air}}) - m\mathbf{g}, \qquad (1.23)$$

and the invariance is now obvious.

For closed systems, the equations of motions are invariant under Galilei transformations; an absolute velocity cannot be defined. Stated differently: only relative motions can be defined and measured (*Galilei's principle of relativity*).

We close this section with two remarks. In all three cases of invariances we had to refer to closed systems; how far do we have to go to get a really closed system? Is our Galaxy sufficient, or have we to take the whole universe? Second, we saw that only relative velocities matter; what about acceleration – why is this absolute?

1.5 Some remarks on the homogeneity of time

How can one check that space and time are really homogeneous? We want to discuss that problem a little bit for the case of time.

We start with the notion 'constant velocity'. How can one check that a mass is moving with constant velocity? Of course by measuring distances and reading clocks. How does one know that the clocks are going uniformly? After some consideration, and looking at standard procedures, one concludes that good clocks are made by taking a periodic process (rotation of the Earth, harmonic oscillator, vibration of a molecule) and dividing that into smaller parts. But how does one know that this fundamental process is really periodic – no clock to measure it is available! The only way out is *to define* that process as being periodic. But which kind of process should one use for that?

Of course, one has to consult Newton's equations of motion

$$m\ddot{\mathbf{r}} = \frac{\mathrm{d}^2\mathbf{r}}{\mathrm{d}t^2} = \mathbf{F} \qquad (1.24)$$

and to take a process, such as the rotation of the Earth around the Sun, which is periodic when these equations hold.

To see that really a definition of the time is hidden here in the equations of motion, consider a transformation

$$T = f(t) \quad \Rightarrow \quad \mathrm{d}T = f'\mathrm{d}t, \ \mathrm{d}/\mathrm{d}t = f'\mathrm{d}/\mathrm{d}T \qquad (1.25)$$

of the time. In the new time variable T the equations of motion (1.24) read

$$f'^2 \frac{\mathrm{d}^2\mathbf{r}}{\mathrm{d}T^2} + f'f'' \frac{\mathrm{d}\mathbf{r}}{\mathrm{d}T} = \mathbf{F}; \qquad (1.26)$$

they no longer have the Newtonian form.

We conclude that the correct, appropriate time coordinate is that in which the equations of motion take the simple form (1.24); the laws of mechanics guarantee that such a time really exists. But it here remains an open question whether this time coordinate, which is derived from planetary motion, is also the appropriate time to describe phenomena in other fields of physics such as light propagation. This questions will be answered by Special Relativity – in the negative.

Exercises

1.1   Is the equation $m\ddot{\mathbf{r}} - k\mathbf{r} = 0$ (harmonic oscillator) invariant under translations?

1.2   Show that a rotation $D^{\alpha'}{}_\beta$ always has one real eigenvector $\mathbf{w}$ with $D^{\alpha'}{}_\beta\, w^\beta = \lambda w^\alpha$, and that $w^\alpha = (1, \mathrm{i}, 0)$ is a complex eigenvector of the rotation (1.15). What are the corresponding eigenvalues?

1.3   Is $m\ddot{\mathbf{r}} = f(x)\mathbf{r}$ rotationally invariant?

1.4   Show that the Laplacian is invariant under rotations, i.e. that $\partial^2/\partial x^\alpha \partial x_\alpha = \partial^2/\partial x^{\alpha'} \partial x_{\alpha'}$ holds.

2

Light propagation in moving coordinate systems and Lorentz transformations

2.1 The Michelson experiment

At the end of the nineteenth century, it was a common belief that light needs and has a medium in which it propagates: light is a wave in a medium called ether, as sound is a wave in air. This belief was shattered when Michelson (1881) tried to measure the velocity of the Earth on its way around the Sun. He used a sensitive interferometer, with one arm in the direction of the Earth's motion, and the other perpendicular to it. When rotating the instrument through an angle of 90°, a shift of the fringes of interference should take place: light propagates in the ether, and the velocity of the Earth had to be added that of the light in the direction of the respective arms. The result was zero: there was no velocity of the Earth with respect to the ether.

This negative result can be phrased differently. Since the system of the ether is an inertial system, and that of the Earth is moving with a (approximately) constant velocity, the Earth's system is an inertial system too. So the Michelson experiment (together with other experiments) tells us that the velocity of light is the same for all inertial systems which are moving with constant velocity with respect to each other (*principle of the invariance of the velocity of light*). The speed of light in empty space is the same for all inertial systems, independent of the motion of the light source and of the observer.

This result does not violate Galilei's principle of relativity as stated at the end of Section 1.4: it confirms that also the ether cannot serve to define an absolute velocity. But of course something is wrong with the transformation law for the velocities: light moving with velocity c in the system of the ether should have velocity $c+v$ in the system of the Earth.

This contradiction can be given a geometric illustration (see Fig. 2.1). Consider two observers Σ (coordinates x, y, z, t) and Σ' (coordinates $x' = x - vt$, $y' = y$, $z' = z$, $t' = t$), moving with constant velocity $\mathbf{v}$ with respect to each other. At $t = 0$, when their coordinate systems coincide, a light signal is emitted at the origin. Since for both of them the light velocity is c, after a time T the light signal has reached the sphere

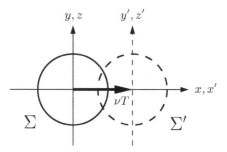

Fig. 2.1. Light propagation as seen by two observers in relative motion; $t = t' = T$.

$x^2 + y^2 + z^2 = c^2T^2$ for Σ, and $(x - vT)^2 + y^2 + z^2 = c^2T^2$ for Σ'. But this a contradiction, the light front cannot be simultaneously at the two spheres!

It will turn out that it is exactly this 'simultaneously' which has to be amended.

2.2 The Lorentz transformations

Coordinates The wave front of light emitted at $t = 0$ at the origin has reached the three-dimensional light sphere

$$x^2 + y^2 + z^2 - c^2t^2 = 0 \tag{2.1}$$

at the time t. Space and time coordinates enter here in a very symmetric way. Therefore we adapt our coordinates to this light sphere and take the time as a fourth coordinate $x^4 = ct$. More exactly, we use

$$x^a = (x, y, z, ct), \quad x_a = (x, y, z, -ct), \quad a = 1, \ldots, 4. \tag{2.2}$$

The two types of coordinates are obviously related by means of a matrix $\boldsymbol{\eta}$, which can be used to raise and lower indices:

$$x_a = \eta_{ab}x^b, \qquad \eta_{ab} = \eta^{ab} = \begin{pmatrix} 1 & & & \\ & 1 & & \\ & & 1 & \\ & & & -1 \end{pmatrix}, \qquad \eta_b^a = \delta_b^a. \tag{2.3}$$

Using these coordinates, (2.1) can be written as

$$x^a x_a = \eta_{ab}\, x^a x^b = x^2 + y^2 + z^2 - c^2t^2 = 0. \tag{2.4}$$

Invariance of light propagation and Lorentz transformations We now determine the coordinate transformations which leave the light sphere (2.4) invariant, thus ensuring that the light velocity is the same in both systems. Unlike the Galilei transformations (1.20), where the time coordinate was kept constant, it too is transformed here: the definition of the time scale will be adjusted to the light propagation, as it is adjusted to the equations of motion in Newtonian mechanics, cp. Section 1.4.

The transformations we are looking for should be one-to-one, and no finite point should go into infinity; they have to be linear. Neglecting translations, they have the form

$$x^{n'} = L^{n'}{}_a x^a, \quad x_{m'} = L_{m'}{}^b x_b, \quad L_{m'}{}^b = \eta_{m'n'}\eta^{ab}L^{n'}{}_a \qquad (2.5)$$

(for the notation, see the remarks after equation (1.10); note that $\eta_{m'n'}$ and η_{ab} have the same numerical components).

To give the light sphere the same form $x_n x^n = 0 = x^{n'} x_{n'}$ in both coordinates, the transformations (2.5) have to satisfy

$$x^{n'} x_{n'} = L^{n'}{}_a L_{n'}{}^b x^a x_b = x^b x_b, \qquad (2.6)$$

which for all x^a is possible only if

$$L^{n'}{}_a L_{n'}{}^b = \delta^b_a, \quad a, b, n' = 1, \ldots, 4. \qquad (2.7)$$

These equations define the *Lorentz transformations*, first given by Waldemar Voigt (1887). The discussion of these transformations will fill the next chapters of this book.

If we also admit translations,

$$x^{n'} = L^{n'}{}_a x^a + c^{n'}, \quad c^{n'} = \text{const.}, \qquad (2.8)$$

we obtain the *Poincaré transformations*.

Lorentz transformations, rotations and pseudorotations Equation (2.7) looks very similar to the defining equation (1.13) for rotations, $D^{v'}{}_\alpha D_{v'}{}^\beta = \delta^\beta_\alpha$, to which it reduces when the time (the fourth coordinate) is kept fixed:

$$L^{n'}{}_a = \begin{pmatrix} D^{v'}{}_\alpha & 0 \\ 0 & 1 \end{pmatrix}. \qquad (2.9)$$

Rotations leave $x^\alpha x_\alpha = x^2 + y^2 + z^2$ invariant, Lorentz transformations $x^a x_a = x^2 + y^2 + z^2 - c^2 t^2$.

We now determine the special Lorentz transformation which corresponds to a motion (with constant velocity) in the x-direction. We start from

$$\begin{aligned} x' &= Ax + Bct, \quad y' = y \\ ct' &= Cx + Dct, \quad z' = z \end{aligned} \quad \Longleftrightarrow \quad L^{n'}{}_a = \begin{pmatrix} A & 0 & 0 & B \\ 0 & 1 & 0 & 0 \\ 0 & 0 & 1 & 0 \\ C & 0 & 0 & D \end{pmatrix}. \qquad (2.10)$$

When we insert this expression for $L^{n'}_a$ into (2.7), we get the three conditions $A^2 - C^2 = 1$, $D^2 - B^2 = 1$, $AB = CD$, which can be parametrically solved by $A = D = \cosh\varphi$, $B = C = -\sinh\varphi$, so that the Lorentz transformation is given by

$$
\begin{aligned}
x' &= x\cosh\varphi - ct\sinh\varphi, & y' &= y, \\
ct' &= -x\sinh\varphi + ct\cosh\varphi, & z' &= z.
\end{aligned}
\tag{2.11}
$$

The analogy with the rotations

$$
\begin{aligned}
x' &= x\cos\varphi - y\sin\varphi, & z' &= z, \\
y' &= -x\sin\varphi + y\cos\varphi, & t' &= t
\end{aligned}
\tag{2.12}
$$

is obvious – but what is the physical meaning of φ in the case of the pseudorotations (2.11)?

To see this, we consider the motion of the origin $x' = 0$ of the moving coordinate system Σ' as seen from Σ. From $x' = 0$ and (2.11) we have

$$
v = \frac{\mathrm{d}x}{\mathrm{d}t} = \frac{c\sinh\varphi}{\cosh\varphi} \quad \Rightarrow \quad \tanh\varphi = \frac{v}{c},
\tag{2.13}
$$

φ is in a simple way related to the velocity v. If we substitute v for φ in the pseudorotations (2.11), we get the well-known form

$$
x' = \frac{x - vt}{\sqrt{1 - v^2/c^2}}, \quad ct' = \frac{ct - vx/c}{\sqrt{1 - v^2/c^2}}, \quad y' = y, \quad z' = z
\tag{2.14}
$$

of the special Lorentz transformation. This transformation describes the transformation between a system Σ and a system Σ' which moves in the x-direction with constant velocity v with respect to Σ.

For small velocities, $v/c \ll 1$, we regain $x' = x - vt$, i.e. the Galilei transformation; we see that Newtonian mechanics is valid for small velocities, discrepancies will appear only if the particles are moving very fast. We shall come back to this question in Chapter 4.

If we solve (2.14) for the x^a, we will get the same equations, with the primed and unprimed coordinates exchanged and v replaced by $-v$.

2.3 Some properties of Lorentz transformations

In this section we shall discuss some of the more mathematical properties of the Lorentz transformations. Many of the physical implications will be dealt with in the following chapters, in particular in Chapter 3.

Group property The Lorentz transformations form a group. To prove this, we remark that matrix multiplication is associative, and see by inspection that the identity $L^{n'}{}_a = \delta^n_a$ is contained. Two successive transformations yield

$$x^{m''} = L^{m''}{}_{n'} x^{n'} = L^{m''}{}_{n'} L^{n'}{}_a x^a = L^{m''}{}_a x^a. \qquad (2.15)$$

This will be a Lorentz transformation if $L^m{}_a$ satisfies (2.7), which is indeed the case:

$$L^{m''}{}_a L_{m''}{}^d = L^{m''}{}_{n'} L^{n'}{}_a L_{m''}{}^{b'} L_{b'}{}^d = \delta^b_n L^{n'}{}_a L_{b'}{}^d = L^{b'}{}_a L_{b'}{}^d = \delta^d_a. \qquad (2.16)$$

In a similar way one can show that the inverse of a Lorentz transformation is again such a transformation.

Classification of Lorentz transformations The 4×4 matrices $L^{n'}{}_a$ which describe Lorentz transformations have 16 parameters which are subject to the ten conditions (2.5); there are six independent Lorentz transformations, corresponding to three motions (e.g. in the direction of the axes) and three rotations. As we shall show now, there are four distinct types of Lorentz transformations.

From the defining equations (2.5) and (2.7) we immediately get

$$\left\|L^{n'}{}_a L_{n'}{}^b\right\| = \left\|\delta^b_a\right\| = 1, \quad \left\|L_{n'}{}^b\right\| = \left\|\eta_{n'm'}\right\| \cdot \left\|\eta^{ab}\right\| \cdot \left\|L^{m'}{}_a\right\| = \left\|L^{m'}{}_a\right\|, \qquad (2.17)$$

so that

$$\left\|L^{n'}{}_a\right\| = \begin{cases} +1 \\ -1 \end{cases} \qquad (2.18)$$

holds. Evaluating the (4,4)-component of (2.5), we obtain (remember that indices are raised and lowered by means of η!)

$$\begin{aligned} 1 &= -\eta_{44} L^{n'}{}_4 L_{n'}{}^4 \\ &= -\eta^{n'm'} L_{n'4} L_{m'4} = (L_{4'4})^2 - (L_{1'4})^2 - (L_{2'4})^2 - (L_{3'4})^2 \end{aligned} \qquad (2.19)$$

and conclude that

$$L^{4'}{}_4 = \begin{cases} \geq +1 \\ \leq -1 \end{cases}. \qquad (2.20)$$

Equations (2.18) and (2.20) show that there are four distinct classes of Lorentz transformations. Those which do not contain reflections have $\|L^n{}_a\| = +1$ and are called proper. Transformations with $L^4{}_4 \geq +1$ are called orthochronous; because of $ct' = L^4{}_4\, ct + \cdots$ they preserve the direction of time.

Normal form of a proper orthochronous Lorentz transformation By using an adapted coordinate system, any proper orthochroneous Lorentz transformation can be written in the form

$$L^{n'}{}_a = \begin{pmatrix} \heartsuit & 0 & 0 & \heartsuit \\ 0 & \times & \times & 0 \\ 0 & \times & \times & 0 \\ \heartsuit & 0 & 0 & \heartsuit \end{pmatrix} \qquad (2.21)$$

of direct product of a special Lorentz transformation (motion) ($\heartsuit$) and a rotation ($\times$) in the plane perpendicular to that motion. We leave the proof to the reader, see Exercise 2.2.

Lorentz transformation for an arbitrarily directed velocity We start with a question: how does a Lorentz transformation between two systems whose spatial axes are parallel, as in Fig. 2.2, look? By 'parallel' we mean that, for a fixed time, x' (for example) does not change if only y and z vary: in

$$x' = L^{1'}{}_a x^a = L^{1'}{}_1 x + L^{1'}{}_2 y + L^{1'}{}_3 z + L^{1'}{}_4 ct \qquad (2.22)$$

the $L^{1'}{}_2$ and $L^{1'}{}_3$ are assumed to be zero, and from the y'- and z'-equations we see that also $L^{2'}{}_1$, $L^{2'}{}_3$, $L^{3'}{}_1$ and $L^{3'}{}_2$ should vanish. There should be at least one component of the velocity, so we assume $L^{1'}{}_4 \neq 0$. Inserting all this into the defining equations (2.7), the result may be a surprise to the reader: the Lorentz transformation necessarily is of the form (2.11) of a motion in the x-direction (which is preferred here because of the assumption $L^{1'}{}_4 \neq 0$). So if the spatial axes of the two systems should be parallel, then the motion must be in the direction of one of the axes! For all other cases, the Lorentz transformations contain also terms which cause a rotation of the spatial system. For rotations the analogous effect is well known: none of the axes of a coordinate system can remain unchanged unless it coincides with the axis of the rotation.

So one should not be surprised that the Lorentz transformation describing the motion of the system Σ' with an arbitrarily directed velocity V^α (with no 'extra' rotation) looks rather complicated:

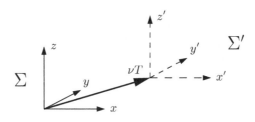

Fig. 2.2. Lorentz transformations between parallel systems.

$$L^{a'}{}_b = \begin{pmatrix} (\gamma - 1)n^\alpha n_\beta + \delta^\alpha_\beta & -v\gamma n^\alpha/c \\ -v\gamma n_\beta/c & \gamma \end{pmatrix}, \quad V^\alpha = vn^\alpha/c,$$ (2.23)

$$n^\alpha n_\alpha = 1, \quad \gamma \equiv (1 - v^2/c^2)^{-1/2}, \quad \alpha, \beta = 1, 2, 3.$$

Note that the rotational part in (2.23), the term $(\gamma-1)n^\alpha n_\beta$, is of second order in v/c.

Velocity addition formula for parallel velocities What is the result if we perform two successive Lorentz transformations, both corresponding to motions in the x-direction? Since the Lorentz transformations form a group, of course again a transformation of that type – but with what velocity?

Lorentz transformations are pseudorotations, i.e. they satisfy

$$\begin{aligned} x' &= x \cosh \varphi_1 - ct \sinh \varphi_1, & ct' &= -x \sinh \varphi_1 + ct \cosh \varphi_1, \\ x'' &= x' \cosh \varphi_2 - ct' \sinh \varphi_2, & ct'' &= -x' \sinh \varphi_2 + ct' \cosh \varphi_2. \end{aligned}$$ (2.24)

To get (x'', ct'') in terms of (x, ct), we observe that one adds rotations about the same axis by adding the angles:

$$\begin{aligned} x'' &= x \cosh \varphi - ct \sinh \varphi, \\ ct'' &= -x \sinh \varphi - ct \cosh \varphi, \end{aligned} \quad \varphi = \varphi_1 + \varphi_2.$$ (2.25)

To translate this relation into one for the velocities, we have to use (2.13), i.e. $\tanh \varphi = v/c$, and the well-known theorem for the hyperbolic tangent,

$$\tanh \varphi = \tanh(\varphi_1 + \varphi_2) = \frac{\tanh \varphi_1 + \tanh \varphi_2}{1 + \tanh \varphi_1 \tanh \varphi_2}.$$ (2.26)

We obtain

$$v = \frac{v_1 + v_2}{1 + v_1 v_2/c^2}.$$ (2.27)

For small velocities, $v_n/c \ll 1$, we get the Galilean addition formula $v = v_1 + v_2$. If we take the velocity of light as one of the velocities (as a limiting case, since the Lorentz transformations (2.14) are singular for $v = c$), we get

$$v = \frac{c + v_2}{1 + v_2/c} = c,$$ (2.28)

the velocity of light cannot be surpassed.

On the other hand, if we take two velocities smaller than that of light, we have, with $v_1 = c - \lambda$, $v_2 = c - \mu$, $\lambda, \mu > 0$,

$$v = \frac{2c - \lambda - \mu}{1 + (c - \lambda)(c - \mu)/c^2} = c\frac{2c - \lambda - \mu}{2c - \lambda - \mu + \lambda\mu/c}$$
$$= \frac{c}{1 + \lambda\mu/[c(2c - \lambda - \mu)]} \leq c, \tag{2.29}$$

it is not possible to reach the velocity of light by adding velocities less than that of light. The velocity addition formula (2.27) seems to indicate that the velocity of light plays the role of a maximum speed; we shall come back to this in the next chapter.

The addition of two non-parallel velocities will be considered in Section 4.4.

Exercises

2.1   Show that the inverse of a Lorentz transformation is again a Lorentz transformation.

2.2   Show by considering the eigenvalue equation $L^{a'}{}_b x_b = \lambda x^a$ that the four eigenvalues λ_a of a proper orthochroneous Lorentz transformation obey $\lambda_1\lambda_2 = 1 = \lambda_3\lambda_4$, and that by using the eigenvectors the Lorentz transformation can be written as indicated in (2.21).

2.3   Show that the transformation (2.23) is indeed a Lorentz transformation, and that origin of the system Σ' obtained from Σ by (2.23) moves with the velocity V^α.

2.4   Show by directly applying (2.14) twice that (2.26) is true.

2.5   In a moving system Σ', a rod is at rest, with an angle φ' with respect to the x'-axis. What is the angle φ with respect to the x-axis?

3

Our world as a Minkowski space

In this chapter we will deal with the physical consequences of the Lorentz transformations. Most of them were first found and understood by Einstein (1905), although most of the more technical properties considered in the last chapter were known before him.

3.1 The concept of Minkowski space

We have seen that the velocity of light is the same for all inertial systems, i.e. for all observers which move with constant velocity with respect to each other. The velocity of light is just one aspect of Maxwell equations, so that in fact the Michelson experiment shows that Maxwell equations are the same in all inertial systems. Since the elements of our world interact not only by electromagnetic fields, but also by gravitation, heat exchange, and nuclear forces, for example, the same must be true for all these interactions. *The laws of physics are the same for all inertial systems* (principle of relativity).

The principle of relativity does not exclude the Galilei transformations of mechanics, if one does not specify the transformations between inertial systems. This can be done by demanding that *the velocity of light is the same for all inertial systems* (principle of the invariance of the velocity of light).

Both principles together characterize Special Relativity. They are most easily incorporated into the laws of physics if one uses the concept of Minkowski space.

The four-dimensional Minkowski space, or world, or space-time, comprises space and time in a single entity. This is done by using Minkowski coordinates

$$x^a = (x^\alpha, ct) = (\mathbf{r}, ct), \quad x_a = \eta_{ab} x^b = (x_\alpha, -ct). \quad (3.1)$$

A point in this space is characterized by specifying space and time; it may be called an event.

The metrical properties of Minkowski space (in Minkowski coordinates) are given by its line element

$$\mathrm{d}s^2 = \mathrm{d}x^2 + \mathrm{d}y^2 + \mathrm{d}z^2 - c^2 \, \mathrm{d}t^2 = \mathrm{d}\mathbf{r}^2 - c^2 \, \mathrm{d}t^2 = \eta_{ab} \, \mathrm{d}x^a \, \mathrm{d}x^b. \quad (3.2)$$

This line element is invariant under Lorentz transformations

$$x^{n'} = L^{n'}{}_a x^a, \quad L^{n'}{}_a L_{n'}{}^b = \delta_a^b \quad (3.3)$$

since $x^n x_n$ is. Note that $\mathrm{d}s^2$ is not positive definite!

3.2 Four-vectors and light cones

A four-vector $a^n = (a^1, a^2, a^3, a^4) = (\mathbf{a}, a^4)$ is a set of four elements which transforms like the components x^n of the position vector,

$$a^{n'} = L^{n'}{}_m a^m. \quad (3.4)$$

An example is the vector connecting two points P_1 and P_2 of Minkowski space, $\overrightarrow{P_1 P_2} = (x^2 - x^1, y^2 - y^1, z^2 - z^1, ct^2 - ct^1)$.

Obviously, a Lorentz transformation mixes the spacelike and the timelike parts of a four-vector, but leaves the 'length' fixed:

$$a^{n'} a_{n'} = L^{n'}{}_m L_{n'}{}^b a^m a_b = a^n a_n = \text{inv.} \qquad (3.5)$$

This invariant can have either sign, or can be zero, depending on the relative size of the spacelike and timelike parts of the vector. This leads to the following invariant classification of four-vectors:

$$a^n a_n = \mathbf{a}^2 - (a^4)^2 \begin{cases} > 0 & \text{spacelike vector} \\ = 0 & \text{null vector} \\ < 0 & \text{timelike vector} \end{cases} . \qquad (3.6)$$

For a given vector a^n, one can always perform a (spatial) rotation of the coordinate system so that $\mathbf{a}$ points in the x-direction: $a^n = (a^1, 0, 0, a^4)$. A special Lorentz transformation (2.14) then yields

$$a^{1'} = \frac{a^1 - va^4/c}{\sqrt{1 - v^2/c^2}}, \quad a^{4'} = \frac{a^4 - va^1/c}{\sqrt{1 - v^2/c^2}}. \qquad (3.7)$$

For $|a^1/a^4| > 1$, one can make $a^{4'}$ vanish by choice of v (note that v has to be smaller than c !), and similarly in the other cases. So one gets the following normal forms of four-vectors.

$$\begin{array}{lll}
& \text{spacelike vector:} & a^n = (a, 0, 0, 0) \\
\text{Normal forms:} & \text{null vector:} & a^n = (a, 0, 0, a) \\
& \text{timelike vector:} & a^n = (0, 0, 0, a).
\end{array} \qquad (3.8)$$

If we have two four-vectors a^n and b^n, then we can define the scalar product of the two by

$$|ab| = a^i b_i = \eta_{in} a^i b^n. \qquad (3.9)$$

This is of course an invariant under Lorentz transformations. When $|ab|$ is zero, the two vectors are called orthogonal, or perpendicular, to each other. Note that in this sense a null vector is perpendicular to itself.

A light wave emanating at $t = 0$ from the origin of the coordinate system will at time t have reached the points $\mathbf{r}$ with

$$\mathbf{r}^2 - c^2 t^2 = 0. \qquad (3.10)$$

If we suppress one of the spatial coordinates, equation (3.10) describes a cone in (x, y, ct)-space. Therefore one calls (3.10) the *light cone*. As Fig. 3.1 shows, the light cone separates timelike vectors inside it from the spacelike vectors outside; null vectors are tangent to it.

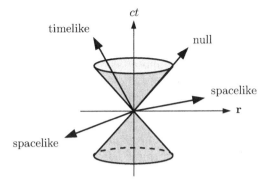

Fig. 3.1. Light cone structure of Minkowski space.

One can attempt to visualize the special Lorentz transformation (2.14) in Minkowski space by drawing the lines $x' = 0$ and $ct' = 0$ as ct'-axis or x'-axis, respectively, for a given value of v/c, see Fig. 3.2. This figure clearly shows that the new ct'-axis always lies inside the light cone (and the new x'-axis outside), that the transformation becomes singular for $v = c$, and that any timelike (spacelike) vector can be given its normal form by a suitable Lorentz transformation. But it does not show that the two coordinate systems are completely equivalent as in fact they are.

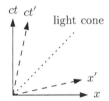

Fig. 3.2. Visualization of a Lorentz transformation.

3.3 Measuring length and time in Minkowski space

The problem   One may argue that the results of any measurement should be independent of the observer who made them. If we admit observers in relative motion, then only invariants with respect to Lorentz transformations will satisfy that condition. So for example (spacelike) distances which occur only as a part of a four-vector do not have an invariant meaning.

In practice one is accustomed to measuring spatial distances and time-intervals separately, and one often insists on using these concepts. But then the results of a measurement depend on the state of motion of the observer, as the components of a three-vector depend on the orientation of the Cartesian coordinate system one uses. The typical question which then arises is the following: suppose two observer Σ and Σ' (in relative motion) make some measurements; how are their results related? The answer to this question leads to some of the most spectacular results of Special Relativity theory.

The notion of simultaneity As a prerequisite, we will consider the meaning of 'same place at different times'. If an observer Σ states this for an object, it means the object is at rest at $x = 0$ (for example). For an observer Σ' moving with respect to Σ and to the object, the object changes its position; from (2.14) one gets

$$x' = \frac{x - vt}{\sqrt{1 - v^2/c^2}}, \quad x = 0 \quad \Rightarrow \quad x' = \frac{-vt}{\sqrt{1 - v^2/c^2}}. \qquad (3.11)$$

There is no *absolute* being at the same place for different times.

This is trivial – but the corresponding result obtained by interchanging the role of space and time is not. If an observer Σ states that two events at different places x_A and x_B are simultaneous (observed at the same time t_0), then the application of a Lorentz transformation gives

$$ct'_A = \frac{ct_0 - vx_A/c}{\sqrt{1 - v^2/c^2}}, \quad ct'_B = \frac{ct_0 - vx_B/c}{\sqrt{1 - v^2/c^2}}, \quad c(t'_A - t'_B) = \frac{(x_B - x_A)v}{c\sqrt{1 - v^2/c^2}}.$$

$$(3.12)$$

For an observer Σ' the two events are no longer simultaneous: *there is no absolute simultaneity at different places.*

This result has been much debated. In the beginning many people objected to that statement, and most of the attempts to disprove Special Relativity rely on the (hidden) assumption of an absolute simultaneity. There seems to be a psychological barrier which makes us refuse to acknowledge that our personal time which we feel passing may be only relative.

We now shall analyze the notion of simultaneity in more detail, just for a single observer. How can we judge and decide that two events at different places A and B happen at the same time? Just to assume 'we know it' is tantamount to assuming that there are signals with an infinite velocity coming from A and B which tell us that events have taken place; also, though not said in those terms, Newtonian physics uses this concept. To get a more precise notion, our first attempt may be to say: two events are simultaneous if two synchronized clocks situated at A and B show the same time. But how can we be sure that the two clocks are synchronized? We cannot transport one of two identical clocks from A to B, since the transport may badly disturb the clock and we have no way of checking that. Nor can we send a signal from A to B, divide the distance $\overline{AB}$ by the signal's velocity V to get the travelling time, and compare thus the clocks: without a clock at B, we cannot know the velocity V!

Considerations like this tell us that we need to *define* simultaneity. As with the definition of time discussed in Section 1.5, simultaneity has to be defined so that the laws of nature become simple, which means here: so that the Lorentz transformations hold. Einstein showed that a possible definition is like this: two events at A and B are simultaneous if light signals emitted simultaneously with those events arrive simultaneously in the middle of the line $\overline{AB}$. Note that here 'simultaneously' has been used only for events occurring at the same place!

By procedures like this, an observer Σ can synchronize his system of clocks in space-times; for a different observer Σ', this system is of course no longer synchronized.

Time dilatation At $\mathbf{r} = 0$, an event takes place between $t_B = 0$ and $t_E = T$; for an observer Σ at rest with that event the corresponding time-interval is of course $\Delta t = T$. Because of the Lorentz transformation (2.14) we then have $ct'_B = 0$, $ct'_E = cT/\sqrt{1 - v^2/c^2}$; for a moving observer Σ' this event lasts

$$\Delta t' = \frac{\Delta t}{\sqrt{1 - v^2/c^2}}. \tag{3.13}$$

A moving clock runs slower than one at rest, any clock runs fastest for an observer who is at rest with respect to it.

Length contraction When we measure the length of a rod at rest, the times t_A and t_B at which we look at the two endpoints $x_A = 0$ and $x_B = L$ are unimportant, its length is always $L = \Delta x$. For a moving observer Σ' this is different: since the rod is moving in his system of reference, he has to take care to determine its two endpoints *simultaneously*! So when using the relations

$$x'_A = \frac{-vt_A}{\sqrt{1 - v^2/c^2}}, \quad ct'_A = \frac{ct_A}{\sqrt{1 - v^2/c^2}},$$
$$x'_B = \frac{L - vt_B}{\sqrt{1 - v^2/c^2}}, \quad ct'_B = \frac{ct_B - vL/c}{\sqrt{1 - v^2/c^2}}, \tag{3.14}$$

he has to set $t'_A = t'_B$. Choosing $t'_A = 0$, this amounts to $t_A = 0$, $t_B = vL/c^2$, and thus to $x'_A = 0$, $x'_B = L\sqrt{1 - v^2/c^2}$, or to

$$\Delta x' = \Delta x \sqrt{1 - v^2/c^2}. \qquad (3.15)$$

A moving rod is shorter than one at rest, a rod is longest for an observer at rest with respect to it.

3.4 Two thought experiments

The two effects explained above, the time dilatation and the length contraction, are experimentally well confirmed. To get a better understanding of them, we will now discuss in some detail two gedanken (thought) experiments.

3.4.1 A rod moving through a tube

We take a rod of length $2L$, and a tube of length L (both measured at rest), see Fig. 3.3.

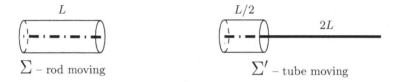

Fig. 3.3. Rod and tube.

System Σ (Tube at rest, rod moving) The length of the tube is L. If the rod moves with velocity $v = c\sqrt{3}/2$, application of (3.15) yields $2L\sqrt{1 - v^2/c^2} = L$ as the length of the rod; if it moves through the tube, it just fits in!

System Σ' (Rod at rest, tube moving) The rod is four times as long as the tube, it never can fit into the tube!

How can the two results both be true? Observer Σ' will state that Σ did not measure the position of the rod's endpoints simultaneously: Σ determined the position of its tip when it had already reached the end of the tube, and then waited until the end of the rod just entered the tube.

3.4.2 The twin paradox

Imagine a pair of twins; one is travelling around in space with a high velocity, the other just stays on Earth.

System Σ (Earth at rest) The travelling twin, assumed to have a constant velocity (except at the turning point), experiences a time dilatation, his biological clock runs slower; when coming back to Earth he is younger.

System Σ′ (Travelling twin at rest) For the travelling twin, the Earth is moving with a (nearly always) constant velocity, and for him the twin staying on Earth remains younger.

So when the two meet again on Earth: who really is younger than the other? It is in the nature of a 'paradox' that the contradiction is only apparent, something wrong has entered. So what is wrong in the above reasoning?

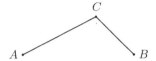

Fig. 3.4. 'Shortest' connection between A and B.

The wrong assumption is that the two systems of reference are equivalent. They are not, since the system (Σ) of the Earth is an inertial system, whereas the system (Σ') of the travelling twin is not: when returning, it must undergo an acceleration. One may argue that the effect of this acceleration can be neglected if the times of constant motion (which cause the time dilatation) are long enough. But that is not true, as a detailed analysis shows. The reasoning saying that the two systems are nearly equivalent, only the effect of the short times of the acceleration needs to be neglected, is on a similar level as saying that – since straight lines are the shortest connection between two points – the connection of A and B via C is the shortest one, the short deviation from a straight line in C does not count, see Fig. 3.4.

For a really reliable answer one has to know how to deal with accelerated systems; here General Relativity is to be asked, and the answer is: yes, travelling (deviating from geodesic motion) keeps you younger.

3.5 Causality, and velocities larger than that of light

Special Relativity denies the existence of an absolute simultaneity. Can it happen that even the temporal order of two events can be changed, i.e. that what for an observer Σ is the cause of an effect, is later than that effect for an observer Σ' ?

To be precise, let us have a cause at $x_1 = 0$, $t_1 = 0$, and its effect at $x_2 = L$, $t_2 = T$. The velocity of that phenomenon obviously is $V = L/T$. An observer Σ', moving with velocity v, sees the cause when $t_1' = 0$, and the effect when $ct'_2 = (cT - vL/c)/\sqrt{1 - v^2/c^2}$. He will state that the effect takes place before the cause if $t'_2 < 0$, i.e. if

$$vV > c^2. \qquad (3.16)$$

This relation can be true only if at least one of the two velocities exceeds that of light. So to avoid this acausal behaviour, one has to demand: *velocities greater than that of light must not and cannot occur.*

This statement is so strict that it invites people to challenge it, and newspapers regularly report on new findings that a velocity exceeding that of light has been observed. Because of its importance, we want to discuss three aspects of the above statement.

First we can say it reflects the logical consistency of Special Relativity. It is supported by the property of Lorentz transformations, discussed in Section 2.3, that by addition of velocities we cannot surpass c, and will be strengthened in the next chapter where we shall show that a material body cannot be accelerated from zero velocity to that of light.

Second, we have to stress that the limitation is only for velocities of material bodies or of processes which can be started voluntarily at the first point and are therefore suitable for transmitting information. To make that clearer, we consider two thought experiments.

Imagine, as in Fig. 3.5, one blade (1) at rest and the other one (2) moving with velocity V_0. The intersection P of the two moves with velocity $V = V_0/\sin\alpha$ to the right. By choosing α small, we can make V arbitrarily large, even exceeding c. Although the arrival of P may cut your finger or your throat, the blade does not carry information, it is always moving, and $V > c$ does not violate the above statement. Blade (2) would carry information, however, if we were able to start it at a prescribed time *simultaneously over its whole length,* acting on it only at P. This is not possible, since no material blade is completely rigid.

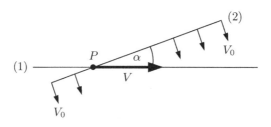

Fig. 3.5. Moving blades.

To turn the argument around: Special Relativity forbids the existence of completely rigid bodies.

The second thought experiment uses a laser pointing to the Moon. By rotating it, the point where it hits the Moon can be made to move on its surface with arbitrary large velocity – but the cause is on the Earth, not at the point hit first on the Moon!

Our third remark is a rather hypothetical one. Special Relativity does not exclude particles with velocities larger than c as long as they cannot be used to carry information, i.e. as long as they cannot be generated or cannot interact with ordinary matter. Such hypothetical particles are called *tachyons*.

So although we can *define* velocities (as in the above thought experiments, or as, for example, phase velocities of waves in dispersive media) which exceed that of light, this does not contradict our statement. For us it is certain that no physical law should permit us to change the past, an acausal behaviour should not take place. But there is no logical way of excluding the possibility that future experiments could disprove the theory by, for example, finding particles faster than light; but so far Special Relativity has always proved to be right.

We close this section by showing how causal relations between points in space-time can be visualized. The light cone at a point P divides space-time into three parts. Points P_1 in the upper part (inside and on the future light cone) are in the (absolute) future of P; they can be influenced by P, but cannot act back. Points P_2 in the lower part (inside and on the past light cone) are in the absolute past; the can act on P, but P cannot react. Points P_3 outside the light cone are in the absolute present; there is no causal connection between them and P.

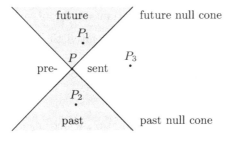

Fig. 3.6. Causal domains in space-time.

Exercises

3.1   Show that the sum of two timelike vectors, both pointing into the future (i.e. with positive timelike components) is again timelike and pointing into the future.

3.2   Show that a vector orthogonal to a timelike vector is spacelike.

3.3   Show that two null vectors which are perpendicular to each other are parallel.

3.4   Show that (as stated in the text) the observer Σ' does not determine the position of the rod's endpoints simultaneously when judged by Σ.

3.5   How fast has a sphere of diameter D to be moved to pass through a circular hole (diameter d) in a plane sheet of paper?

4

Mechanics of Special Relativity

Lorentz transformations were derived from the properties of light propagation. So naturally one would expect that we now start to discuss Electrodynamics and Optics. But since for that we need tensor analysis, we will reverse the logical order and treat Classical Mechanics first.

4.1 Kinematics

Newtonian mechanics is Galilei invariant; its foundations have to be changed to make it Lorentz invariant. The best way to avoid mistakes when doing this is to use only invariants and four-vectors.

World lines and proper time In non-relativistic mechanics, the motion of a point particle is described by giving its position as a function of time, $\mathbf{r} = \mathbf{r}(t)$. Sometimes one uses also the arclength l defined by

$$\mathrm{d}l^2 = \mathrm{d}x^2 + \mathrm{d}y^2 + \mathrm{d}z^2, \ \ x = x(l), \ y = y(l), \ z = z(l), \ t = t(l). \quad (4.1)$$

Here we rather want to have a notation which incorporates space and time on an equal level; this is done by choosing the *proper time* τ defined by

$$-c^2 \, \mathrm{d}\tau^2 = \mathrm{d}s^2 = \mathrm{d}x^n \, \mathrm{d}x_n = \mathrm{d}x^2 + \mathrm{d}y^2 + \mathrm{d}z^2 - c^2 \, \mathrm{d}t^2 \qquad (4.2)$$

as our parameter, and by describing the *world line* of the particle as $x^n(\tau) = \big(\mathbf{r}(\tau), ct(\tau)\big)$. Writing (4.2) as

$$-c^2 \, \mathrm{d}\tau^2 = -c^2 \, \mathrm{d}t^2(1 - \mathrm{d}^2\mathbf{r}/c^2 \, \mathrm{d}t^2) = -c^2 \, \mathrm{d}t^2(1 - v^2/c^2), \qquad (4.3)$$

we see that

$$\mathrm{d}\tau = \mathrm{d}t \sqrt{1 - v^2/c^2} \qquad (4.4)$$

holds, where $v = v(\tau)$ is the – in general non-constant – velocity of the particle. For a particle at rest, proper time and time coincide. Fig. 4.1 shows some typical examples of world lines.

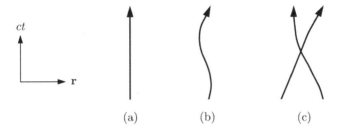

(a) (b) (c)

Fig. 4.1. Typical examples of world lines of particles: (a) particle at rest, (b) accelerated particle, (c) impact of two particles.

Four-velocity Using only the world line representation $x^n = x^n(\tau)$ and the parameter τ as ingredients, there is only one four-vector generalizing the three-dimensional velocity, namely the four-velocity u^n defined by

$$u^n(\tau) = \frac{\mathrm{d}x^n}{\mathrm{d}\tau} = \frac{\mathrm{d}x^n}{\mathrm{d}t \sqrt{1 - v^2/c^2}} = \left(\frac{\mathbf{v}}{\sqrt{1 - v^2/c^2}}, \frac{c}{\sqrt{1 - v^2/c^2}} \right). \qquad (4.5)$$

Because of the definition (4.2) of the proper time τ, the four-velocity obeys

$$u^n u_n = \frac{\mathrm{d}x^n}{\mathrm{d}\tau} \frac{\mathrm{d}x_n}{\mathrm{d}\tau} = -c^2. \qquad (4.6)$$

This equation shows that the four components of u^n are not independent of each other, and that u^n generalizes the tangent vector $\mathrm{d}\mathbf{r}/\mathrm{d}l$ (which is always of length 1).

Four-acceleration Similarly we define the four-acceleration $\dot{u}^n$ by

$$\dot{u}^n = \frac{\mathrm{d}u^n}{\mathrm{d}\tau} = \frac{\mathrm{d}^2x^n}{\mathrm{d}\tau^2}. \qquad (4.7)$$

Because of (4.6) one has

$$d(u^n u_n)/d\tau = 0 = \dot{u}^n u_n, \qquad (4.8)$$

four-velocity and four-acceleration are orthogonal to each other.

Rest system of a particle   Since in general $v(\tau)$ is not constant, an accelerated particle cannot be transformed to rest by a Lorentz transformation. This is possible only for a single instant of time $\tau = \tau_0$, where we can have $u^n(\tau_0) = (0, 0, 0, c)$.

4.2 Equations of motion

In the foregoing section, we introduced some new notations. Now we come to a more difficult part: since Newtonian mechanics contradicts Lorentz invariance, we have to find a new law of nature! To do this, we imitate Newtonian mechanics by demanding that the equations of motion should have the form 'acceleration is proportional to force', i.e. by demanding the form

$$m_0 \frac{d^2 x^n}{d\tau^2} = f^n. \qquad (4.9)$$

The constant m_0 is called *proper mass*, or *rest mass*, and the four-vector f^n the *four-force*, or simply force.

Since force and acceleration are four-vectors, and m_0 is an invariant, this equation is certainly Lorentz invariant. But is it correct? Does it really describe nature? One possible way of testing this is to study the Newtonian limit $c \to 0$. As one can see from (4.5), the Newtonian equations of motion (1.1) should be contained in the spatial components of (4.9). But how is the three-force $\mathbf{F}$ contained in f^n? Here we need a result from electrodynamics (which we will give later in Section 7.4) which says that f^n has the form

$$f^n = \left(\frac{\mathbf{F}}{\sqrt{1 - v^2/c^2}}, f^4 \right), \qquad (4.10)$$

with a not yet determined fourth component f^4.

Spacelike components of the equations of motion   Inserting the expression (4.10) for f^n into the equation of motion (4.9), and using (4.5) and (4.4), we obtain

$$\frac{d}{dt} m\mathbf{v} = \mathbf{F}, \quad m = \frac{m_0}{\sqrt{1 - v^2/c^2}}. \qquad (4.11)$$

For $v \ll c$, this is indeed the Newtonian equation of motion, with mass

m and proper mass m_0 coinciding. In general, however, it only has the same form, but with a velocity-dependent mass m. It is a kind of a surprise that the Newtonian form is still valid in Special Relativity!

For large velocities v, the mass m is growing, tending to infinity if v approaches c; this shows that a finite force $\mathbf{F}$ cannot accelerate a point mass so that its velocity equals that of light in a finite time.

Timelike component of the equations of motion The Newtonian equations of motion do not have a timelike component – so what new law is hidden here? The answer is simple, but nevertheless surprising. If we write the equations of motion as

$$m_0 \dot{u}^n = f^n \tag{4.12}$$

and multiply this equation by u_n, we see that, because of (4.8),

$$f^n u_n = 0 \tag{4.13}$$

holds, so the four-force must be orthogonal to the four-velocity. This relation can be used to determine f^4; the simple calculation gives

$$f^n = \left(\frac{\mathbf{F}}{\sqrt{1 - v^2/c^2}}, \frac{\mathbf{F}\mathbf{v}}{c\sqrt{1 - v^2/c^2}} \right). \tag{4.14}$$

Using this result, the timelike component of the equation of motion (4.9) reads

$$\frac{\mathrm{d}}{\mathrm{d}t} mc^2 = \mathbf{F}\mathbf{v}. \tag{4.15}$$

A relation similar to this appears in Newtonian mechanics if we multiply $m\,\mathrm{d}\mathbf{v}/\mathrm{d}t = \mathbf{F}$ by $\mathbf{v}$ and write the result as

$$\frac{\mathrm{d}}{\mathrm{d}t} \left(\tfrac{1}{2} m\mathbf{v}^2 \right) = \mathbf{F}\mathbf{v}. \tag{4.16}$$

It tells how the kinetic energy $mv^2/2$ changes with time. So by analogy we conclude that also (4.15) is the balance equation for the kinetic energy of a particle. This interpretation is supported by the fact that the development of mc^2 with respect to v/c yields

$$mc^2 = \frac{m_0}{\sqrt{1 - v^2/c^2}} = m_0 c^2 + \frac{1}{2} m v^2 + \cdots. \tag{4.17}$$

Energy E and (inertial) mass m are – except for the factor c^2 – just two names for the same thing,

$$E = mc^2 = \frac{m_0 c^2}{\sqrt{1 - v^2/c^2}}. \tag{4.18}$$

When this was first formulated by Einstein, it was a very surprising result, with remarkable consequences, so that the above relation became a trademark of Einstein's. A consequence of (4.18) is that, besides a part due to the motion of the particle, there is an energy m_0c^2 due to its rest mass. Kinetic energy can be transformed into other forms of energy, for example into heat, and so it was argued that the same should be the case for the rest energy m_0c^2; we know nowadays that this is true. Conversely, any energy of a system contributes to its rest mass, and so photons (electromagnetic radiation) in a box also have a rest mass; since any mass is a source for the gravitational field, photons therefore also must have a gravitational field.

Energy-momentum four-vector The fact that energy and momentum of a particle are closely related is best seen from the energy-momentum four-vector p^n, which just the four-velocity multiplied by the rest mass:

$$p^n = m_0 u^n = \left(\frac{m_0 \mathbf{v}}{\sqrt{1 - v^2/c^2}}, \frac{m_0 c}{\sqrt{1 - v^2/c^2}} \right)$$

$$= (m\mathbf{v}, E/c) = (\mathbf{p}, E/c), \quad p_n p^n = -m_0^2 c^2. \tag{4.19}$$

The equations of motion (4.9) are the balance equations for this vector.

4.3 Hyperbolic motion

Simple examples for relativistic mechanics are rare; most of the daily life examples cannot be generalized meaningfully as (for example) a Lorentz invariant gravitational field does not exist. An interesting example is, however, the generalization of the one-dimensional motion under a constant force (which in analogy to the homogeneous gravitational field we name $m_0 g$). The corresponding x-component of the equation of motion reads

$$\frac{\mathrm{d}}{\mathrm{d}t} mv = F = m_0 \frac{\mathrm{d}}{\mathrm{d}t} \frac{v}{\sqrt{1 - v^2/c^2}} = m_0 g = \text{const.} \tag{4.20}$$

The question whether this is really a correct relativistic generalization of a motion under a constant force is left to the reader, see Exercise 4.2.

Taking the initial value $v(0) = 0$, the straightforward integration of (4.20) yields

$$\frac{v}{\sqrt{1 - v^2/c^2}} = gt, \quad v = \frac{\mathrm{d}x}{\mathrm{d}t} = \frac{gt}{\sqrt{1 + (gt/c)^2}},$$

$$x(t) = c^2 g^{-1}\sqrt{1 + (gt/c)^2} + b, \quad v(t) = c^2 t/(x - b), \tag{4.21}$$

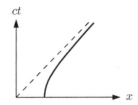

Fig. 4.2. Hyperbolic motion.

from which

$$(x - b)^2 - c^2 t^2 = c^4/g^2. \tag{4.22}$$

follows. This equation describes a hyperbola in x-t-space, which explains the name hyperbolic motion; see Fig. 4.2. A photon emitted at $t = 0$ from a point to the left of $x = b$ can never reach the particle.

To obtain x and t as functions of the proper time τ, we have to evaluate

$$d\tau = dt \sqrt{1 - v^2/c^2} = dt / \sqrt{1 + (gt/c)^2}. \tag{4.23}$$

The result is

$$\tau = \frac{c}{g} \operatorname{arcsinh}\frac{gt}{c}, \quad ct(\tau) = \frac{c^2}{g} \sinh \frac{g\tau}{c}, \quad x(\tau) - b = \frac{c^2}{g} \cosh \frac{g\tau}{c}. \tag{4.24}$$

For $v \ll c$, we regain with $v = gt$, $x-$ const. $= gt^2/2$, the well-known results of Newtonian mechanics.

The simplicity of the hyperbolic motion admits some interesting insights into relativistic effects.

Take two rockets, let them start at the same time $t = 0$ at $x = x_0$ and $x = x_0 + L$, respectively, and let them have the *same* acceleration. Imagine now a rope spanned between them. What happens to the rope when the two rockets are accelerating? From the viewpoint of an observer at rest, the distance L between the rockets remains the same

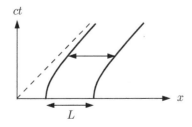

Fig. 4.3. Two identical rockets in hyperbolic motion.

for all times. For him a material rope spanned between them thus also has the same length for all times. But since that rope is moving, this can only be so if in the rope's rest system its length L' is growing according to $L' = L(1 - v^2/c^2)^{-1/2}$, so just compensating the length contraction (3.15) due to its motion; the rope expands forever, until it finally tears into pieces.

On the other hand, take two rockets with *different* accelerations $g = c^2/X$, but both their world lines being described by hyperbolae

$$x^2 - c^2 t^2 = X^2. \tag{4.25}$$

For either of the rockets (for any fixed value of X), its velocity v can be obtained from $d(x^2 - c^2 t^2)$ as $v = dx/dt = c^2 t/x$; it obeys $\sqrt{1 - v^2/c^2} = X/x$. Assume now the two rockets start at $t = 0$ with an infinitesimal distance $\Delta x_0 = \Delta X$. Because of (4.25), for any (constant) time we have $x\,dx = X\,dX$, which amounts to $\Delta x = X\Delta X/x = dX\sqrt{1 - v^2/c^2}$. So we have $\Delta x = \Delta x_0 \sqrt{1 - v^2/c^2}$, which is exactly the formula for the length contraction of a moving rod! If two rockets of the family (4.25), now at a finite distance from each other, are connected by a rod, then this rod undergoes different velocities at each of its parts, but as a whole it is neither contracted nor expanded by the motion of the differently accelerated rockets.

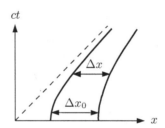

Fig. 4.4. Two rockets in hyperbolic motion, with different acceleration.

4.4 Systems of particles

Actio = reactio? In Newtonian mechanics, for a system of, say, two particles, the equations of motion read

$$m_1 \frac{d^2 \mathbf{r}_1}{dt^2} = \mathbf{F}_{12} + \mathbf{F}_1^{\text{ext}}, \quad m_2 \frac{d^2 \mathbf{r}_2}{dt^2} = \mathbf{F}_{21} + \mathbf{F}_2^{\text{ext}}, \tag{4.26}$$

where F_{AB} denotes the force exerted from particle A on particle B, and

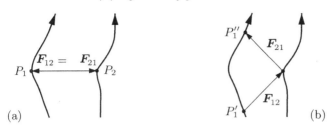

Fig. 4.5. Interaction between two particles: (a) Newtonian physics, (b) Special Relativity.

the $\mathbf{F}_A^{\text{ext}}$ are forces from outside the system. It is one of the fundamental laws that *actio = reactio* holds, i.e. the forces between the two particles are equal in magnitude, but opposite in direction,

$$\boldsymbol{F}_{12} = -\boldsymbol{F}_{21}. \tag{4.27}$$

The two forces act – 'naturally' – at the same time, see Fig. 4.5.

In Special Relativity, causality has to be obeyed. This means that the force $\boldsymbol{F}_{12}$, which has its cause in particle 1 at some instant of time, can have its effect on particle 2 only at a later time such that the action propagates at most with the velocity of light. Whereas in Newtonian mechanics $\boldsymbol{F}_{12}$ and $\boldsymbol{F}_{21}$ refer to the same track of interaction $P_1 P_2$, this is different in relativity: the action on particle 2 at P_2 comes from P_1', and it acts back on particle 1 at P_1''. In general, *actio = reactio* no longer holds, which makes the treatment of interacting particles much more difficult. The deeper reason for these difficulties is the fact that the field by which the interaction is mediated (e.g. the electromagnetic field) has its own momentum which contributes to the momenta of the particles which are balanced in the equations of motion (4.26).

As a consequence of this, the equations and calculations are simple only if the interaction takes place at the same point in space-time: this happens for a collision, or a decay, of particles, and we shall consider only those from now on.

Collision and decay of particles – energy-momentum balance and centre of mass There are no external forces $\mathbf{F}_A^{\text{ext}}$, and interaction between the particles takes place only when they are at the same point in space-time. We therefore can assume $\mathbf{v}_A = \mathbf{v}_B$ for all particles participating in the process, and moreover $\mathbf{F}_{AB} = -\mathbf{F}_{BA}$. In generalizing the relativistic equations of motion (4.11) and (4.15), or (4.9), and the Newtonian equations (4.26), we conclude that

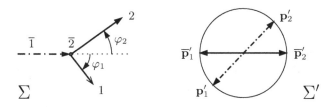

Fig. 4.6. Elastic collision of two particles.

$$\sum_A m_A \mathbf{v}_A = \text{const.}, \quad \sum_A m_A c = \sum_A E_A/c = \text{const.} \qquad (4.28)$$

or

$$P^n \equiv \sum_A p_A^n = \text{const.} \qquad (4.29)$$

holds: three-momentum and energy, or four-momentum, are conserved.

The vector P^n, being the sum of timelike vectors, all pointing into the future, is again timelike and pointing into the future, see Exercise 4.1. So there is a coordinate system in which it has the normal form

$$P^n = (0, P^4). \qquad (4.30)$$

This system is called the centre of mass system. Since the four-momentum of a particle satisfies $p^n p_n = -m_0^2 c^2$, we attribute a rest mass $M_0 = P^4/c$ to the system of particles; but this rest mass is *not* the sum of the particles' rest masses m_{0A}, since

$$M_0 = P^4/c = \sum_A m_A(v_A) = \sum_A m_{0A}(1 - v_A^2/c^2)^{-1/2} \geq \sum_A m_{0A} \qquad (4.31)$$

holds. The kinetic energy of the particles contributes to the system's rest mass.

Elastic collision of two particles   A collision is called elastic if the two rest masses remain unchanged during the collision. We consider two particles. In the observer's system Σ, the first is moving in the x-direction, the second is at rest; a bar denotes the respective values before the collision. The conservation laws (4.28) yield

$$\overline{\mathbf{p}}_1 = \mathbf{p}_1 + \mathbf{p}_2, \quad \overline{E}_1 + \overline{E}_2 = E_1 + E_2 = E. \qquad (4.32)$$

What are the deflection angles φ_1 and φ_2? To calculate them, we use the centre of mass system Σ'. Because of (4.32), it has $P^n = (MV, 0, 0, Mc) = (\overline{m}_1 \overline{v}_1, 0, 0, E_1 + E_2)$, so its velocity V (in the x-direction) is

$$V = \overline{m}_1 \overline{v}_1 c^2 / (\overline{E}_1 + \overline{E}_2), \qquad (4.33)$$

and the conservation laws (4.32) read

$$\overline{\mathbf{p}}_1' + \overline{\mathbf{p}}_2' = 0 \quad = \mathbf{p}_1' + \mathbf{p}_2', \qquad E/c = \sqrt{\mathbf{p}^2 + m_0^2 c^2}$$
$$\overline{E}_1' + \overline{E}_2' = E' = E_1' + E_2', \qquad \text{for each particle.} \qquad (4.34)$$

If we now substitute the momenta $\mathbf{p}$ in the energy conservation (and use $\mathbf{p}_1^2 = \mathbf{p}_2^2$ etc.), we obtain

$$E'/c = \sqrt{\overline{\mathbf{p}}_1'^2 + m_{01}^2 c^2} + \sqrt{\overline{\mathbf{p}}_1'^2 + m_{02}^2 c^2}$$
$$= \sqrt{\mathbf{p}_1'^2 + m_0^2 c^2} + \sqrt{\mathbf{p}_1'^2 + m_0^2 c^2}. \qquad (4.35)$$

To get $\overline{\mathbf{p}}_1'$ in terms of $\mathbf{p}_1'$, we observe that an equation of the form $A - \sqrt{a+x} = \sqrt{b+x}$ has a unique solution for x. Since $\overline{\mathbf{p}}_1'^2 = \mathbf{p}_1'^2$ is a solution of (4.35), it is the only one: the three-momenta of the two particles before and after the collision are of equal magnitude, they have only been turned around, see Fig. 4.6, and the energies before and after the collision are the same.

All that holds in the centre of mass system Σ'. To translate this into the observer's system Σ, we make use of the fact that the particles' four-momenta $\overline{p}_A^n$ and p_A^n (before and after the collision, respectively) and the four-velocity V^n of the centre of mass obey

$$(p_A^n - \overline{p}_A^n) V_n = 0, \quad A = 1, 2, \qquad (4.36)$$

in any system of reference, since this relation is Lorentz invariant, and correct in the centre of mass system. If we apply it in the observer's system, where $V_n = (1 - V^2/c^2)^{-1/2}(V, 0, 0, c)$ holds, and write the momenta after the collision as $p_A^n = (p_a \cos \varphi_A, \cdot, \cdot, E_A/c)$ – the dots indicate terms not entering (4.36) – then we obtain after a short calculation

$$\cos \varphi_1 = [E_1(\overline{E}_1 + \overline{E}_2) - \overline{E}_1 \overline{E}_2 - m_{01}^2 c^4]/p_1 \overline{p}_1 c^2,$$
$$\cos \varphi_2 = (E_1 + \overline{E}_2)(E_2 - \overline{E}_2)/p_1 \overline{p}_1 c^2. \qquad (4.37)$$

Velocities in the observer system Σ and in the centre of mass system Σ'
If the system Σ' moves with the (constant) velocity V with respect to Σ in the x-direction, we can apply the inverse of the Lorentz transformation (2.14), with $v = V$, and obtain for the differentials

$$dx = \frac{dx' + V dt'}{\sqrt{1 - V^2/c^2}}, \, dy = dy', dz = dz', \, dt = \frac{dt' + V dx'/c^2}{\sqrt{1 - V^2/c^2}}. \qquad (4.38)$$

From this we get for the velocity (with $v_x = dx/dt$ etc.)

$$v_x = \frac{v'_x + V}{1 + v'_x V/c^2}, \; v_y = \frac{v'_y \sqrt{1 - V^2/c^2}}{1 + v'_x V/c^2}, \; v_z = \frac{v'_z \sqrt{1 - V^2/c^2}}{1 + v'_x V/c^2}. \quad (4.39)$$

These relations describe a special case of the addition of non-parallel velocities; for $v_x = 0 = v_y$ we regain (2.27).

Exercises

4.1   Show that (4.14) follows from (4.13) and (4.5)!

4.2   Compute for (4.20) the components of the acceleration $\dot{u}^n$ and show that $\dot{u}^n \dot{u}_n = m_0^2 g^2$ holds!

4.3   Transform the two-dimensional line element $ds^2 = dx^2 - c^2 dt^2$ by introducing coordinates $x = X \cosh cT$, $ct = X \sinh cT$ adapted to the hyperbolae (4.25), and relate the lines $T = $ const. to the velocity of the rockets flying on those hyperbolae!

4.4   An elastic rod tears when expanded to twice its rest length. Such a rod undergoes from $t = 0$ a constant acceleration in its length direction; when will it tear?

4.5   Apply (4.37) to the case where particle 1 is a photon, with $\bar{p}_1 = h/\bar{\lambda}$, $p_1 = h\lambda$, where λ is the photon's wavelength and h is Planck's constant, and give λ in terms of $\bar{\lambda}$ (Compton scattering).

5
Optics of plane waves

5.1 Invariance of phase and null vectors

Optics is a part of Maxwell's theory; but for most of its daily-life applications, the description of light by a scalar function suffices. This scalar function $u(x^n) = u(\mathbf{r}, ct)$ is a solution of the wave equation

$$\left(\frac{\partial^2}{\partial x^2} + \frac{\partial^2}{\partial y^2} + \frac{\partial^2}{\partial z^2} - \frac{\partial^2}{c^2 \partial t^2} \right) u = \eta^{nm} \frac{\partial^2}{\partial x^n x^m} u = \Box u = 0. \quad (5.1)$$

A simple solution to this equation is the general plane wave

$$u(\mathbf{r}, ct) = u_0 \, e^{i\Phi}, \quad \Phi = \mathbf{kr} - \omega t, \quad \mathbf{k}^2 = \omega^2/c^2. \quad (5.2)$$

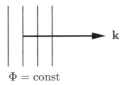

$\Phi = \text{const}$

Fig. 5.1. Plane wave and wave vector in three-space.

The three-vector $\mathbf{k}$ is called the *wave vector*; it points in the direction orthogonal to the planes $\Phi = \text{const.}$ and is thus tangent to the light rays, see Fig. 5.1. Its magnitude $k = 2\pi/\lambda$ gives the number of wavelengths λ per length π, and $\omega = 2\pi\nu$ is related to the frequency ν of the wave.

To see that the above description is in fact already Lorentz invariant, we first observe that for many optical experiments the phase Φ is what matters. Interference is due to differences in the phase of waves, and the fact that at a given space-time point there is light, or is none, cannot depend on the velocity of the observer looking there. So we conclude that *the phase* $\Phi = \mathbf{k}\mathbf{r} - \omega t$ *is a Lorentz invariant*. Since it obviously can be written as

$$\Phi = k^n x_n, \quad k^n = (\mathbf{k}, \omega/c) = (\mathbf{k}, 2\pi\nu/c), \tag{5.3}$$

we further conclude that k^n *is a four-vector*. Because of (5.2), $\mathbf{k}^2 = \omega^2/c^2$, it is a null vector. So we have

$$k^{n'} = L^{n'}{}_m k^m, \quad k^n k_n = 0. \tag{5.4}$$

Most of the relativistic effects in optics are due to, and can be explained in terms of, the Lorentz transformation of the vector k^n.

5.2 The Doppler effect – shift in the frequency of a wave

Suppose there is a plane wave travelling in the x-direction of an observer's rest frame Σ. It is characterized by its vector $k^n = (k, 0, 0, \omega/c)$, with $k = \omega/c$.

Longitudinal Doppler effect For an observer Σ' moving with velocity v in the direction of the wave (in the direction of $\mathbf{k}$), application of a Lorentz transformation to k^n yields

$$k^{4'} = \frac{\omega'}{c} = \frac{k^4 - vk^1/c}{\sqrt{1 - v^2/c^2}} = \frac{\omega/c - vk/c}{\sqrt{1 - v^2/c^2}} = \frac{\omega}{c}\frac{1 - v/c}{\sqrt{1 - v^2/c^2}}, \tag{5.5}$$

and because of $\omega = 2\pi\nu$ one gets for the frequencies

$$\nu' = \nu \sqrt{\frac{1 - v/c}{1 + v/c}}. \tag{5.6}$$

Whereas most of the relativistic effects are of order v^2/c^2, here terms of order v/c are included, which indicates that there may be a classical part in it. Indeed one has

$$\text{Relativ.:} \quad \nu' = \nu(1 - v/c + v^2/2c^2 + \cdots),$$

$$\text{Class., moving observ.:} \quad \nu' = \nu(1 - v/c), \tag{5.7}$$

$$\text{Class., moving source:} \quad v' = \nu(1 + v/c)^{-1} = \nu(1 - v/c + v^2/c^2 - \cdots).$$

Transversal Doppler effect For an observer Σ' moving with velocity v in the y-direction orthogonal to the direction of the wave (orthogonal to $\mathbf{k}$), application of a Lorentz transformation to k^n yields

$$k^{4'} = \frac{\omega'}{c} = \frac{k^4}{\sqrt{1 - v^2/c^2}}, \quad \nu' = \frac{\nu}{\sqrt{1 - v^2/c^2}}. \tag{5.8}$$

In non-relativistic optics there is no such effect.

5.3 Aberration – change in the direction of a light ray

In a system Σ, the wave vector has the form (see Fig. 5.2)

$$k^n = (-k\cos\alpha, -k\sin\alpha, 0, k), \quad \tan\alpha = k_y/k_x. \tag{5.9}$$

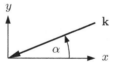

Fig. 5.2. Incident wave.

Seen from a system Σ', which moves in the x-direction with velocity v with respect to Σ, one then has

$$k'_x = \frac{-k\cos\alpha - vk/c}{\sqrt{1 - v^2/c^2}}, \; k'_y = k_y, \tag{5.10}$$

and therefore, with (5.9) and $\tan\alpha' = k'_y/k'_x$,

$$\tan\alpha' = \frac{\sqrt{1 - v^2/c^2}\sin\alpha}{\cos\alpha + v/c}. \tag{5.11}$$

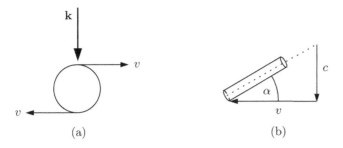

Fig. 5.3. Aberration of (a) star light and (b) rain drops.

The best-known non-relativistic application of this effect is that a telescope monitoring the position of a star has to be turned during the year due to the motion of the Earth on its orbit around the Sun, see Fig. 5.3. In this case, one has

$$\alpha = \pi/2, \quad \tan\alpha' = \frac{\sqrt{1 - v^2/c^2}}{v/c} = \frac{c}{v}\left(1 - \frac{v^2}{2c^2} + \cdots\right). \qquad (5.12)$$

In the non-relativistic limit, the aberration angle α' is exactly that angle by which one has to incline a tube (moving with velocity v) to permit rain drops (falling vertically with velocity c) to reach its bottom.

The relation (5.11) is rather asymmetric in the angles α and α'. This can be remedied by taking the tangent of $\alpha/2$ instead that of α. Using $\tan(\alpha/2) = \tan\alpha/(1 + \sqrt{1 + \tan^2\alpha})$, a straightforward calculation gives

$$\tan\frac{\alpha'}{2} = \sqrt{\frac{c - v}{c + v}}\tan\frac{\alpha}{2}. \qquad (5.13)$$

5.4 The visual shape of moving bodies

Penrose (1959) taught us to understand what equation (5.11) 'really' means – about 50 years after that equation was first written down.

Consider a sphere moving at a large distance with velocity v. What is its shape as judged by an observer at rest (in monocular, non-stereoscopic vision)? The naive answer is that because of the length contraction, the diameter of the sphere will be contracted in the direction of the motion, but not altered orthogonal to that direction: instead of a circle, one will see an ellipse. But in this answer one forgot that 'seeing' means 'following light rays', and light rays are subject to aberration. A correct answer can be obtained as follows.

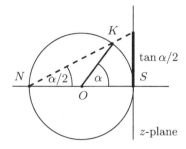

Fig. 5.4. Interpretation of the aberration formula.

When an observer Σ at O sees and locates a far away object at K, this can be understood as saying that he marks its position on a unit sphere surrounding him, using spherical coordinates α and φ (φ is the angle around the axis, not shown in the figure). This sphere can be projected onto a plane which it touches at its south pole S, see Fig. 5.4, a procedure well-known from the theory of complex functions. In that plane, the complex coordinate $z = \tan(\alpha/2) \exp(\mathrm{i}\,\varphi)$ is used. The angle $\alpha/2$ then naturally appears as the angle under which K is seen from the north pole N, and $\tan \alpha/2$ as the projection onto the plane.

The aberration formula (5.13) now tells us that the change to a moving system Σ' induces a simple scale transformation $z' = \mathrm{const}.\ z$. From the theory of holomorphic functions one knows that such a transformation always maps circles (and as their limits also straight lines) onto circles, not only at the plane, but also on the sphere. So what is a circle to Σ will remain a circle to Σ': the moving sphere has the apparent shape of a circle, only its diameter has changed. Similarly, since the projection of a straight rod onto the sphere is a part of a circle for Σ, it is again a part of a circle for Σ'.

For a general body, at a large distance, the application of the aberration formula gives again a surprising result, see Exercise 5.2. Here we shall use a different approach, see Fig. 5.5. Neglecting the z-extension, we ask which parts of a cube (square) can be seen by an observer far away at $y \to -\infty$, i.e. by parallel projection onto the plane $y = 0$. When the square is at rest, obviously only the front side BC (length l) is visible. When the square is moving with velocity v to the right, light arriving simultaneously with that from B and C is that from A emitted when A was at the position A', $l' = vl/c$ left of A. Moreover, because of the length contraction the front side BC has the length $l'' = l\sqrt{1 - v^2/c^2}$.

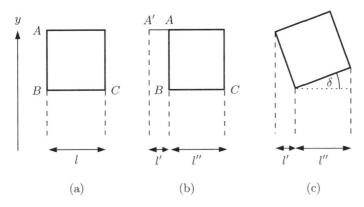

Fig. 5.5. Image of a moving square: (a) at rest, (b) moving, and (c) rotated.

The projection is the same as if the square has been turned around by the angle δ, with

$$\sin \delta = v/c. \tag{5.14}$$

Note that in both these examples a rather simplified model (monocular seeing, parallel projection) has been used. A realistic description would be much more difficult.

5.5 Reflection at a moving mirror

To treat the reflection at a moving mirror, we take the solution for a mirror at rest (in Σ'), and transform it into the system Σ in which the mirror moves with constant velocity v.

Classical optics shows that for the reflection at a mirror at rest, the two wave vectors have to satisfy (see Fig. 5.6)

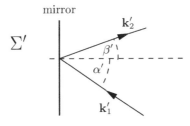

Fig. 5.6. Reflection at a mirror.

$$k_1^{n'} = (-k_1' \cos \alpha', k_1' \sin \alpha', 0, k_1'), \qquad \sin \alpha' = \sin \beta',$$
$$\text{with} \qquad\qquad\qquad (5.15)$$
$$k_2^{n'} = (k_2' \cos \beta', k_2' \sin \beta', 0, k_2'), \qquad k_1' = k_2' = k;$$

the two angles are equal, and the frequency $\omega = ck$ is not changed by the reflection.

In the system Σ, the mirror is moving (receding) with velocity $-v$ in the x-direction. Application of a Lorentz transformation to the wave vectors gives

$$
\begin{aligned}
k_1^n &= \left(\frac{-k'(\cos \alpha' + v/c)}{\sqrt{1 - v^2/c^2}}, \; k' \sin \alpha', \; 0, \; \frac{k'(1 + v[\cos \alpha']/c}{\sqrt{1 - v^2/c^2}} \right) \\
&= \quad (-k_1 \cos \alpha, \qquad k_1 \sin \alpha, \; 0, \qquad k_1), \\
k_2^n &= \left(\frac{k'(\cos \alpha' - v/c)}{\sqrt{1 - v^2/c^2}}, \; k' \sin \alpha', \; 0, \; \frac{k'(1 - [\cos \alpha']/c)}{\sqrt{1 - v^2/c^2}} \right) \\
&= \quad (-k_2 \cos \beta, \qquad k_2 \sin \beta, \; 0, \qquad k_2).
\end{aligned}
\qquad (5.16)
$$

One can easily read off that both frequencies $\nu_A = ck_A/2\pi$ are changed (Doppler shifted), and that the angles obey

$$\frac{\sin \alpha}{\sin \beta} = \frac{\nu_2}{\nu_1}. \qquad (5.17)$$

To eliminate the angles still hidden in the ν_A, one best introduces $\alpha/2$ and $\beta/2$, as in the discussion of the aberration formula. A straightforward calculation yields

$$\tan \frac{\alpha}{2} = \sqrt{\frac{1 - v/c}{1 + v/c}} \tan \frac{\alpha'}{2}, \quad \frac{\tan(\alpha/2)}{\tan(\beta/2)} = \frac{c - v}{c + v}. \qquad (5.18)$$

This result can also be understood in terms of photons which are reflected less hard when the mirror is moving in the x-direction.

5.6 Dragging of light within a fluid

In fluid with refractive index n, the velocity of light is not c, but only $V = c/n$. If that fluid is moving with velocity v, then an observer Σ' at rest will measure a velocity as obtained by applying (2.27) to V and v. The result is

$$V' = \frac{V + v}{1 + vV/c^2} = V + v\left(1 - \frac{1}{n^2}\right) + \cdots = \frac{c}{n'}. \qquad (5.19)$$

Two interpretations of this experiment due to Fizeau are possible: the fluid drags the light with it and changes its velocity, or a moving fluid has a different refractive index n'.

Exercises

5.1 Show that the wave equation operator $\square$ from (5.1) is Lorentz invariant.

5.2 Show that the aberration formula (5.11), applied to $\alpha = \pi/2$, gives just the result (5.14) with $\delta = \pi/2 - \alpha'$.

5.3 Show that a motion of the mirror in its plane does affect the reflection angles and the frequencies.

5.4 A rocket flies through a large ring (radius R) orthogonal to its orbit. How far back is the plane of the ring with respect to the rocket when the pilot sees it as exactly to his left?

6
Four-dimensional vectors and tensors

Before we can treat Electrodynamics and Fluid Mechanics, we need some more tools from tensor algebra and analysis which we will supply now. We shall do that in some detail, since they are very similar to those used in General Relativity.

6.1 Some definitions

We repeat and list here the basic definitions, with only short comments. *Lorentz transformations* are transformations of Minkowski space,

$$x^{n'} = L^{n'}{}_a\, x^a, \quad x_{m'} = L_{m'}{}^b\, x_b, \tag{6.1}$$

which satisfy

$$L_{m'}{}^b = \eta_{m'n'}\eta^{ab} L^{n'}{}_a, \quad L^{n'}{}_a L_{m'}{}^a = \delta^n_m. \tag{6.2}$$

Scalars, or tensors of rank 0, are functions φ which remain invariant under Lorentz transformations,

$$\varphi' = \varphi. \tag{6.3}$$

Examples of scalars are the rest mass m_0, the proper time τ, and the phase Φ of a wave.

Four-vectors, or tensors of rank 1, are objects with four components which transform as the components of x^n,

$$a^{n'} = L^{n'}{}_m a^m, \quad a_{m'} = L_{m'}{}^n a_n. \tag{6.4}$$

Examples are the force f^n, the four-velocity u^n, and the wave vector k^n. *Indices* can be *raised*, or *lowered*, with the help of the matrix η,

$$a_n = \eta_{nm} a^m, \quad a^n = \eta^{nm} a_m,$$
$$\eta^{nm} = \mathrm{diag}\,(1,1,1,-1) = \eta_{nm}, \quad \eta^m_n = \delta^m_n. \tag{6.5}$$

The a^n are called the *contravariant*, and the a_n the *covariant* components of the vector.

A quadratic 4×4 matrix T^{nm} is a *tensor of rank 2* if – under Lorentz transformations – it transforms like the product of two four-vectors,

$$T^{n'm'} = L^{n'}{}_a L^{m'}{}_b T^{ab}. \tag{6.6}$$

T^{nm} are the *contravariant*, $T^n{}_a = \eta_{am} T^{nm}$ the *mixed*, and $T_{ab} = \eta_{an} \eta_{bm} T^{nm}$ the *covariant components* of that tensor.

The quantities $T^{a_1}{}_{a_2 a_3}{}^{a_4 \cdots}{}_{\cdots a_n}$ are the components of a *tensor of rank n* if they transform like a contravariant vector with respect to each contravariant (upper) index, and like a covariant vector with respect to each covariant (lower) index.

Pseudotensors transform similarly to tensors, but additional signs enter into the transformation law which depend on the components of the transformation matrix $L^{n'}{}_a$. The most important example is the so-called ε-tensor ε^{abcd}. It is so defined that under interchange of two arbitrary indices its sign changes (it is completely antisymmetric), and that

$$\varepsilon^{1234} = 1 \tag{6.7}$$

always holds (i.e. also after a Lorentz transformation has been performed). Note that for a Riemannian space equation (6.7) has to be replaced by (17.21)!

If we apply a Lorentz transformation to the ε-tensor and erroneously treat it as a real tensor, we first see that $\varepsilon^{a'b'c'd'} = L^{a'}{}_n L^{b'}{}_m L^{c'}{}_p L^{d'}{}_q \varepsilon^{abcd}$ is again completely antisymmetric, and because of (6.7) we have

$$\varepsilon^{1'2'3'4'} = L^{1'}{}_n L^{2'}{}_m L^{3'}{}_p L^{4'}{}_q \varepsilon^{nmpq} = \|L^{n'}{}_a\| = \pm 1 \tag{6.8}$$

(the middle term of this equation is the rule for calculating the determinant of a matrix). We see that only for proper Lorentz transformations with $\|L^{n'}{}_a\| = +1$ do we get the correct result $\varepsilon^{1'2'3'4'} = 1$; in the general case, we have to add a factor $\|L^{n'}{}_a\|$ to the transformation law.

6.2 Tensor algebra

All rules for handling tensors have to guarantee that the results are again tensors. The proof of the tensor property will only occasionally be given in the text – we recommend the reader to fill that gap.

Addition   One adds tensors of the same rank and the same index form by adding their components,

$$T^{ab} + S^{ab} = R^{ab}. \tag{6.9}$$

Structures of the form $T^a + S^{ab}$, or $T_a{}^b + S^{ab}$, are forbidden.

Multiplication   Multiplication of the components of an nth rank tensor by those of an mth rank tensor produces an $(n + m)$th rank tensor, for example

$$S^a{}_b{}^c T^{np}{}_q = N^{acnp}{}_{bq}. \tag{6.10}$$

Contraction   Summing over a covariant and a contravariant index of a tensor gives another tensor, whose rank is reduced by 2:

$$T^{ab}{}_{nm} \rightarrow T^{ab}{}_{am} = S^b{}_m. \tag{6.11}$$

The simplest example of a contraction is the *trace* $T = T^b{}_b$ of a second-rank tensor.

Inner product, raising and lowering of indices, scalar product   The multiplication of two tensors with simultaneous contraction over indices of the two factors is called taking the inner product:

$$S^a{}_b T^{nb}{}_q = N^{an}{}_q. \tag{6.12}$$

Important examples are the raising and lowering of indices by which one can interchange covariant and contravariant components:

$$T^n = \eta^{na} T_a, \quad T^{nr}{}_{pq} = \eta_{mp} \eta^{rs} T^n{}_s{}^m{}_q. \tag{6.13}$$

Another example is the *scalar product* of two vectors:

$$a^n b_n = \eta_{nm} a^n b^m = \eta^{nm} a_n b_m, \tag{6.14}$$

with the special case

$$a^2 = a_n a^n. \tag{6.15}$$

The quotient law A structure $N^{nm\cdots}{}_{pq\cdots}$ is a tensor if, and only if, the contraction with every tensor $T^{pq\cdots}{}_{nm\cdots}$ is an invariant,

$$N^{nm\cdots}{}_{pq\cdots} T^{pq\cdots}{}_{nm\cdots} = \text{inv.} \tag{6.16}$$

The proof (for a simple example) is left to the reader, see Exercise 6.2.

Formulae for products of ε-tensors If during a calculation a product of two ε-tensors occurs, then it can be expressed in terms of Kronecker symbols $\eta^a_b = \delta^a_b$ as follows:

$$
\begin{aligned}
\varepsilon_{abcd}\varepsilon^{pqnm} =&- \delta^p_a\delta^q_b\delta^n_c\delta^m_d + \delta^q_a\delta^n_b\delta^m_c\delta^p_d - \delta^n_a\delta^m_b\delta^p_c\delta^q_d + \delta^m_a\delta^p_b\delta^q_c\delta^n_d \\
&+ \delta^q_a\delta^p_b\delta^n_c\delta^m_d - \delta^p_a\delta^n_b\delta^m_c\delta^q_d + \delta^n_a\delta^m_b\delta^p_c\delta^q_d - \delta^m_a\delta^q_b\delta^p_c\delta^n_d \\
&+ \delta^n_a\delta^q_b\delta^p_c\delta^m_d - \delta^q_a\delta^p_b\delta^n_c\delta^m_d + \delta^p_a\delta^m_b\delta^n_c\delta^q_d - \delta^m_a\delta^n_b\delta^p_c\delta^q_d \\
&+ \delta^m_a\delta^q_b\delta^n_c\delta^p_d - \delta^q_a\delta^n_b\delta^p_c\delta^m_d + \delta^n_a\delta^p_b\delta^m_c\delta^q_d - \delta^p_a\delta^m_b\delta^q_c\delta^n_d \\
&+ \delta^p_a\delta^n_b\delta^q_c\delta^m_d - \delta^n_a\delta^q_b\delta^m_c\delta^p_d + \delta^q_a\delta^m_b\delta^p_c\delta^n_d - \delta^m_a\delta^p_b\delta^n_c\delta^q_d \\
&+ \delta^p_a\delta^q_b\delta^m_c\delta^n_d - \delta^q_a\delta^m_b\delta^n_c\delta^p_d + \delta^m_a\delta^n_b\delta^p_c\delta^q_d - \delta^n_a\delta^p_b\delta^q_c\delta^m_d,
\end{aligned}
\tag{6.17}
$$

$$
\begin{aligned}
\varepsilon_{abcd}\varepsilon^{aqnm} =&- \delta^q_b\delta^n_c\delta^m_d - \delta^n_b\delta^m_c\delta^q_d - \delta^m_b\delta^q_c\delta^n_d \\
&+ \delta^q_b\delta^m_c\delta^n_d + \delta^m_b\delta^n_c\delta^q_d + \delta^n_b\delta^q_c\delta^m_d,
\end{aligned}
\tag{6.18}
$$

$$\varepsilon_{abcd}\varepsilon^{abnm} = -2(\delta^n_c\delta^m_d - \delta^m_c\delta^n_d), \tag{6.19}$$

$$\varepsilon_{abcd}\varepsilon^{abcm} = -6\delta^m_d \tag{6.20}$$

$$\varepsilon_{abcd}\varepsilon^{abcd} = -24. \tag{6.21}$$

Formula (6.17) is a consequence of (6.7) and the symmetry properties of the ε-tensor. In particular, the components of this pseudotensor only differ from zero when the four indices have different values, so that products fail to vanish only when the indices coincide pairwise. The remaining formulae follow from (6.17) by contraction, noticing that $\delta^n_n = 4$.

6.3 Symmetries of tensors

A tensor is called *symmetric* with respect to two indices n, m which are either both contravariant or both covariant if its components do not alter under the interchange of these indices:

$$T^{pq\cdots}{}_{nm\cdots} = T^{pq\cdots}{}_{mn\cdots}. \tag{6.22}$$

It is called *antisymmetric* if its sign changes under this interchange:

$$T^{pq\cdots}{}_{nm\cdots} = -T^{pq\cdots}{}_{mn\cdots} \qquad (6.23)$$

These symmetries remain preserved under Lorentz transformations.

The symmetric part with respect to the two indices a, m of an arbitrary tensor is the sum of the component and its permutation,

$$T_{(a|bc|m)} = \tfrac{1}{2}(T_{abcm} + T_{mbca}). \qquad (6.24)$$

The antisymmetric part is obtained analogously:

$$T_{[a|bc|m]} = \tfrac{1}{2}(T_{abcm} - T_{mbca}). \qquad (6.25)$$

Here we have used the convention of *Bach brackets*: round brackets denote symmetrization, square brackets antisymmetrization. Indices in the brackets not touched by the procedure are to be set between vertical lines. This convention is especially useful when one symmetrizes or antisymmetrizes with respect to several indices. One symmetrizes by forming the sum of the tensor components with all permutations of the indices and dividing by the number of permutations (when antisymmetrizing, one adds the even permutations and subtracts the odd permutations). For example

$$T_{(n_1 n_2 \cdots n_\nu)} = (T_{n_1 n_2 \cdots n_\nu} + T_{n_2 n_1 \cdots n_\nu} + \cdots + T_{n_\nu n_1 n_2 \cdots}) / \nu!, \qquad (6.26)$$

$$T_{[abc]} = (T_{abc} - T_{bac} + T_{bca} - T_{cba} + T_{cab} - T_{acb}) / 3!. \qquad (6.27)$$

One sees the advantage of this formulation when applying it to formula (6.17), which can be written simply as

$$\varepsilon_{abcd}\varepsilon^{pqnm} = -24\delta_a^{[p}\delta_b^q\delta_c^n\delta_d^{m]}. \qquad (6.28)$$

Tensors which are symmetric or antisymmetric with respect to all indices are called completely symmetric or antisymmetric, respectively.

A completely antisymmetric third rank tensor T_{abc} has exactly four essentially different components, for example $T_{123}, T_{124}, T_{134}$ and T_{234}, and therefore precisely the same number as a vector. One can exploit this fact to map it to a pseudovector T^n with the aid of the ε-tensor:

$$\varepsilon^{abcn}T_{abc} = T^n, \qquad T_{abc} = \tfrac{1}{3!}\varepsilon_{nabc}T^n, \qquad (6.29)$$

in analogy to the mapping of an antisymmetric second-rank tensor (e.g. the vector product of two vectors) to a pseudovector in three-dimensional Euclidean space.

A completely antisymmetric tensor of the fourth rank has essentially only one component, and, with the aid of the ε-tensor, can be mapped onto a pseudoscalar T,

$$\varepsilon_{abcd}T^{abcd} = T;\qquad(6.30)$$

it is proportional to the ε-tensor.

In four-dimensional Minkowski space there are no completely anti-symmetric tensors of rank higher than four.

6.4 Algebraic properties of second rank tensors

An arbitrary second rank tensor can be decomposed into its antisymmetric and symmetric parts, and the latter further into a trace-free term and a term proportional to η_{ab}:

$$T_{ab} = T_{[ab]} + T_{(ab)} = T_{[ab]} + \{T_{(ab)} - \tfrac{1}{4}T^n_n\eta_{ab}\} + \tfrac{1}{4}T^n_n\eta_{ab}. \qquad(6.31)$$

The physically important second-rank tensors often belong to one of the symmetry classes, or at least their constituent parts have different physical meanings. Thus, for example, the electromagnetic field tensor is antisymmetric, and its energy-momentum tensor is symmetric and trace-free.

Because of the particular importance of symmetric and antisymmetric tensors of second rank we will examine more closely their algebraic properties (eigenvectors, eigenvalues, normal forms).

The tensor η_{ab}   The defining equation

$$\eta_a{}^b w_b = \lambda w_a \qquad(6.32)$$

for an eigenvector is trivially satisfied (with $\lambda = 1$) for every vector w_a. Every vector is an eigenvector, η_{ab} singles out no direction in space-time.

Symmetric tensors   The eigenvector equation for a symmetric tensor T_{ab},

$$T_a{}^b w_b = \lambda w_a \longleftrightarrow (T_{ab} - \lambda\eta_{ab})w^b = 0, \qquad(6.33)$$

can be regarded as a linear system of equations for the w^b. The condition for the existence of a solution is the *secular equation*

$$\|T_{ab} - \lambda\eta_{ab}\| = 0. \qquad(6.34)$$

The eigenvalues λ can be determined from this equation. Under a Lorentz transformation the secular equation is only multiplied by the square of the determinant of the $L^{a'}{}_n$ which because of (2.19) equals unity. So the eigenvalues are invariant, and with them also the coefficients α_A of the equation

$$\lambda^4 + \alpha_1\lambda^3 + \alpha_2\lambda^2 + \alpha_3\lambda + \alpha_4 = 0, \qquad(6.35)$$

which follows from (6.34). As the α_A are derived from the components of T_{ab} by algebraic operations, they are algebraic invariants of that tensor. One can show that all other algebraic invariants can be constructed out of the α_A. The invariance property is also recognizable directly in, for example,

$$\alpha_1 = T_n^n, \quad \alpha_4 = -\|T_{ab}\|. \tag{6.36}$$

Equation (6.35) gives in general four different eigenvalues λ from which the eigenvectors can be determined. We will not go into the details here, but instead indicate an important property. Whilst in three-dimensional Euclidean space one can always transform symmetric tensors by orthogonal transformations to principle axes, that is no longer possible in Minkowski space by Lorentz transformations. This is intimately connected with the occurrence of null vectors, as can be seen by the example

$$T_{ab} = k_a k_b = \begin{pmatrix} 0 & & & \\ & 0 & & \\ & & 1 & 1 \\ & & 1 & 1 \end{pmatrix}, \quad k_a = (0, 0, 1, 1), \tag{6.37}$$

in which T_{ab} is constructed from the null vector k_a.

Antisymmetric tensors An antisymmetric tensor F_{ab} can, of course, never be brought to diagonal form, but nevertheless the question of eigenvalues and eigenvectors is again significant. For an antisymmetric tensor F_{ab}, the eigenvalue equation

$$F_a{}^b w_b = \lambda w_a \tag{6.38}$$

implies, by contraction with w^a, the relation

$$\lambda w^b w_b = 0. \tag{6.39}$$

That is, either the eigenvalue λ is zero, or the eigenvector w^a is a null vector, or both.

The antisymmetry of F_{ab} also implies that

$$\|F_{ab} - \lambda \eta_{ab}\| = \|-F_{ba} - \lambda \eta_{ba}\| = \|F_{ba} + \lambda \eta_{ba}\|, \tag{6.40}$$

and since one can interchange rows and columns in the determinant it follows that

$$\|F_{ab} - \lambda \eta_{ab}\| = \|F_{ab} + \lambda \eta_{ab}\|. \tag{6.41}$$

The secular equation $\|F_{ab} - \lambda \eta_{ab}\| = 0$ therefore transforms into itself when λ is replaced by $-\lambda$, and hence contains only even powers of λ:

$$\lambda^4 + \beta_2 \lambda^2 + \beta_4 = 0. \tag{6.42}$$

It thus furnishes only two invariants, β_2 and β_4.

Every antisymmetric tensor F_{ab} can be *dualized*; that is with the aid of the ε-tensor its associated (pseudo-) tensor can be constructed:

$$\tilde{F}^{ab} = \tfrac{1}{2}\varepsilon^{abcd}F_{cd}. \tag{6.43}$$

Because of the property (6.19) of the ε-tensor, a double application of the duality operation yields the original tensor, apart from a sign:

$$\tilde{\tilde{F}}_{nm} = \tfrac{1}{2}\varepsilon_{nmab}\tilde{F}^{ab} = \tfrac{1}{4}\varepsilon_{nmab}\varepsilon^{abcd}F_{cd} = -F_{nm}. \tag{6.44}$$

One can show that the two invariants β_2 and β_4 can be simply expressed in terms of F_{ab} and $\tilde{F}^{ab}$. In fact

$$\beta_2 = \tfrac{1}{2}F_{ab}F^{ab}, \quad \beta_4 = -\tfrac{1}{16}\left(F_{ab}\tilde{F}^{ab}\right)^2. \tag{6.45}$$

6.5 Tensor analysis

As with algebraic manipulations with tensors, we demand that differentiation of tensors (of tensor components) should result in tensors again. This demand can easily be satisfied. Take for example a scalar φ. Since it is an invariant, the same is true for its differential $d\varphi = (\partial\varphi/\partial x^n)\, dx^n$. Using the quotient law from above (or Exercise 6.2), we can conclude that

$$\varphi_{,n} = \frac{\partial\varphi}{\partial x^n} \tag{6.46}$$

is a tensor. Here and in the following we shall use the comma followed by an index as an abbreviation for a partial derivative.

Similarly we can conclude that the partial derivatives of any tensor of rank n give a tensor of rank $n+1$,

$$T^{ab}{}_{mn} \longrightarrow T^{ab}{}_{mn,p} = \frac{\partial T^{ab}{}_{mn}}{\partial x^p}. \tag{6.47}$$

The operations (6.46) and (6.47) are generalizations of the three-dimensional gradient.

In a similar way, second and higher derivatives can be constructed, as in

$$T^{ab}{}_{mn,pq} = \frac{\partial^2 T^{ab}{}_{mn}}{\partial x^p \partial x^q}. \tag{6.48}$$

Three differential operators are of particular importance in physics. The first is the generalized *divergence*, the partial differentiation of a vector or a tensor followed by a contraction,

$$T^n \longrightarrow T^n{}_{,n} = \frac{\partial T^n}{\partial x^n}. \tag{6.49}$$

The second is the generalized curl, the partial differentiation of a vector followed by antisymmetrization,

$$T_n \longrightarrow \tfrac{1}{2}\left(T_{n,m} - T_{m,n}\right), \tag{6.50}$$

and the third the generalized Δ-operator, applied to a scalar or a tensor, e.g.

$$\Box T^a = \eta^{mn} \frac{\partial^2}{\partial x^n \partial x^m} T^a = \eta^{mn} T^a{}_{,mn}. \tag{6.51}$$

Although partial derivatives always result in a covariant index, this last equation is sometimes also written as

$$\eta^{mn} \frac{\partial^2}{\partial x^n \partial x^m} T^a = T^a{}_{,n}{}^{,n}. \tag{6.52}$$

We end this chapter with a remark concerning tensor equations, and why it is so important to use only tensors in the formulation of physical laws. If one knows that an equation is a tensor equation, then it suffices to check its validity in any special coordinate system. For if $T_{ab} = S_{ab}$, or $B_{ab} \equiv T_{ab} - S_{ab} = 0$, in some special system, and if B_{ab} is a tensor, then because of the homogeneous transformation law (6.6) B_{ab} vanishes in any system.

Exercises

6.1   Show that $S^b{}_m$ as defined by (6.11) is indeed a tensor. Hint: start with the transformation law for $T^{ab}{}_{nm}$.

6.2   Show that B_n are the components of a vector if $B_n\,\mathrm{d}x^n = \mathrm{inv.}$ for any choice of the $\mathrm{d}x^n$.

6.3   Show that $\eta^n{}_m$ is a second rank tensor.

6.4   Prove that the symmetry property $T_{ab} = T_{ba}$ is preserved under Lorentz transformations.

6.5   Show by direct transformations of the differentials that $T_{,n}$ is a vector.

7

Electrodynamics in vacuo

7.1 The Maxwell equations in three-dimensional notation

It is clear from the very beginning that Classical Electrodynamics need not be changed to make it Lorentz invariant; the Lorentz transformations were found by studying Maxwell's theory! So what we have to do here is to make that invariance explicit and visible, and then to use it for answering interesting questions.

In a beginners' course on electrodynamics, Maxwell's equations are usually given in a three-dimensional notation. We therefore will start from such a formulation. Unfortunately, the people agreeing on the MSKA-system of units in electrodynamics neglected the needs of relativists, as for example in the equation

$$\text{curl } \mathbf{E} = -\partial \mathbf{B}/\partial t \qquad (7.1)$$

shows: clearly a 'c' is missing here at the ∂t! There are several ways of solving this dilemma. The easiest seems to be to set $c = 1$ by choice of units. We shall take a different approach and start from Maxwell's equations in the rational Gauss system (called 'rational' since there are no factors 4π in the field equations).

Maxwell's equations then read as follows (below we sometimes use a dot to denote the partial derivative with respect to t):

$$\text{curl } \mathbf{E} + \partial \mathbf{B}/c\partial t = 0, \qquad \text{div } \mathbf{B} = 0, \qquad (7.2)$$

$$\text{div } \mathbf{D} = \rho, \qquad \text{curl } \mathbf{H} = (\partial \mathbf{D}/\partial t + \mathbf{j})/c, \qquad (7.3)$$

$$\text{In vacuo:} \quad \mathbf{E} = \mathbf{D}, \qquad \mathbf{B} = \mathbf{H}. \qquad (7.4)$$

(Electric field $\mathbf{E}$, displacement $\mathbf{D}$, magnetic field $\mathbf{H}$, magnetic induction $\mathbf{B}$, current density $\mathbf{j}$, charge density ρ. In this chapter, from now on we will set $\mathbf{E} = \mathbf{D}$ and $\mathbf{H} = \mathbf{B}$.)

As an integrability condition for Maxwell's equations the equation of continuity (conservation of charge) has to be satisfied:

$$\text{div } \mathbf{j} + \dot{\rho} = 0. \qquad (7.5)$$

Because of the system (7.2), a scalar potential U and a vector potential $\mathbf{A}$ can be introduced by

$$\mathbf{B} = \mathrm{curl}\ \mathbf{A}, \quad \mathbf{E} = -\ \mathrm{grad}\ U - \partial \mathbf{A}/c\partial t. \tag{7.6}$$

If these potentials satisfy the Lorentz condition (Lorentz gauge)

$$\mathrm{div}\ \mathbf{A} + \partial U/c\partial t = 0, \tag{7.7}$$

the second set (7.3) of Maxwell's equations reduces to the inhomogeneous wave equations

$$\Delta \mathbf{A} - \frac{\partial^2}{c^2\partial t^2}\mathbf{A} = -\frac{1}{c}\mathbf{j}, \quad \Delta U - \frac{\partial^2}{c^2\partial t^2}U = -\rho. \tag{7.8}$$

The Lorentz force exerted on a point charge q is

$$\mathbf{F} = q\left(\mathbf{E} + \mathbf{v} \times \mathbf{B}/c\right), \tag{7.9}$$

and the Poynting vector is given by

$$\mathbf{S} = c\left(\mathbf{E} \times \mathbf{B}\right). \tag{7.10}$$

7.2 Current four-vector, four-potential, and the retarded potentials

We now start to translate the above equations into a four-dimensional language, in which we shall use only manifestly covariant equations, that is, only tensor equations. For this we have to find out, for example, which three-vectors can be upgraded to four-vectors, and how.

First we remember that a moving charge is a current; a charge at rest is a current for a moving observer. So obviously j and ρ are to be tied together. If we try $j^n = (\mathbf{j}, a\rho)$, with some unknown constant a, then the equation of continuity (7.5) shows that this constant has to equal c, and we end up with

$$j^n = (\mathbf{j}, c\rho), \quad j^n_{,n} = 0. \tag{7.11}$$

If the current has its origin in a moving charge density (convection current), then in three-dimensional notation we have $\mathbf{j} = \rho \mathbf{v}$, which obviously translates into $j^n = (\rho \mathbf{v}, \rho c)$. Using $u^n = (\mathbf{v}, c)/\sqrt{1 - v^2c^2}$, we end up with

$$j^n = \rho_0 u^n, \quad \rho = \frac{\rho_0}{\sqrt{1 - v^2c^2}}, \tag{7.12}$$

where ρ_0 is the rest-charge density.

Does (7.12) really indicate that the charge grows when it is moving? No, ρ is the charge *density*, and (7.12) – together with the volume contraction – just guarantees charge conservation.

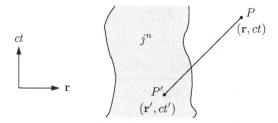

Fig. 7.1. Causal structure of the retarded potentials.

When $\mathbf{j}$ and ρ form a four-vector, then equations (7.8) indicate that the same is true for $\mathbf{A}$ and U; there is a four-potential A^n,

$$A^n = (\mathbf{A}, U), \quad A_n = (\mathbf{A}, -U), \tag{7.13}$$

which satisfies the Lorentz convention

$$A^n_{,n} = 0 \tag{7.14}$$

and the inhomogeneous wave equations

$$\Box A^n = \eta^{mn}\frac{\partial^2}{\partial x^n \partial x^m}A^n = \Delta A^n - \frac{\partial^2}{c^2\partial^2 t^2}A^n = -\frac{1}{c}j^n. \tag{7.15}$$

All these equations are invariant under the gauge transformations

$$\overline{A}_n = A_n + \varphi_{,n}, \tag{7.16}$$

where φ is an arbitrary function of all four coordinates.

The solution to the above inhomogeneous wave equations is usually given in terms of the retarded potentials as

$$A^n(\mathbf{r}, t) = \frac{1}{4\pi c}\int \frac{j^n(\mathbf{r}', t - |\mathbf{r} - \mathbf{r}'|/c)}{|\mathbf{r} - \mathbf{r}'|}\mathrm{d}^3\mathbf{r}'. \tag{7.17}$$

A manifestly covariant formulation of this relation is of course possible, but we will not pursue this here. We stress, however, that the retardation inherent in these integrals just says that electromagnetic fields propagate on light cones, since $t' = t - |\mathbf{r} - \mathbf{r}'|/c$ is the equation for the light cone between the two points P and P', cp. Fig. 7.1.

7.3 Field tensor and the Maxwell equations

Since there are no obvious candidates which might be added to the three-vectors $\mathbf{E}$ and $\mathbf{B}$ to make four-vectors out of them, the essential idea is to start with the relations between fields and potentials. Equations (7.6) show that $\mathbf{E}$ and $\mathbf{B}$ are given in terms of the derivatives of A^n. To make these equations tensor equations, we have to construct a tensor from the derivatives of A_n. There are three preferred candidates: the full set $A_{m,n}$, its symmetric part $(A_{m,n} + A_{n,m})$, and its antisymmetric part $(A_{m,n} - A_{n,m})$. Counting the number of components of these tensors, we get 16, or 10, or 6, respectively. Since the two three-vectors have altogether 6 components, the choice is clear: the *field-strength tensor*, or for short *field tensor*

$$F_{mn} = A_{n,m} - A_{m,n} \qquad (7.18)$$

incorporates the components of both $\mathbf{E}$ and $\mathbf{B}$.

When checking this assertion by calculating the components of F_{mn}, we first note that F_{mn} is of course antisymmetric. Taking now $A_n = (\mathbf{A}, -U)$ we obtain for example

$$F_{\mu 4} = A_{4,\mu} - A_{\mu,4} = -U_{,\mu} - \partial A_\mu/\partial ct = E_\mu, \quad \mu = 1, 2, 3, \qquad (7.19)$$

or, with $x^n = (x, y, z, ct)$,

$$F_{12} = A_{2,1} - A_{1,2} = \frac{\partial A_y}{\partial x} - \frac{\partial A_x}{\partial y} = B_z = B_3. \qquad (7.20)$$

Putting all the pieces together, we find that

$$F_{mn} = \begin{pmatrix} 0 & B_z & -B_y & E_x \\ -B_z & 0 & B_x & E_y \\ B_y & -B_x & 0 & E_z \\ -E_x & -E_y & -E_z & 0 \end{pmatrix}, \quad F^{mn} = \begin{pmatrix} 0 & B_z & -B_y & -E_x \\ -B_z & 0 & B_x & -E_y \\ B_y & -B_x & 0 & -E_z \\ E_x & E_y & E_z & 0 \end{pmatrix}$$

$$(7.21)$$

holds. The field tensor F_{mn} indeed comprises $\mathbf{E}$ and $\mathbf{B}$ in a single entity. The three components of $\mathbf{B}$ can be extracted from F_{mn} by

$$B^\alpha = \varepsilon^{\alpha\mu\nu} F_{\mu\nu}, \quad F^{\mu\nu} = \varepsilon^{\mu\nu\alpha} B_\alpha, \quad \alpha, \mu, \nu = 1, 2, 3, \qquad (7.22)$$

where $\varepsilon^{\alpha\mu\nu}$ are the components of the completely antisymmetric three-dimensional ε-tensor, with $\varepsilon^{123} = 1$.

There is a different, although equivalent, way of constructing a tensor out of $\mathbf{E}$ and $\mathbf{B}$: one uses the dualized field tensor $\widetilde{F}_{mn}$ instead, which is defined as

$$\widetilde{F}_{mn} = \tfrac{1}{2}\varepsilon_{mnab}F^{ab}. \qquad (7.23)$$

So for example we have

$$\begin{aligned} \widetilde{F}_{12} &= \tfrac{1}{2}\varepsilon_{12ab}F^{ab} = \tfrac{1}{2}(\varepsilon_{1243}F^{43} + \varepsilon_{1234}F^{34}) \\ &= \tfrac{1}{2}(F^{43} - F^{34}) = F^{43} = E_z, \end{aligned} \qquad (7.24)$$

and the final result is

$$\tilde{F}_{mn} = \begin{pmatrix} 0 & E_z & -E_y & -B_x \\ -E_z & 0 & E_x & -B_y \\ E_y & -E_x & 0 & -B_z \\ B_x & B_y & B_z & 0 \end{pmatrix}, \quad \tilde{F}^{mn} = \begin{pmatrix} 0 & E_z & -E_y & B_x \\ -E_z & 0 & E_x & B_y \\ E_y & -E_x & 0 & B_z \\ -B_x & -B_y & -B_z & 0 \end{pmatrix}.$$

$$(7.25)$$

Comparing (7.21) and (7.25) we see that dualization is the transition

$$\mathbf{E} \to -\mathbf{B}, \quad \mathbf{B} \to \mathbf{E}. \tag{7.26}$$

Up to now all these are rather formal definitions; we still have to see whether they fit into the framework of Maxwell's equations. The answer will be a clear 'yes'. If we start with the equation div $\mathbf{E} = \rho$, we see that we can write it as

$$\frac{\partial F^{41}}{\partial x} + \frac{\partial F^{42}}{\partial y} + \frac{\partial F^{43}}{\partial z} + \frac{\partial F^{44}}{c \partial t} = F^{4n}{}_{,n} = \rho = \frac{1}{c} j^4 \tag{7.27}$$

(note that $F^{44} = 0$!). This is the fourth component of a tensor equation; so we may guess that one full set of equations is given by

$$F^{mn}{}_{,n} = j^m / c. \tag{7.28}$$

If that is correct, we can dualize this equation: the first set (7.2) follows – in vacuo – from the second set (7.3) by a dualization (7.2), together with $j^n = 0$, and we thus get

$$\tilde{F}^{mn}{}_{,n} = 0. \tag{7.29}$$

We still have to prove that (7.28) is correct. We do this by inserting $F_{mn} = A_{n,m} - A_{m,n}$ into this equation. Exchanging the order of partial differentiations and making use of the Lorentz convention $A^n{}_{,n} = 0$ and of the inhomogeneous wave equations (7.15), we obtain

$$F^{mn}{}_{,n} = \left(A^{n,m} - A^{m,n} \right)_{,n} = \left(A^n{}_{,n} \right)^{,m} - A^{m,n}{}_{,n} = -\Box A^n = j^n / c. \tag{7.30}$$

The system (7.29) admits another representation. We can write it as

$$F_{\langle ab,c \rangle} \equiv F_{ab,c} + F_{bc,a} + F_{ca,b} = 0, \tag{7.31}$$

where $\langle abc \rangle$ denotes, as explicated, the rule for a cyclic permutation of the indices together with a summation of the components. We leave the proof of this form of (7.29) to the reader.

We close with a simple observation: if there are no charges and currents, $j^n = 0$, then Maxwell's equations are equivalent to

$$\Phi^{mn}{}_{,n} = 0, \quad \Phi^{mn} \equiv F^{mn} + \mathrm{i}\,\tilde{F}^{mn}. \tag{7.32}$$

7.4 Poynting's theorem, Lorentz force, and the energy-momentum tensor

Poynting's theorem is the energy-balance equation for the electromagnetic field. As an immediate consequence of Maxwell's equations one obtains for the Poynting vector $\mathbf{S}$ the equation

$$\text{div } \mathbf{S} + \partial w / \partial t = -\mathbf{j}\mathbf{E}, \qquad (7.33)$$

where $w = (\mathbf{E}^2 + \mathbf{B}^2)/2$ is the energy density of the field.

Here it is not so easy to identify the four-dimensional structure of this equation. The right hand side is the product of a vector and tensor components, and after some thought one sees that it can be written as $F^{4n}j_n$. So Poynting's theorem (7.33) is the fourth component of a vector equation, with $F^{mn}j_n$ on its right hand side. This reminds us of mechanics, where we learned that the energy is the fourth component of the energy-momentum four-vector, and we may guess that Poynting's theorem is just the energy-momentum balance for the electromagnetic field. This guess is supported, and justified, by the spatial components of $F^{mn}j_n$ which can be written as

$$F^{\mu n}j_n = F^{\mu\nu}j_\nu + F^{\mu 4}j_4 = \varepsilon^{\mu\nu\alpha}B_\alpha j_\nu + E^\mu c\rho \sim \mathbf{j} \times \mathbf{B} + c\rho\mathbf{E}. \quad (7.34)$$

This is – except for a factor c – the Lorentz force density, and with $\mathbf{j} = q\mathbf{v}$, $\rho = qc$ we regain (7.9).

The left hand side of (7.33) clearly is a divergence, not of a vector, but of a second-rank tensor, which is quadratic in the fields. It turns out that this tensor, called the *energy-momentum tensor*, is given by

$$T^{mn} = F^{am}F_a{}^n - \tfrac{1}{4}\eta^{mn}F_{ab}F^{ab}. \qquad (7.35)$$

It satisfies

$$T^{mn}{}_{,n} = -F^{mn}j_n/c. \qquad (7.36)$$

We leave the proof to the reader, see Exercise 7.3.

The energy-momentum tensor T^{mn} is symmetric, and has a vanishing trace, $T^n{}_n = 0$. To get more insight into its structure, we study its components. With

$$F^{ab}F_{ab} = F^{12}F_{12} + F^{21}F_{21} + F^{14}F_{14} + F^{41}F_{41} + \cdots = 2(\mathbf{B}^2 - \mathbf{E}^2), \quad (7.37)$$

we immediately get

$$T^{44} = F^{4a}F_a{}^4 + \tfrac{1}{2}(\mathbf{E}^2 - \mathbf{B}^2) = \tfrac{1}{2}(\mathbf{E}^2 + \mathbf{B}^2) = w, \qquad (7.38)$$

the T^{44}-component is the energy density. Similarly we have

$$T^{4\nu} = F^{a4}F_a{}^{\nu} = F^{\alpha 4}F_\alpha{}^{\nu} = \varepsilon^{\alpha\nu\mu}F^4{}_\alpha B_\mu = (E \times B)^{\nu} = S^{\nu}/c; \quad (7.39)$$

here the Poynting vector appears. The $T^{\alpha\beta}$-components are less easy to discuss. The result of a short calculation is

$$T^{\alpha\beta} = -E^\alpha E^\beta - B^\alpha B^\beta + \tfrac{1}{2}(E^2 + B^2)\delta^{\alpha\beta}. \quad (7.40)$$

These are, up to a sign, the components of Maxwell's stress tensor $\sigma^{\alpha\beta}$, an object often not very much discussed and seldom beloved by students. Collecting all these pieces, we have found that T^{mn} has the following structure:

$$T^{mn} = \begin{pmatrix} -\sigma^{\mu\nu} & S^\mu/c \\ \\ S^\nu/c & w \end{pmatrix}, \quad \text{with} \quad \begin{array}{l} \sigma^{\mu\nu} : \text{Maxwell's} \\ \qquad \text{stress tensor,} \\ S^\nu : \text{Poynting vector,} \\ w : \text{energy density.} \end{array} \quad (7.41)$$

In mechanics, we have learned that momentum and energy of a particle are tied together. Here the same is true for energy, Poynting vector, and stress tensor. The relation

$$T^{\alpha n}{}_{,n} = -\sigma^{\alpha\beta}{}_{,\beta} + \frac{1}{c^2}\frac{\partial S^\alpha}{\partial t} = \frac{1}{c}F^{\alpha n}j_n = \frac{1}{c}(\mathbf{j} \times \mathbf{B} + c\rho\mathbf{E})^\alpha, \quad (7.42)$$

with the Lorentz force density on the right hand side, shows that – since force equals the time derivative of momentum – S^n/c^2 is to be interpreted as the momentum density, and the equations (7.36) are the energy-momentum balance of the electromagnetic field.

We close this section with the remark that the Lorentz force density, which enters the above equation, is part of the four-vector $F^{mn}j_n$. We can therefore conclude that the force density, and not the force itself, is part of a four-vector – a fact we already used when formulating the equations of motions in Section 4.2.

7.5 The variational principle for the Maxwell equations

Most of the fundamental equation of physics can be derived from variational principles of the form

$$W = \int L\,\mathrm{d}t = \frac{1}{c}\int \mathcal{L}\,\mathrm{d}^4x = \int \mathcal{L}\,\mathrm{d}x\,\mathrm{d}y\,\mathrm{d}z\,\mathrm{d}t = \text{extremum}. \quad (7.43)$$

The action W is either given in terms of a Lagrange function L (as for example in mechanics) or in terms of a Lagrange density $\mathcal{L}$ (as will be the case here). Since the volume element is invariant under Lorentz trans-formations, the Lorentz-invariant nature of the action W is guaranteed provided $\mathcal{L}$ itself is an invariant.

There are not so many candidates for the $\mathcal{L}$ of the Maxwell field; it should be at most quadratic in the field variables F_{ab}, A_n and j_n, and the possible constitutive parts should be tensors (or vectors). Out of the antisymmetric tensor F_{mn} only two invariants can be constructed, see Section 6.4 and equation (6.45); these are

$$F_{ab}F^{ab} = 2(\mathbf{B}^2 - \mathbf{E}^2), \quad \tilde{F}_{ab}F^{ab} = 4\mathbf{EB}. \tag{7.44}$$

The other two fields just combine to $A_n j^n$.

It turns out that the correct choice, with all factors adjusted, is

$$\mathcal{L} = -\tfrac{1}{4}F^{ab}F_{ab} + A_n j^n/c, \tag{7.45}$$

which leads to

$$\delta W = \delta \frac{1}{c} \int \left[A_n j^n/c - \tfrac{1}{4}\eta^{na}\eta^{mb}(A_{n,m} - A_{m,n})(A_{a,b} - A_{b,a}) \right] \mathrm{d}^4x = 0. \tag{7.46}$$

For a given source j^n, out of all possible fields $A_n(x^i)$ only those for which W is stationary, $\delta W = 0$, are realized in nature. The variations δA_n (the difference between the 'real' A_n and those admitted for com-parison) are small and must vanish at the surface of the four-dimensional volume of integration.

We shall now derive Maxwell's equations from the above action prin-ciple. When varying the A_n, we obtain for a general $\mathcal{L}$

$$\begin{aligned}
\delta W &= \frac{1}{c} \int \delta \mathcal{L} \, \mathrm{d}^4 x = \frac{1}{c} \int \left[\frac{\partial \mathcal{L}}{\partial A_n} \delta A_n + \frac{\partial \mathcal{L}}{\partial A_{n,m}} \delta A_{n,m} \right] \mathrm{d}^4 x \\
&= \frac{1}{c} \int \left[\frac{\partial \mathcal{L}}{\partial A_n} \delta A_n - \left(\frac{\partial \mathcal{L}}{\partial A_{n,m}} \right)_{,m} \delta A_n + \left(\frac{\partial \mathcal{L}}{\partial A_{n,m}} \delta A_n \right)_{,m} \right] \mathrm{d}^4 x.
\end{aligned} \tag{7.47}$$

The last term under the integral sign is a four-dimensional divergence; using the Gauss law, we can transform it into a surface integral which vanishes since, by assumption, $\delta A_n = 0$ on the surface. The action principle then reads

$$\delta W = \frac{1}{c} \int \left[\frac{\partial \mathcal{L}}{\partial A_n} - \left(\frac{\partial \mathcal{L}}{\partial A_{n,m}} \right)_{,m} \right] \delta A_n \, \mathrm{d}^4 x = 0, \tag{7.48}$$

which for arbitrary functions δA_n can be satisfied only if

$$\frac{\delta \mathcal{L}}{\delta A_n} \equiv \frac{\partial \mathcal{L}}{\partial A_n} - \left(\frac{\partial \mathcal{L}}{\partial A_{n,m}} \right)_{,m} = 0. \tag{7.49}$$

For the $\mathcal{L}$ given by (7.45) we get $\partial\mathcal{L}/\partial A_n = j^n/c$ and

$$\frac{\partial\mathcal{L}}{\partial A_{n,m}} = \frac{\partial\mathcal{L}}{\partial F_{ab}}\frac{\partial F_{ab}}{\partial A_{n,m}} = -\frac{1}{2}F^{ab}\frac{\partial F_{ab}}{\partial A_{n,m}}$$

$$= -\frac{1}{2}F^{ab}(\delta_b^n\delta_a^m - \delta_a^n\delta_b^m) = F^{nm}, \tag{7.50}$$

and we correctly have

$$\frac{\delta\mathcal{L}}{\delta A_n} = j^n/c - F^{nm}{}_{,m} = 0. \tag{7.51}$$

Maxwell's equations follow indeed from the action principle with $\mathcal{L}$ given by (7.45). The second set (7.29) of Maxwell's equations is identically satisfied by the introduction of the four-potential A_n.

Exercises

7.1   Show that $F_{\langle ab,c\rangle} = 0$ holds by writing F_{ab} in terms of the four-potential A_n.

7.2   The field equations $\Phi^{mn}{}_{,n} = 0$ are invariant under the substitution $\Phi^{mn} \rightarrow e^{i\alpha}\Phi^{mn}$, $\alpha =$ const. ('duality rotation'). How do the three-vectors $\mathbf{E}$ and $\mathbf{B}$ transform?

7.3   Show that $T^{mn}{}_{,n} = -\frac{1}{c}F^{mn}j_n$ is true by using the definition of T^{mn}, Maxwell's equations (7.27), and equations (7.32) to substitute for $F^{ab}F_{ab,n}$.

7.4   Derive the field equations for a Lagrange density of the form $\mathcal{L} = a_1 F^{mn}F_{mn} + a_2 \widetilde{F}^{mn}F_{mn}$, $a_1, a_2 =$ const.

8

Transformation properties of electromagnetic fields: examples

So far we have treated the electromagnetic field only in a rather formal way. Here we want to give some examples of how the different quantities are transformed under the special Lorentz transformations

$$x' = \frac{x - vt}{\sqrt{1 - v^2/c^2}}, \quad ct' = \frac{ct - vx/c}{\sqrt{1 - v^2/c^2}}, \quad y' = y, \quad z' = z. \tag{8.1}$$

8.1 Current and four-potential

In case of the current density $j^n = (\mathbf{j}, c\rho)$, the transformation law (6.4) of a four-vector yields

$$j'_x = \frac{j_x - v\rho}{\sqrt{1 - v^2/c^2}}, \quad c\rho' = \frac{c\rho - vj_x/c}{\sqrt{1 - v^2/c^2}}, \quad j'_x = j_y, \quad j'_z = j_z. \tag{8.2}$$

Except for the relativistic correction $1/\sqrt{1 - v^2/c^2}$, these relations say that moving charges contribute to the current, and part of the current becomes charge if you move with it.

The four-potential undergoes the same transformation law,

$$A'_x = \frac{A_x - vU/c}{\sqrt{1 - v^2/c^2}}, \quad U' = \frac{U - vA_x/c}{\sqrt{1 - v^2/c^2}}, \quad A'_x = A_y, \quad A'_z = A_z. \tag{8.3}$$

In both cases one has to be careful when discussing physical applications. Although the transformation law

$$A^{n'} = L^{n'}_{\ a} A^a, \tag{8.4}$$

is correct, it does not show explicitly that the arguments of the functions $A^n(x^i)$ have to be transformed, too, to get the $A^{n'}$ in terms of the new coordinates $x^{n'}$:

$$A^{n'}[x^{i'}] = L^{n'}_{\ a} A^a[x^b(x^{i'})] = L^{n'}_{\ a} A^a[(L^{-1})^b_{\ i'} x^{i'}], \tag{8.5}$$

where $(L^{-1})^i_{\ b}$ is the inverse of the Lorentz transformation $L^{n'}_{\ b}$, $x^b = (L^{-1})^b_{\ i'} x^{i'}$.

Take, for example, a point charge Q at rest in the origin of the system Σ. Its potentials are

$$\mathbf{A} = 0, \quad U = \frac{Q}{4\pi r} = \frac{Q}{4\pi \sqrt{x^2 + y^2 + z^2}}. \tag{8.6}$$

For the observer Σ', this simple electric field has changed into a superposition of a magnetic field (due to the motion of the charge) and an electric field,

$$A'_x = \frac{-vU/c}{\sqrt{1 - v^2/c^2}}, \quad U' = \frac{U}{\sqrt{1 - v^2/c^2}}, \quad A'_x = 0 = A'_z. \tag{8.7}$$

In detail, the potential U is given by

$$\begin{aligned} U'(r', ct') &= \frac{Q}{4\pi\sqrt{1 - v^2/c^2}} \frac{1}{\sqrt{(x' + vt')^2/(1 - v^2/c^2) + y'^2 + z'^2}} \\ &= \frac{Q}{4\pi} \frac{1}{\sqrt{(x' + vt')^2 + (1 - v^2/c^2)(y'^2 + z'^2)}}. \end{aligned} \tag{8.8}$$

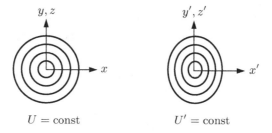

Fig. 8.1. Lines of equal potential U and U', respectively.

Instead of spheres, the lines of equal potential U are now rotationally symmetric ellipsoids, flattened in the direction of the motion (Fig. 8.1).

8.2 Field tensor and energy-momentum tensor

To get the explicit transformation laws for the field components, one has to take the transformation law

$$F^{n'm'} = L^{n'}{}_a L^{m'}{}_b F^{ab} \qquad (8.9)$$

and to insert into it the matrices

$$L^{n'}{}_a = \begin{pmatrix} \Gamma & & & -v\Gamma/c \\ 0 & 1 & 0 & 0 \\ 0 & 0 & 1 & 0 \\ -v\Gamma/c & 0 & 0 & \Gamma \end{pmatrix}, \quad F^{mn} = \begin{pmatrix} 0 & B_z & -B_y & -E_x \\ -B_z & 0 & B_x & -E_y \\ B_y & -B_x & 0 & -E_z \\ E_x & E_y & E_z & 0 \end{pmatrix},$$

$$(8.10)$$

where $\Gamma = (1 - v^2/c^2)^{-1/2}$.

To calculate the new E'_x, for example, one may first write down only the non-zero components of $L^{n'}{}_a$, and then insert F^{mn}:

$$E'_x = F^{4'1'} = L^{4'}{}_a L^{1'}{}_b F^{ab}$$

$$= L^{4'}{}_1 L^{1'}{}_1 F^{11} + L^{4'}{}_1 L^{1'}{}_4 F^{14} + L^{4'}{}_4 L^{1'}{}_1 F^{41} + L^{4'}{}_4 L^{1'}{}_4 F^{44} \qquad (8.11)$$

$$= (1 - v^2/c^2)^{-1} \left(v^2 F^{14}/c^2 + F^{41} \right) = F^{41} = E_x.$$

Proceeding in a similar way for the other components, we finally obtain

$$E'_x = E_x, \quad E'_y = \frac{E_y - vB_z/c}{\sqrt{1 - v^2/c^2}}, \quad E'_z = \frac{E_z + vB_y/c}{\sqrt{1 - v^2/c^2}},$$

$$B'_x = B_x, \quad B'_y = \frac{E_y + vB_z/c}{\sqrt{1 - v^2/c^2}}, \quad B'_z = \frac{B_z - vE_y/c}{\sqrt{1 - v^2/c^2}}.$$

$$(8.12)$$

With respect to a velocity $\mathbf{v}$ *in the direction of one of the spatial axes*, this can be written as

$$\mathbf{E}'_\perp = \frac{\mathbf{E}_\perp + (\mathbf{v} \times \mathbf{B})/c}{\sqrt{1 - v^2/c^2}}, \quad \mathbf{B}'_\perp = \frac{\mathbf{B}_\perp - (\mathbf{v} \times \mathbf{E})/c}{\sqrt{1 - v^2/c^2}}, \quad \mathbf{E}'_\| = \mathbf{E}_\|, \ \mathbf{B}'_\| = \mathbf{B}_\|.$$
$$(8.13)$$

These equations clearly show that $\mathbf{E}$ and $\mathbf{B}$ do not transform as the spatial components of four-vectors. They also show that 'electric' and 'magnetic' is a distinction which depends on the motion of the observer, as 'charge' and 'current' is. Equations (8.13) are invariant under a duality transformation $\mathbf{E} \to -\mathbf{B}, \mathbf{B} \to \mathbf{E}$.

We apply the above transformations (8.13) to the field of an infinite charged wire, with charge line density η, which extends in the x-direction. In its rest system Σ we have only $\mathbf{E}_\perp$ non-zero,

$$E_x = 0 = E_\varphi, \quad E_r = \eta/2\pi r, \quad \mathbf{B} = 0 \qquad (8.14)$$

(here φ is the angle around the x-axis, and $r^2 = y^2 + z^2$). For an observer Σ' moving along the x-axis, we obtain as the only nonzero components

$$E'_r = \frac{\eta}{2\pi r \sqrt{1 - v^2/c^2}}, \quad B'_\varphi = -\frac{v}{c} \frac{\eta}{2\pi r \sqrt{1 - v^2/c^2}}, \qquad (8.15)$$

cp. Fig. 8.2. For small velocities, the magnetic field of moving charges is rather small compared to its electric field: magnetic fields are a typical relativistic effect! That they are so big when we have a current in a metallic (non-charged) wire, with only slowly moving electrons, is due to the enormous number of these electrons.

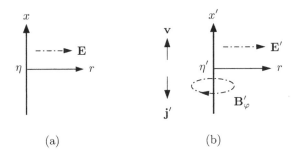

(a)   (b)

Fig. 8.2. Field of a charged wire (a) at rest and (b) moving.

This last example shows that a purely electric field becomes electric *and* magnetic for a moving observer. Can it conversely happen that an arbitrary electromagnetic field is purely electric or purely magnetic to a special observer? Since both **EB** and $\mathbf{E}^2 - \mathbf{B}^2$ are invariant, cp. (7.44), this is possible only when **EB** = 0. If $\mathbf{E}^2 - \mathbf{B}^2 \neq 0$, this condition is also sufficient. To prove this, take the directions of **E** and **B** as the y- and z-directions, respectively. The transformation law (8.12) then gives

$$E' = \frac{E - vB/c}{\sqrt{1 - v^2/c^2}}, \qquad B' = \frac{B - vE/c}{\sqrt{1 - v^2/c^2}}, \tag{8.16}$$

$$\mathbf{E} = (0, E, 0), \quad \mathbf{E}' = (0, E', 0), \quad \mathbf{B} = (0, 0, B), \quad \mathbf{B}' = (0, 0, B').$$

Depending on the relative size of E and B, one can find a velocity $v < c$ so that either $\mathbf{E}' = 0$ (for $E/B < 1$) or $\mathbf{B}' = 0$ (for $E/B > 1$); $E = B$ violates the assumption $\mathbf{E}^2 - \mathbf{B}^2 \neq 0$.

When written explicitly, the transformation law for the components of the energy momentum tensor (7.41) becomes rather clumsy; so we shall not discuss it in detail. What we want to study is the behaviour of the Poynting vector **S** and the energy density w. These components are more intimately related than one may guess. To see this, we observe that because of

$$w^2 - \mathbf{S}^2/c^2 = \tfrac{1}{4}(\mathbf{E}^2 + \mathbf{B}^2)^2 - (\mathbf{E} \times \mathbf{B})^2 = (\mathbf{EB})^2 + \tfrac{1}{4}(\mathbf{E}^2 - \mathbf{B}^2)^2. \tag{8.17}$$

$w^2 - \mathbf{S}^2/c^2 = T^{4n}T_{4n}$ is an invariant – although $(\mathbf{S}/c, w)$ is not a four-vector!

An immediate consequence of (8.17) is that the energy always dominates, $w^2 \geq \mathbf{S}^2/c^2$; only if both field invariants vanish, $\mathbf{EB} = 0 = \mathbf{E}^2 - \mathbf{B}^2$, field momentum and field energy are of equal magnitude. We will come back to those fields later in Section 9.4.

Exercises

8.1 How does the constant electric field inside a condenser look to an observer moving (a) parallel or (b) orthogonal to it?

8.2 Use equations (8.12) to show that there is a coordinate system in which **E** and **B** are parallel unless both invariants of the field tensor vanish at that point.

9

Null vectors and the algebraic properties of electromagnetic field tensors

9.1 Null tetrads and Lorentz transformations

The occurrence of null vectors is a typical effect of the indefinite metric of Minkowski space. In this chapter we shall show that null vectors can be used in a systematic way for the representation of Lorentz transformations and for the study of the electromagnetic field.

In Minkowski space the preferred coordinates are the (quasi-)Cartesian coordinates $x^n = (x, y, z, ct)$, in which we have

$$ds^2 = \eta_{ab}\,dx^a\,dx^b = dx^2 + dy^2 + dz^2 - c^2\,dt^2, \quad \eta_{ab} = \begin{pmatrix} 1 & & & \\ & 1 & & \\ & & 1 & \\ & & & -1 \end{pmatrix}. \quad (9.1)$$

Linked up with the four coordinates is a 'tetrad' of four unit vectors z_a, w_a, v_a and u_a/c, which are orthogonal to the hypersurfaces $x^a = $ const. and thus to each other,

$$
\begin{aligned}
z_a &= x_{,a} = (1,0,0,0), & v_a &= z_{,a} = (0,0,1,0), \\
w_a &= y_{,a} = (0,1,0,0), & u_a/c &= -ct_{,a} = (0,0,0,-1).
\end{aligned}
\quad (9.2)
$$

$$z^a z_a = w^a w_a = v^a v_a = -u^a u_a/c^2 = 1, \quad \text{all other products zero.} \quad (9.3)$$

These four vectors form a complete system, any vector or tensor can be expressed using their linear combinations and products, for example

$$\eta_{ab} = z_a z_b + w_a w_b + v_a v_b - u_a u_b/c^2. \quad (9.4)$$

We now want to use a tetrad of four null vectors instead of the above tetrad of one timelike and three spacelike vectors. Since there are only two *real* linearly independent null vectors in Minkowski space, which we may take as

$$k_a = (u_a/c + v_a)/\sqrt{2}, \quad l_a = (u_a/c - v_a)/\sqrt{2}, \quad (9.5)$$

we have to add two *complex* vectors

$$m_a = (z_a - \mathrm{i}\,w_a)/\sqrt{2}, \quad \overline{m}_a = (z_a + \mathrm{i}\,w_a)/\sqrt{2}, \quad (9.6)$$

which are complex conjugates of each other. The system $(k_a, l_a, m_a, \overline{m}_a)$ of four null vectors is called a *null tetrad*, or a *Sachs tetrad*, or a *Newman–Penrose* tetrad (the reader should be aware of different sign conventions

63

in the literature). Only two of the products of these null vectors are non-zero,

$$k^a l_a = -1, \quad m^a \overline{m}_a = 1, \quad \text{all other products zero.} \qquad (9.7)$$

Instead of (9.4), we now have

$$\eta_{ab} = m_a \overline{m}_b + \overline{m}_a m_b - k_a l_b - l_a k_b. \qquad (9.8)$$

The four real basic vectors are of course not uniquely defined: it is exactly the Lorentz transformations which mediate between two systems $(z_a, w_a, v_a, u_a/c)$ and $(z'_a, w'_a, v'_a, u'_a/c)$ both satisfying (9.3)–(9.4). The same is true for the system $(k_a, l_a, m_a, \overline{m}_a)$ of null vectors. To find the transformations between two such systems of null vectors explicitly, we first observe that the multiplication of k^a by some constant (or even function) A does not affect its being null; then to keep $k^a l_a$ fixed also (and k^a pointing into the future), we have to take

$$k^{a'} = Ak^a, \quad l^{a'} = A^{-1} l^a, \quad A > 0. \qquad (9.9)$$

The transformations keeping k^a fixed turn out to be

$$k^{a'} = k^a, \quad l^{a'} = l^a + B\overline{B}k^a + \overline{B}m^a + B\overline{m}^a, \quad m^{a'} = e^{i\Theta}(m^a + Bk^a),$$
$$\Theta \text{ real}, B \text{ complex}, \qquad (9.10)$$

and a special transformation keeping l^a fixed is

$$k^{a'} = k^a + E\overline{E}l^a + \overline{E}m^a + E\overline{m}^a, \quad l^{a'} = l^a, \quad m^{a'} = m^a + El^a,$$
$$E \text{ complex}. \qquad (9.11)$$

We also may simply interchange k^a and l^a,

$$k^{a'} = l^a, \quad l^{a'} = l^a. \qquad (9.12)$$

The general transformation between two null tetrads contains six real parameters and thus precisely the same number as a general Lorentz transformation. And indeed we are dealing here with a particularly simple representation of the Lorentz transformations. The parameter Θ on its own produces a rotation in the m^a-$\overline{m}^a$- (x-y-) plane. To see this, we use (9.2) and (9.6) to write $m_a = (x + iy)_{,a}/\sqrt{2}$, and infer from $m^{a'} = e^{i\Theta}m^a$ that

$$(x + iy)' = e^{i\Theta}(x + iy) \qquad (9.13)$$

holds. This splits into $x' = x \cos \Theta - y \sin \Theta$, $y' = x \sin \Theta + y \cos \Theta$. The parameter A gives a special Lorentz transformation in the z-ct-plane, and the parameters B and E describe so-called null rotations.

9.2 Self-dual bivectors and the electromagnetic field tensor

Like every tensor, the antisymmetric field tensor F_{ab} can be given in terms of the null tetrad introduced above by an expansion of the kind

$$F_{ab} = \alpha_1(k_a l_b - k_b l_a) + \alpha_2(k_a m_b - k_b m_a) + \cdots, \tag{9.14}$$

but the combinations

$$U_{ab} = \overline{m}_a l_b - \overline{m}_b l_a, \quad V_{ab} = k_a m_b - k_b m_a,$$
$$W_{ab} = m_a \overline{m}_b - m_b \overline{m}_a - k_a l_b + k_b l_a \tag{9.15}$$

play a special rôle. These antisymmetric tensors, or bivectors, are self-dual in the sense that under dualization they reproduce themselves up to a factor $-\,\mathrm{i}$:

$$\tilde{U}_{ab} = \tfrac{1}{2}\varepsilon_{abpq}U^{pq} = -\,\mathrm{i}\,U_{ab}, \quad \tilde{V}_{ab} = -\,\mathrm{i}\,V_{ab}, \quad \tilde{W}_{ab} = -\,\mathrm{i}\,W_{ab}. \tag{9.16}$$

One can verify these relations by using, for example, the fact that the antisymmetric tensor $\varepsilon_{abpq}l^p m^q$ yields zero upon contraction with l_a or m_a, and therefore because of (9.3) must be constructed from the vectors l_a and m_a, and that for the vectors of the null tetrad (verifiable with $\sqrt{2}k^a = (0,0,1,1)$, $\sqrt{2}l^a = (0,0,-1,1)$, and $\sqrt{2}m^a = (1,-\mathrm{i},0,0)$), we have

$$\varepsilon_{abpq}k^a l^b m^p \overline{m}^q = -\,\mathrm{i}. \tag{9.17}$$

Because of their definition (9.15) and the properties (9.7) of the null vectors, the self-dual bivectors have the 'scalar products'

$$W_{ab}W^{ab} = -4, \quad U_{ab}V^{ab} = 2,$$
$$W_{ab}W^{ab} = W_{ab}U^{ab} = V_{ab}V^{ab} = U_{ab}U^{ab} = 0. \tag{9.18}$$

The non-self-dual field tensor F_{ab} of a Maxwell field cannot of course be expanded in terms of the self-dual bivectors (9.15). Instead we have to use the complex field tensor

$$\Phi_{ab} = F_{ab} + \mathrm{i}\,\tilde{F}_{ab} \tag{9.19}$$

already introduced in Section 7.3, which is self-dual in the above sense,

$$\tilde{\Phi}_{ab} = -\,\mathrm{i}\,\Phi_{ab}. \tag{9.20}$$

Hence it can be expanded with respect to the bivectors U, V, and W:

$$\Phi_{ab} = \varphi_0 U_{ab} + \varphi_1 W_{ab} + \varphi_2 V_{ab}. \tag{9.21}$$

Corresponding to the six independent components of a second-rank antisymmetric tensor there occur three complex coefficients φ_i. They can be calculated from the field tensor, because of (9.18), according to

$$\varphi_0 = \tfrac{1}{2}\Phi_{ab}V^{ab} \quad = B_y - E_x + \mathrm{i}\,(E_y + B_x),$$

$$\varphi_1 = -\tfrac{1}{4}\Phi_{ab}W^{ab} = E_z - \mathrm{i}\,B_z, \tag{9.22}$$

$$\varphi_2 = \tfrac{1}{2}\Phi_{ab}U^{ab} \quad = E_x + B_y + \mathrm{i}\,(E_y - B_x).$$

9.3 The algebraic classification of electromagnetic fields

First formulation Symmetric tensors become particularly simple when one carries out a transformation to principal axes. Setting up the analogous problem for antisymmetric tensors consists of simplifying the expansion (9.21) by choice of the direction k^a of the null tetrad; that is, by adapting the null tetrad to the antisymmetric tensor under consideration.

The self-dual bivectors transform under a rotation (9.11) according to

$$W'_{ab} = W_{ab} - 2EU_{ab}, \quad U'_{ab} = U_{ab}, \quad V'_{ab} = V_{ab} - EW_{ab} + E^2 U_{ab}. \tag{9.23}$$

Hence we have for the expansion coefficients occurring in (9.21)

$$\varphi_0 = \varphi'_0 - 2E\varphi'_1 + E^2\varphi'_2, \quad \varphi_1 = \varphi'_1 - E\varphi'_2, \quad \varphi_2 = \varphi'_2. \tag{9.24}$$

We can therefore make one of the two coefficients φ_0 or φ_1 vanish by suitable choice of E, that is, by suitable choice of the new direction k^a, and thereby simplify the expansion (9.21). Since only φ_0 remains invariant under the transformations (9.10), which leave k^a fixed but alter l^a and m^a, and since we seek the most invariant choice possible, we demand that φ_0 vanishes:

$$\varphi'_0 - 2\varphi'_1 E + \varphi'_2 E^2 = 0. \tag{9.25}$$

According to the number of roots E of this equation (and taking into account a few special cases) one can divide the electromagnetic fields into two classes.

Fields for whose field tensor the inequality

$$\varphi'^2_1 - \varphi'_0\varphi'_2 \neq 0 \tag{9.26}$$

holds are called *non-degenerate*. They possess two different directions k^a for which φ_0 vanishes, since in general (9.25) has two distinct roots E. For $\varphi'_2 = 0$ only one root exists, but then l^a is one of the null directions singled out and one can again obtain $\varphi_0 = 0$ through the interchange of labels $l^a \leftrightarrow -k^a$, $U_{ab} \leftrightarrow \overline{V}_{ab}$.

Fields whose field tensor satisfies

$$\varphi'^2_1 - \varphi'_0\varphi'_2 = 0 \tag{9.27}$$

are called *degenerate* or *null* fields. They possess only one null direction k^a with $\varphi'_0 = 0$. If one has achieved $\varphi_0 = 0$, then because of (9.27) φ_1 also vanishes.

The relation

$$\varphi_1'^2 - \varphi_0' \varphi_2' = -4\Phi^{ab}\Phi_{ab} = -2(F^{ab}F_{ab} + \mathrm{i}\,\tilde{F}^{ab}F_{ab}), \tag{9.28}$$

which follows from (9.18) and (9.21), shows particularly clearly that the classification of electromagnetic fields introduced above is independent of the choice of the null tetrad and of the interpretation through the bivector expansion.

Equation (9.28) implies a simple prescription for establishing the type of an electromagnetic field: a Maxwell field is degenerate, or a null field, if and only if its two invariants vanish, that is if and only if

$$F^{ab}F_{ab} = 0 = \tilde{F}^{ab}F_{ab}. \tag{9.29}$$

Second formulation One can also translate the classification just set up into the more usual language of eigenvalue equations and eigenvectors. As one can deduce from (9.21) and (9.15), $\varphi_0 = 0$ is equivalent to

$$\Phi_a{}^b k_b = (F_a{}^b + \mathrm{i}\,\tilde{F}_a{}^b)k_b = \varphi_1 k_a. \tag{9.30}$$

Non-degenerate fields thus possess two distinct null eigenvectors k^a for which (9.30) holds, or for which

$$k_{[c}F_{a]b}k^b = 0 = k_{[c}\tilde{F}_{a]b}k^b. \tag{9.31}$$

Degenerate fields (null fields) for which φ_0 *and* φ_1 vanish possess only one null eigenvector k^a with

$$F_{ab}k^b = 0 = \tilde{F}_{ab}k^b. \tag{9.32}$$

Their field tensor has the simple structure $\Phi_{ab} = \varphi_2 V_{ab}$, that is

$$F_{ab} = k_a p_b - k_b p_a, \quad p_a k^a = 0 = p_a l^a \tag{9.33}$$

(p^a is spacelike).

9.4 The physical interpretation of electromagnetic null fields

The simplest example of an electromagnetic null field is a plane wave,

$$\begin{aligned} A_n &= \mathrm{Re}\,[\hat{p}_n\, \mathrm{e}^{\mathrm{i}\,k_r x^r}], \quad \hat{p}_n k^n = 0, \quad k^n k_n = 0, \\ F_{nm} &= \mathrm{Re}\,[(\hat{p}_m k_n - \hat{p}_n k_m)\mathrm{i}\,\mathrm{e}^{\mathrm{i}\,k_r x^r}]. \end{aligned} \tag{9.34}$$

One easily verifies that the necessary and sufficient condition (9.29) for the vanishing of the invariants is satisfied.

Plane waves (null fields) also occur as far fields of isolated charge and current distributions. If one starts from the representation (7.17) of the four-potential, i.e. from

$$A^n(\mathbf{r}, t) = \frac{1}{4\pi c} \int \frac{j^n(\mathbf{r}', t - |\mathbf{r} - \mathbf{r}'|/c)}{|\mathbf{r} - \mathbf{r}'|} \mathrm{d}^3 r', \qquad (9.35)$$

and expands the corresponding field tensor in powers of $1/r$,

$$F_{nm} = \overset{1}{F}_{nm}/r + \overset{2}{F}_{nm}/r^2 + \cdots, \qquad r^2 = \mathbf{r}^2, \qquad (9.36)$$

then one sees that $\overset{1}{F}_{nm}$ has the structure

$$\overset{1}{F}_{nm} = p_m k_n - p_n k_m, \qquad (9.37)$$

with

$$p_m = \partial\left[\int j^n(\mathbf{r}', t - |\mathbf{r} - \mathbf{r}'|/c)\mathrm{d}^3 r'\right]/4\pi c \partial t,$$

$$k_n = -(t - |\mathbf{r} - \mathbf{r}'|/c)_{,n} = \left(\frac{x_n - x_n'}{c|\mathbf{r} - \mathbf{r}'|}, -\frac{1}{c}\right) \approx (x_n/rc, -1/c), \qquad (9.38)$$

$$k^n k_n = 0, \qquad A^n_{,n} \approx p^n k_n = 0.$$

A related example is the far field of an accelerated charged particle, see Section 10.4 and Exercise 10.1.

The energy-momentum tensor (7.35) of a general null field has the simple form

$$T^{mn} = F^{am} F_a{}^n = p_a p^a k^m k^n = \lambda^2 k^m k^n. \qquad (9.39)$$

Between the energy flux density (Poynting vector) $S^\alpha = cT^{4\alpha}$ and the energy density $w = T^{44}$ the relation $|S^\alpha| = wc$ holds; the energy flux density is as large as if the whole field energy moves with the velocity of light: electromagnetic null fields are pure radiation fields.

Exercises

9.1   How is the parameter A in (9.9) related to the velocity between the two systems?

9.2   Show that Φ_{ab} as defined in (9.19) is in fact self-dual.

9.3   Show that degenerate fields are characterized by $(\mathbf{B} + \mathrm{i}\,\mathbf{E})^2 = 0$.

9.4   Show that the energy-momentum tensor can be written as $T^{mn} = \frac{1}{2}\Phi^{am}\bar{\Phi}_a{}^n$.

10
Charged point particles and their field

10.1 The equations of motion of charged test particles

In (non-relativistic) electrodynamics, the equation of motion for a particle of mass m and charge e reads

$$m\ddot{\mathbf{r}} = \frac{\mathrm{d}(m\mathbf{v})}{\mathrm{d}t} = \mathbf{F} = e\left(\mathbf{E} + \frac{\mathbf{v}}{c} \times \mathbf{B}\right). \qquad (10.1)$$

The particle is a test particle in that the electromagnetic field generated by the particle itself does not act back. From the considerations on relativistic mechanics and the Lorentz force in Sections 4.2 and 7.4, it is obvious that the relativistic equation of motion reads

$$m_0 \frac{\mathrm{d}^2 x^a}{\mathrm{d}\tau^2} = m_0 \frac{\mathrm{d}u^a}{\mathrm{d}\tau} = f^a = \frac{e}{c} F^{ab} u_b. \qquad (10.2)$$

Its constitutive parts are invariants (m_0, τ, e), four-vectors (f^a, u^a), and a tensor (F^{ab}).

As in Section 4.2, we split this equation into its spatial part and the rest. With (7.34) and (4.11) we immediately obtain

$$\frac{\mathrm{d}}{\mathrm{d}t}(m\mathbf{v}) = e\left(\mathbf{E} + \frac{\mathbf{v}}{c} \times \mathbf{B}\right), \quad m = \frac{m_0}{\sqrt{1 - v^2/c^2}}, \qquad (10.3)$$

which differs from (10.1) only in the velocity-dependence of m. A redefinition of e, which one may have expected, is not necessary. For the timelike component we get from (4.15) and $\mathbf{v}e(\mathbf{E} + \mathbf{v} \times \mathbf{B}/c) = e\mathbf{E}\mathbf{v}$

$$\mathrm{d}(mc^2)/\mathrm{d}t = e\mathbf{E}\mathbf{v}. \qquad (10.4)$$

The magnetic field does not change the particle's energy.

As an example we consider the motion of a particle in the x-y-plane under the influence of a constant electric field $\mathbf{E} = (E, 0, 0)$, with the initial conditions $mv_x = 0$, $mv_y = p_0$ at $t = 0$. From the equations of motion

$$\mathrm{d}(mv_x)/\mathrm{d}t = eE, \quad \mathrm{d}(mv_y)/\mathrm{d}t = 0 \qquad (10.5)$$

one immediately gets

$$mv_x = eEt, \quad mv_y = p_0. \qquad (10.6)$$

Instead of solving the timelike component of the equations of motion, we use the identity $m^2 v^2 - m^2 c^2 = -m_o^2 c^2$, cp. (4.19), which gives with

$$m^2 c^2 = (eEt)^2 + A^2, \quad A^2 \equiv m_0^2 c^2 + p_0^2, \qquad (10.7)$$

the time-dependence of m. With that (10.6) now reads

$$v_x = \frac{dx}{dt} = \frac{ecEt}{\sqrt{(eEt)^2 + A^2}}, \quad v_y = \frac{dy}{dt} = \frac{p_0 c}{\sqrt{(eEt)^2 + A^2}}, \qquad (10.8)$$

and we can integrate it by

$$x - x_0 = \frac{c}{eE}\sqrt{(eEt)^2 + A^2}, \quad y - y_0 = \frac{p_0 c}{eE} \operatorname{arc\,sinh} (eEt/A). \quad (10.9)$$

To see the geometric form of this orbit, we eliminate t and get

$$x - x_0 = \frac{cA}{eE} \cosh \left(\frac{eE(y - y_0)}{p_0 c} \right). \qquad (10.10)$$

The orbit is a catenary. Its non-relativistic limit ($p_0 = m_0 v_0$, $A = m_0 c$, $\cosh \alpha = 1 + \alpha^2/2$) is the parabola

$$x - x_0 = \text{const.} + eE(y - y_0)^2/2m_0 v_0^2. \qquad (10.11)$$

For $p_0 = 0$ (no initial velocity in the y-direction) we regain from (10.9) the hyperbolic motion of Section 4.3.

10.2 The variational principle for charged particles

As for the Maxwell equations in Section 7.5 we need invariants to construct the Lagrange function $\widehat{L}$ for a variational principle of the form typically used in mechanics,

$$W = \int \widehat{L}\, d\tau = \int L\, dt = \text{extremum}. \qquad (10.12)$$

$\widehat{L}$ should have the dimension of an energy. Out of the combinations one can build from the entities (m_0, x^n, u^n) describing the motion of the particle, only $m_0 u^n u_n = -m_0 c^2$ meets this condition, and if we admit for a Maxwell field, $u^n A_n$ is the only additional candidate. So we make the ansatz

$$W = \int (-m_0 c^2 + e A_n u^n/c)d\tau = \int [-m_0 c^2 \sqrt{1 - v^2/c^2} + e\mathbf{A}\mathbf{v}/c - eU]\, dt. \qquad (10.13)$$

In the non-relativistic limit $v \ll c$, the term $-m_0 c^2 \sqrt{1 - v^2/c^2}$ just gives – up to a constant – the kinetic energy $m_0 v^2/2$.

To prove that this Lagrange function L is correct, we have to show that the Euler–Lagrange equations

$$\frac{\mathrm{d}}{\mathrm{d}t}\frac{\partial L}{\partial \dot{x}^\alpha} - \frac{\partial L}{\partial x^\alpha} = 0 \qquad (10.14)$$

give the equations of motion (10.3). This is easily done, since – with $v^2 = \dot{x}^\alpha \dot{x}_\alpha$ – one gets

$$\frac{\partial L}{\partial \dot{x}^\alpha} = \frac{m_0 \dot{x}_\alpha}{\sqrt{1 - v^2/c^2}} + \frac{e}{c}A_\alpha = mv_\alpha + \frac{e}{c}A_\alpha, \qquad \frac{\partial L}{\partial x^\alpha} = \frac{e}{c}A_{\beta,\alpha}\dot{x}^\beta - eU_{,\alpha}, \qquad (10.15)$$

and from this

$$\frac{\mathrm{d}}{\mathrm{d}t}\frac{\partial L}{\partial \dot{x}^\alpha} - \frac{\partial L}{\partial x^\alpha} = \frac{\mathrm{d}}{\mathrm{d}t}(mv_\alpha) + \frac{e}{c}\frac{\partial A_\alpha}{\partial t} + eU_{,\alpha} + \frac{e}{c}(A_{\alpha,\beta} - A_{\beta,\alpha})v^\beta = 0, \qquad (10.16)$$

which agrees with (10.3).

From the Maxwell field F_{nm}, the four-potential A_n is determined only up to gauge transformations (7.16), i.e. up to

$$\overline{A}_n = A_n + \varphi_{,n}; \qquad (10.17)$$

the potential itself does not have an immediate physical meaning. How does it happen that it appears in the action W? If we calculate the change in W induced by such a gauge transformation, we obtain

$$c(\overline{W} - W) = \int_1^2 e\varphi_{,n}u^n \,\mathrm{d}\tau = \int_1^2 e\varphi_{,n}\frac{\mathrm{d}x^n}{\mathrm{d}\tau}\,\mathrm{d}\tau = \int_1^2 e\varphi_{,n}\,\mathrm{d}x^n = \int_1^2 e\,\mathrm{d}\varphi$$

$$= e\varphi\Big|_1^2 = \text{const.} \qquad (10.18)$$

Although W changes, it changes only by a constant which does not give any contribution to the variation.

So far we have encountered two variational principles, one for the Maxwell field in Section 7.5, and one for a charged particle just above. An inspection of both actions shows that for a charged point particle, where we can set

$$\int A_n j^n \,\mathrm{d}^4 x = \int eA_n u^n \,\mathrm{d}\tau, \qquad (10.19)$$

they jointly can be written as

$$\begin{aligned} W &= -\tfrac{1}{4}c^{-1}\int F_{nm}F^{nm}\,\mathrm{d}^4 x + ec^{-1}\int A_n u^n \,\mathrm{d}\tau - \int m_0 c^2 \,\mathrm{d}\tau \\ &= \int L_{\text{Maxw}}\,\mathrm{d}^4 x + \int L_{\text{Interaction}}\,\mathrm{d}\tau + \int \widehat{L}_{\text{Mech}}\,\mathrm{d}\tau. \end{aligned} \qquad (10.20)$$

The source-free Maxwell field, and the force-free particle motion, are described by L_{Maxw} and $\widehat{L}_{\text{Mech}}$, respectively. The interaction works on both sides. This kind of approach for describing interaction by adding a term to the action is quite common in theoretical physics.

10.3 Canonical equations

From the Lagrange function

$$L = -m_0 c^2 \sqrt{1 - \dot{x}^\alpha \dot{x}_\alpha / c^2} + e\mathbf{A}\mathbf{v}/c - eU \qquad (10.21)$$

we construct the Hamilton function H by first calculating the canonical momenta $\widehat{p}_\alpha$,

$$\widehat{p}_\alpha = \frac{\partial L}{\partial \dot{x}^\alpha} = mv_\alpha + \frac{e}{c} A_\alpha, \qquad (10.22)$$

cp. (10.15). These canonical momenta $\widehat{p}_\alpha$ are different from the usual momenta $p_\alpha = mv_\alpha$ (so that we used a different symbol). For the Hamilton function $H = \dot{x}^\alpha \widehat{p}_\alpha - L$ we obtain, after a short calculation,

$$H = mc^2 + eU. \qquad (10.23)$$

We still have to substitute the velocities hidden in m by the canonical momenta. For this we make use of

$$m^2 \mathbf{v}^2 - m^2 c^2 = (\widehat{\mathbf{p}} - e\mathbf{A}/c)^2 - m^2 c^2 = -m_0^2 c^2, \qquad (10.24)$$

and arrive at

$$(H - eU)^2/c^2 = (\widehat{\mathbf{p}} - e\mathbf{A}/c)^2 + m_0^2 c^2. \qquad (10.25)$$

This relation can be read as saying that there is a four-vector $\widehat{p}_n = (\widehat{\mathbf{p}}, -H/c)$ with $\widehat{p}^n \widehat{p}_n = -m_0^2 c^2$. This indicates that there may be a truly four-dimensional canonical formalism, with the usual Hamilton function as a fourth component of a vector, and a still unknown generalized Hamilton function. Although it is in principle possible to formulate such an approach (see for example Sundermeyer 1982), we will not follow this line of thought. Rather we shall give below an example – there are not so many simple but truly relativistic problems!

Equation (10.25) can easily be solved with respect to H,

$$H = eU + \sqrt{m_0 c^4 + c^2 (\widehat{\mathbf{p}} - e\mathbf{A}/c)^2}. \qquad (10.26)$$

For completeness, we add the system of canonical equations, which reads

$$\frac{\mathrm{d}}{\mathrm{d}t} x^\alpha = \frac{\partial H}{\partial \widehat{p}^\alpha}, \quad \frac{\mathrm{d}}{\mathrm{d}t} \widehat{p}_\alpha = -\frac{\partial H}{\partial x^\alpha}, \quad \frac{\mathrm{d}H}{\mathrm{d}t} = \frac{\partial H}{\partial t}. \qquad (10.27)$$

We now want to apply this formalism to the motion of a charge (charge e, rest mass m_0) in a plane electromagnetic wave, see Fig. 10.1. We assume that the charge is at rest at the beginning, and that the linearly polarized plane wave hits the charge at $t = 0$,

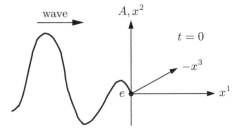

Fig. 10.1. Motion of a charge in a plane electromagnetic wave.

$$A^n = \left(0, A(x^1 - ct), 0, 0\right) \quad \leftrightarrow \quad \mathbf{E} = (0, A', 0), \quad \mathbf{B} = (0, 0, A'), \tag{10.28}$$

at $t = 0$: $x^1 = x^2 = x^3 = 0 = \dot{x}^1 = \dot{x}^2 = \dot{x}^3$. $A = 0$ for $x^1 - ct \geq 0$.

The Hamilton function for this example is

$$H = \sqrt{m_0 c^4 + c^2 \widehat{p}_1^2 + c^2 (\widehat{p}_2 - eA/c)^2 + c^2 \widehat{p}_3^2} = m(\widehat{p}_\alpha, A)c^2, \tag{10.29}$$

and the canonical equations read

$$\widehat{p}_1 = m\dot{x}_1, \quad \widehat{p}_2 = m\dot{x}_2 + eA/c, \quad \widehat{p}_3 = m\dot{x}_3,$$

$$\frac{d\widehat{p}_1}{dt} = \frac{(\widehat{p}_2 - eA/c)ecA'}{H} = \frac{1}{c}\frac{dH}{dt}, \quad \frac{d\widehat{p}_2}{dt} = 0 = \frac{d\widehat{p}_3}{dt}. \tag{10.30}$$

Making use of the initial conditions, one immediately gets

$$\widehat{p}_2 = 0, \quad \widehat{p}_3 = 0 \quad \rightarrow \quad H = \sqrt{m_0 c^4 + c^2 \widehat{p}_1^2 + e^2 A^2}. \tag{10.31}$$

Combining the second and third equations of (10.30) gives

$$d(H - c\widehat{p}_1)/dt = 0 \quad \rightarrow \quad H - c\widehat{p}_1 = m_0 c^2, \tag{10.32}$$

and together with (10.31)

$$\widehat{p}_1 = \frac{e^2 A^2}{2m_0 c^3}, \quad H = \frac{e^2 A^2}{2m_0 c^2} + m_0 c^2 = mc^2. \tag{10.33}$$

We still have to integrate the first system of (10.30), that is

$$m\dot{x}_1 = m_0 \frac{dx^1}{d\tau} = \widehat{p}_1 = \frac{e^2 A^2}{2m_0 c^3}, \quad m_0 \frac{dx^2}{d\tau} = \frac{e}{c} A, \quad m_0 \frac{dx^3}{d\tau} = 0. \tag{10.34}$$

Since

$$m_0 \, dx_1/d\tau = H/c - m_0 c = mc - m_0 c = m_0 c \, dt/d\tau - m_0 c \tag{10.35}$$

we have $d(x_1 - ct)/d\tau = -c$ and from this

$$c\tau = ct - x_1, \quad A = A(x^1 - ct) = A(\tau). \tag{10.36}$$

The potential A just depends on the proper time of the charge, and that makes the final integration easy:

$$x_1(\tau) = \frac{e^2}{2m_0^2 c^3} \int_0^{\tau} A^2(-c\tau')\mathrm{d}\tau', \qquad t(\tau) = \tau + x_1(\tau),$$

$$x_2(\tau) = -\frac{e}{m_0 c} \int_0^{\tau} A(-c\tau')\mathrm{d}\tau', \qquad x_3(\tau) = 0.$$

(10.37)

The charge oscillates in the x_2-direction, due to the electric field. This motion induces a Lorentz force in the x_1-direction, and the charge is driven in the direction of the (Poynting vector) of the wave. When the wave has passed by ($A = A_E = $ const. for some $\tau \geq \tau_E$), the charge moves with constant (zero for $A_E = 0$) velocity in the x_1-direction.

10.4 The field of a charged particle in arbitrary motion

If one gets the problem of finding the electromagnetic field of an accelerated charged particle, the first idea may be to use the formula (7.17) for the retarded potentials. Although it is possible, by means of δ-functions, to apply this representation to the current of a point particle, there is a shorter way which we will take now.

From the retarded potentials we take the knowledge of the causal structure: only those points of the particle's world line $a^n(\tau)$ which lie on the light cone with vertex in x^a contribute to the potentials A^n at a point x^a, i.e. only those points which satisfy (see Fig. 10.2)

$$R^n R_n = [x^n - a^n(\tau)][x_n - a_n(\tau)] = 0. \qquad (10.38)$$

We now use the Lorentz invariance of Maxwell's theory by transforming – for a given point (x^n) and its counterpart $(a^n(\tau))$ – to the system

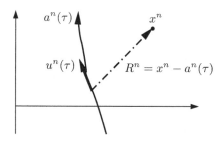

Fig. 10.2. Field of a charged particle. For explanation, see text.

in which the particle is at rest, $u^n = da^n/d\tau = (0, 0, 0, c)$. In that system, only the scalar potential U survives and is given by the well-known expression

$$U = \frac{e}{4\pi\hat{r}} = \frac{e}{4\pi(R^\alpha R_\alpha)^{1/2}}, \quad \mathbf{A} = 0. \tag{10.39}$$

Here $\hat{r}$ is the spatial distance between (x^n) and the point on the world line; because of (10.38), it can also be written as $\hat{r} = R^4$. This can be used to rewrite (10.39) as

$$A^n(x^m) = -\frac{eu^m}{4\pi R^n u_n}\bigg|_{\text{ret}}, \tag{10.40}$$

where $|_{\text{ret}}$ indicates here and later that for the particle's data (e.g. $a^n(\tau)$, $u^n(\tau)$) those on the past light cone through (x^n) have to be taken (with $R^4 = ct - a^4 > 0$).

As derived, equation (10.40) is valid only for the particle's rest system. But it clearly is a tensor equation, and so we can conclude that it is valid in any system!

The equations (10.40) are often given in a three-space notation. With $A^n = (\mathbf{A}, U)$, $u^n = (\mathbf{v}, c)/\sqrt{1 - v^2/c^2}$ and $R^n = (\mathbf{R}, R)$ they read

$$\mathbf{A}(\mathbf{r}, t) = \frac{e\mathbf{v}}{4\pi c(R - \mathbf{R}\mathbf{v}/c)}\bigg|_{\text{ret}}, \quad U = \frac{e}{4\pi(R - \mathbf{R}\mathbf{v}/c)}\bigg|_{\text{ret}} \tag{10.41}$$

and are called 'Liénard–Wiechert potentials'.

For calculating the field tensor from these potentials still another form will be used. To get this, we split the vector R^n into its parts parallel and orthogonal to the four-velocity, respectively:

$$R^a = \rho(n^a + u^a/c), \quad \rho = -R^a u_a/c|_{\text{ret}}, \quad \text{with } u^a n_a = 0, \ n^a n_a = 1 \tag{10.42}$$

(ρ is the R from above, the spatial distance between source and (x^n) in the rest system of the charge). With (10.42) we can write the four-potential as

$$A^a(x^n) = \frac{eu^a}{4\pi c\rho}\bigg|_{\text{ret}}. \tag{10.43}$$

To calculate the field tensor F_{ab} we have to differentiate the potential with respect to the x^n. This has to be done carefully: since the point (x^n) and the position τ of the particle are tied together by the light cone prescription $|_{\text{ret}}$, a shift in the x^n induces a shift in τ. In detail, we get from (10.38)

$$R_n \, dR^n = R_n(dx^n - u^n \, d\tau) = 0, \tag{10.44}$$

or – with (10.42) – $(n_n + u_n/c)\mathrm{d}x^n = (n_n + u_n/c)u^n\,\mathrm{d}\tau = -c\,\mathrm{d}\tau$, i.e.

$$c\partial\tau/\partial x^n = -(n_n + u_n/c). \tag{10.45}$$

To calculate the derivatives of the potential,

$$A_{a,b} = \left[-\frac{eu_a}{4\pi c\rho^2}\rho_{,b} + \frac{e\dot{u}_a}{4\pi c\rho}\tau_{,b}\right]\Big|_{\text{ret}}, \tag{10.46}$$

we still need the derivatives of ρ. We get them from

$$\frac{\partial\rho}{\partial x^b} = \frac{\mathrm{d}}{\mathrm{d}\tau}\big(-R^a u_a c\big)\frac{\partial\tau}{\partial x^b} = (-u^a u_a + R^a \dot{u}_a)(n_b + u_b/c)/c^2$$
$$= n_b + u_b/c + \rho n_a \dot{u}^a[n_b + u_b/c]/c^2. \tag{10.47}$$

Putting all the pieces together, and after some reshuffling, we finally obtain

$$F^{ab}(x^n) = \frac{e}{4\pi c^2\rho}\left[\frac{\dot{u}^a u^b - u^a \dot{u}^b}{c} - n^a \dot{u}^b + n^b \dot{u}^a\right.$$
$$\left.-(n^a u^b - n^b u^a)\frac{n^m \dot{u}_m}{c}\right]\Big|_{\text{ret}} + \frac{e(u^a n^b - u^b n^a)}{4\pi c\rho^2}\Big|_{\text{ret}}. \tag{10.48}$$

This formula gives the field at a point (x^n) generated from a point charge which at the retarded time has four-velocity u^a and four-acceleration $\dot{u}^a$, where ρ is the spatial distance in the particle's rest frame, and n^a the spatial vector from the particle to (x^n) in that frame.

The terms without acceleration $\dot{u}^a$ are exactly those with ρ^{-2}; they correspond to the Coulomb field of a charge, and can be transformed into a pure Coulomb field if the acceleration vanishes identically.

The terms containing the acceleration are those with a ρ^{-1}. One knows from Maxwell's theory that fields which go as r^{-1} for large r (with $r^2 = x^2 + y^2 + z^2$) usually have a Poynting vector which goes with r^{-2} so that there is a net energy flux through a sphere at large distances: these fields are radiative. The same is true here: accelerated charges radiate. To support this assertion, we calculate the Poynting vector with respect to the particle's rest frame. We do this by first calculating the electric and magnetic field vectors of the radiation field in that rest frame,

$$E^a = F^{ab}u_b/c = \frac{e}{4\pi c^2\rho}\left[n^a(n^m\dot{u}_m) - \dot{u}^a\right]\Big|_{\text{ret}},$$
$$B^a = -\widetilde{F}^{ab}u_b/c = -\tfrac{1}{2}\varepsilon^{abmn}F_{mn}u_b/c = \frac{e}{4\pi\rho c^3}\varepsilon^{abmn}n_m\dot{u}_n u_b\Big|_{\text{ret}}. \tag{10.49}$$

Note $u^b\dot{u}_b = 0 = u^b n_b$ and $\varepsilon^{abnm}u_b u^a = 0$; both four-vectors E^a and B^a

are in fact spacelike three-vectors in the space orthogonal to u^b. The same is true for the Poynting vector

$$S^b = -\varepsilon^{bpqh} E_p B_q u_h = \frac{e^2 [\dot{u}^a \dot{u}_a - (n^a \dot{u}_a)^2]}{16\pi^2 c^3 \rho^2} n^b \Big|_{\text{ret}}. \tag{10.50}$$

The Poynting vector points outwards, in the n^a-direction. The radiation field described by (10.49)–(10.50) displays some very special properties. Both E^a and B^a are orthogonal to n^a, and they are orthogonal to each other; the three vectors S^a, E^a and B^a form an pairwise orthogonal set of vectors (all orthogonal to u^a), which is a typical property of a plane electromagnetic wave. The wave emitted by a charged accelerated particle is locally a plane wave.

By using the above Poynting vector, one can calculate the total momentum and energy loss of the particle due to the radiation. The result is

$$\frac{\mathrm{d}P^a}{\mathrm{d}\tau} = \frac{2e^2}{3c^5} (\dot{u}^n \dot{u}_n) u^a, \tag{10.51}$$

which gives in the rest system for the radiated energy

$$\frac{\mathrm{d}W}{\mathrm{d}t} = \frac{2e^2}{3c^3} \dot{u}^n \dot{u}_n. \tag{10.52}$$

10.5 The equations of motion of charged particles – the self-force

As shown in the preceding section, an accelerated charge radiates and loses energy and momentum. This should cause a back-reaction to the particle and should be reflected in its equation of motion. The equation of motion should therefore be of the form

$$m_0 \frac{\mathrm{d}^2 x^n}{\mathrm{d}\tau^2} = \underset{\substack{\text{external electro-}\\\text{magnetic force}}}{\frac{e}{c} F^{nm}_{\text{ext}} u_m} + \underset{\text{self-force}}{\Gamma^n} + \underset{\substack{\text{other}\\\text{forces}}}{\tilde{f}^n} \tag{10.53}$$

(we shall neglect the 'other forces' from now on). What is this self-force Γ^n? The first guess may be that it is equal to the loss term discussed above, i.e. $\Gamma^n = \dot{P}^n = 2e^2 (\dot{u}^a \dot{u}_a) u^n / 3c^5$. But this cannot be true; every force four-vector has to be orthogonal to the four-velocity, and $\dot{P}^n$ is not: $\dot{P}^n u_n = -2e^2 (\dot{u}^a \dot{u}_a) / 3c^3$. We cannot discuss here in detail the possible remedy, but only give a plausible argument. Because of

$$u^a u_a = -c^2, \quad \dot{u}^a u_a = 0, \quad \ddot{u}^a u_a + \dot{u}^a \dot{u}_a = 0 \quad \text{etc.}, \tag{10.54}$$

.

one can simply add an additional force term proportional to $\ddot{u}^n$ to $\dot{P}^n$ to make it orthogonal to u_n. This is done by writing

$$m_0 \frac{d^2 x^n}{d\tau^2} = \frac{e}{c} F_{\text{ext}}^{nm} u_m + \frac{2e^2}{3c^3}\left[\ddot{u}^n - (\dot{u}^a \dot{u}_a)\frac{u^n}{c^2}\right]. \tag{10.55}$$

One may wonder whether this additional term does not lead to an extra irreversible loss of energy. But when writing (10.55) as

$$\frac{d}{d\tau}\left(m_0 u^n - \frac{2e^2}{3c^3}\dot{u}^n\right) = \frac{e}{c}F_{\text{ext}}^{nm}u_m - \frac{2e^2}{3c^3}(\dot{u}^a\dot{u}_a)\frac{u^n}{c^2}, \tag{10.56}$$

one sees that it is a reversible term (which may be interpreted as the momentum of the radiation field).

Equation (10.55), the so-called 'Lorentz–Abraham–Dirac equation', has been much debated. Its most unfortunate feature is the appearance of third derivatives (second derivatives of u^a), which contradicts the fundamental assumptions of Classical Mechanics.

To get an impression of the difficulties, we shall consider the simple example of a one-dimensional motion $x = x(\tau)$,

$$\frac{d}{d\tau}m_0\dot{x} = \frac{2e^2}{3c^3}\left[\dddot{x} - (\ddot{x}^2 - c^2\ddot{t}^2)\frac{\dot{x}}{c^2}\right] + K, \tag{10.57}$$

where K is an external force. If we parametrize the four-velocity by

$$\dot{x} = c\sinh q, \quad c\dot{t} = c\cosh q \longrightarrow \ddot{x} = c\dot{q}\cosh q, \quad c\ddot{t} = c\dot{q}\sinh q, \tag{10.58}$$

we obtain from (10.57)

$$\dot{q} - \tau_0\ddot{q} = \widehat{K}, \quad \tau_0 \equiv 2e^2/3m_0c^3, \quad \widehat{K} \equiv K/m_0c\cosh q. \tag{10.59}$$

For $\widehat{K} = 0$, this second order differential equation for the velocity admits the solution

$$\dot{q} = a_0\, e^{\tau/\tau_0}. \tag{10.60}$$

This is a 'run-away' solution: there is no external force, but the particle is always accelerated (it 'borrows' the energy for this from its own field).

To exclude those unphysical solutions, one has to impose some initial conditions at, for example, $\tau = \pm\infty$. If we have a force depending only on τ, $\widehat{K} = \widehat{K}(\tau)$, this can be done by taking the solution in the form

$$q(\tau) = \int_{-\infty}^{\tau} \widehat{K}(\tau')d\tau' + \int_{\tau}^{\infty} e^{(\tau-\tau')/\tau_0}\widehat{K}(\tau')d\tau', \tag{10.61}$$

where the particle is at rest ($q = 0$) for $\tau \to -\infty$. If we take a force that acts on the particle from $\tau = 0$ on, and is zero before, then we have

$$\tau \le 0: \quad q(\tau) = \int_\tau^\infty e^{(\tau-\tau')/\tau_0} \widehat{K}(\tau')d\tau', \quad \widehat{K}(\tau') = 0 \text{ for } \tau' < 0. \quad (10.62)$$

But this means that $q(\tau)$ is non-zero for $\tau \le 0$, the charge starts moving before the force acts! This can be interpreted as an advanced action, or by saying that its own field makes the point charge extended; for an electron, the time τ_0 is exactly the time the light needs to pass the classical electron radius $2e^2/3m_0c^2 \approx 2 \cdot 10^{-13}$cm. For practical purposes it may be a relief to note that at such small distances classical theory becomes obsolete in any case, and has to be replaced by quantum theory.

It has been proposed (Rohrlich 2001) to replace the Lorentz–Abraham–Dirac equation (10.55) by the equation

$$m_0 \frac{d^2 x^n}{d\tau^2} = \frac{e}{c} F_{ext}^{nm} u_m + \frac{2e^4}{3m_0^2 c^5} \left[\frac{m_0 c}{e} F_{ext,b}^{nm} u^b u_m + F_{ext}^{nm} F_{mp\,ext} u^p \right. \\ \left. + F_{ext}^{am} F_{mp\,ext} u^n u_a u^p \right]. \quad (10.63)$$

Although this equation excludes run-away solutions, it still does not exclude the pre-acceleration described above.

All these attempts show that it is not simply possible to press the degrees of freedom inherent in the point charge *and* its radiation field into a simple particle model.

Exercises

10.1 Show that both the Coulomb part and the radiative part of the field (10.48) can be written as $F^{ab} = k^a p^b - k^b p^a$. How do the vectors p^b in the two cases differ? Hint: use equation (10.38).

10.2 Verify the expression (10.50) for the Poynting vector by direct calculation of T^{ab}.

10.3 Use the result of Exercise 10.1 to calculate the invariants of the field.

10.4 Show that one obtains (10.63) by repeatedly substituting $\dot{u}^n$ by $eF_{ext}^{nm} u_m/m_0 c$ on the right hand side of (10.55), i.e. of

$$m_0 \dot{u}^n = cF_{ext}^{nm} u_m/c + 2e^2(\delta_a^n + u^n u_a/c^2)\ddot{u}^a/3c^3.$$

Further reading for Chapter 10

Rohrlich (1965).

11

Pole-dipole particles and their field

11.1 The current density

Besides a charge, particles may also have a magnetic and an electric dipole moment (and even higher moments). If we want to calculate, for example, the force on a moving magnetic dipole, we first have to find a suitable way of describing that dipole. This is done by giving the current density j^n in terms of four dimensional δ-functions and their derivatives. A thorough discussion of δ-functions requires some knowledge of the mathematical theory of distributions. What we will need, and give now, is a physicist's manual on how to deal with them.

For any function $\varphi(x^a)$ which vanishes with all its derivatives at the boundary of the domain of integration, the following rules hold:

$$\delta^4(x^m) = \delta(x)\delta(y)\delta(z)\delta(ct) = \delta^3(\mathbf{r})\delta(ct),$$

$$\int \delta^4(x^m - c^m)\varphi(x^a)\mathrm{d}x^4 = \varphi(c^m),$$

$$\int \delta^4(x^m)_{,n}\varphi(x^a)\mathrm{d}x^4 = -\int \delta^4(x^m)\varphi(x^a)_n\,\mathrm{d}x^4 = -\varphi_{,n}(0), \quad (11.1)$$

$$\int \delta(\tau - b)\varphi(\tau)\mathrm{d}\tau = \varphi(b),$$

$$\delta[f(\tau)] = \sum_i \delta(\tau - \tau_i)/|f'(\tau_i)|, \quad f(\tau_i) = 0,$$

where τ_i are simple roots of $f(\tau) = 0$, and f' is $\mathrm{d}f/\mathrm{d}\tau$.

Pole particle A simple point charge has the property that the current (charge) density is zero except at the particle's world line

$$x^m = a^m(\tau), \quad \text{with } u^m = \mathrm{d}a^m/\mathrm{d}\tau, \quad u^m u_m = -c^2. \quad (11.2)$$

This property can be ensured by using $\delta^4[x^m - a^m(\tau)]$; this distribution in itself would say that the charge is present not only at one point, but also only for one instant of time, which is obviously wrong. This is amended by performing an integration and writing

$$j^n = \int w^n(\tau)\delta^4[x^m - a^m(\tau)]\mathrm{d}\tau. \quad (11.3)$$

For a particle at rest, proper time and time coincide, one has $c\tau = a^4 = ct'$, and (11.3) gives

$$j^n = \int w^n(t')\delta^3[x^\mu - a^\mu]\delta(ct - ct')\mathrm{d}t' = w^n(t)\delta^3[x^\mu - a^\mu]/c, \quad (11.4)$$

which is the required behaviour. To get the correct expression for w^n, one could demand the Lorentz invariance of the current; we will go a different way and exploit the equation of continuity instead.

From $j^n_{,n} = 0$, (11.3) and the properties (11.1) of δ-functions one obtains

$$0 = \int w^n(\tau)\delta^4_{,n}[x^m - a^m(\tau)]d\tau\varphi(x^b)dx^4 = -\int w^n(\tau)\varphi_{,n}[a^b(\tau)]d\tau. \quad (11.5)$$

Here we encounter a problem. At first glance one may think that since the test functions $\varphi(x^a)$ are arbitrary, so are their derivatives occurring above. But this is not quite true: if φ is chosen along the world line $a^b(\tau)$ of the particle, then its derivatives in the direction of the world line are fixed, and only the derivatives orthogonal to it are free. To evaluate equation (11.5), we therefore split the vector w^n into its projections along and orthogonal to the tangent vector u^n, respectively:

$$w^n = A(\tau)u^n + b^n(\tau), \quad b^n u_n = 0. \quad (11.6)$$

With $d\varphi/d\tau = \varphi_{,n}u^n$ we then get from (11.5)

$$\begin{aligned}0 &= \int[Au^n + b^n]\varphi_{,n}\,d\tau = \int b^n\varphi_{,n}d\tau + \int A(d\varphi/d\tau)d\tau \\ &= \int b^n\varphi_{,n}\,d\tau - \int \varphi(dA/d\tau)d\tau.\end{aligned} \quad (11.7)$$

Since φ and its derivatives in the directions of b^n are arbitrary, (11.7) can be true only for $b^n = 0 = dA/d\tau$, and if we set $A = ec$, we obtain the final result that the current for a charged pole-particle necessarily is

$$j^n = ec \int u^n(\tau)\delta^4[x^m - a^m(\tau)]d\tau. \quad (11.8)$$

Pole-dipole particle For the pole-dipole particle one starts with the ansatz

$$j^n = c \int \left(w^n(\tau)\delta^4[x^m - a^m(\tau)] + w^{ns}(\tau)\delta^4_{,s}[x^m - a^m(\tau)]\right) d\tau, \quad (11.9)$$

where the functions $w^m(\tau)$ and $w^{ns}(\tau)$ are still to be determined from the equation of continuity. First of all, we can assume that $w^{ns}u_s = 0$, since if there was a component in the direction of u^s, for example $f^n u^s$, this term would lead via partial integration,

$$\begin{aligned}\int f^n u^s\{\partial\,\delta^4[x^m - a^m(\tau)]/\partial x^s\}d\tau &= -\int f^n u^s\{\partial\,\delta^4/\partial a^s\}d\tau \\ &= -\int f^n(d\delta^4/d\tau)d\tau = \int(df^n/d\tau)\delta^4 d\tau,\end{aligned} \quad (11.10)$$

to a contribution to the not yet determined vector w^n.

As before, we perform a projection in the direction of u^n, and orthogonal to it, by writing

$$w^n - dp^n/d\tau = eu^n + b^n, \quad w^{ns} = m^{ns} - u^n p^s, \tag{11.11}$$

with $b_a u^a = m^{ns} u_s = p^s u_s = 0$. If we use this in

$$0 = j^n_{,n} = c \int [w^n \delta^4_{,n} + w^{ns} \delta^4_{,sn}] d\tau \tag{11.12}$$

and perform an integration over an arbitrary test function φ, we obtain

$$\begin{aligned}
0 &= \int [-w^n \varphi_{,n} + w^{ns} \varphi_{,ns}] d\tau \\
&= \int [(-dp^n/d\tau - cu^n - b^n)\varphi_{,n} + (m^{ns} - u^n p^s)\varphi_{,ns}] d\tau \tag{11.13} \\
&= \int [c(de/d\tau)\varphi - b^n \varphi_{,n} + m^{ns} \varphi_{,ns}] d\tau.
\end{aligned}$$

In this relation, φ and all its derivatives are arbitrary functions, and we can conclude that the corresponding coefficients have to vanish:

$$e = \text{const.}, \quad b^n = 0, \quad m^{ns} = -m^{sn}. \tag{11.14}$$

The results of all this that the current density of a pole-dipole particle can be written in either of the forms

$$j^n(x) = c \int \left[(eu^n + dp^n/d\tau)\delta^4[x^m - a^m(\tau)] \right. \tag{11.15}$$
$$\left. + (m^{ns} - u^n p^s)\delta^4_{,s}[x^m - a^m(\tau)] \right] d\tau, \quad m^{ns} = -m^{sn},$$

or, using (11.10) to shift the $dp^n/d\tau$-term to the m^{ns},

$$j^n(x) = c \int \left[eu^n \delta^4[x^m - a^m(\tau)] + w^{ns}\delta^4_{,s}[x^m - a^m(\tau)] \right] d\tau, \tag{11.16}$$
$$w^{ns} = m^{sn} - u^n p^s + u^s p^n = -w^{sn},$$

where e is a constant, and m^{ns}, w^{ns}, p^n and u^n are arbitrary functions of the proper time τ.

11.2 The dipole term and its field

The antisymmetric tensor w^{ns} has six independent components, and it is a fair guess that the three components of m^{ns} represent the magnetic dipole moment, and the three components of p^n the electric dipole, cp. also the structure of the polarization tensor (12.12).

To really prove that, one could determine the electromagnetic field produced by the above current, and compare the results with the hopefully known field of dipoles. We will do this here only for the simple case of a static electric dipole at rest. For such a dipole, with $p^n = (\mathbf{p}, 0)$, located at $a^n = (0, 0, 0, ct)$, $\tau = t$, one obtains as the four-potential

$$A^n(\mathbf{r}) = \frac{1}{4\pi c} \int \frac{j^n(\mathbf{r}')\mathrm{d}^3 x'}{|\mathbf{r} - \mathbf{r}'|} = -\frac{1}{4\pi} \int u^n p^{s'} \delta^4_{,s'} \frac{\mathrm{d}^3 x'}{|\mathbf{r} - \mathbf{r}'|} \mathrm{d}\tau$$

$$= \frac{1}{4\pi} \int u^n \frac{\partial}{\partial x^{s'}} \left(\frac{1}{|\mathbf{r} - \mathbf{r}'|} \right) p^{s'} \delta^4(x^i - a^i) \mathrm{d}^3 x' \, \mathrm{d}\tau \qquad (11.17)$$

$$= \frac{1}{4\pi c} u^n p^{s'} \frac{\partial}{\partial x^{s'}} \left(\frac{1}{|\mathbf{r} - \mathbf{r}'|} \right) \Big|_{r'=0},$$

which has as its only non-zero component the well-known potential

$$U = A^4 = \mathbf{p}\mathbf{r}/4\pi r^3. \qquad (11.18)$$

In the general case, the potential of the dipole term can be calculated using the retarded Green's function of the wave equation as

$$cA^m(x^n) = \int j^m(x^{n'})G_{\mathrm{ret}}(x^n - x^{n'})\mathrm{d}^4 x',$$
$$2\pi G_{\mathrm{ret}}(x^n - x^{n'}) = \delta[(x^n - x^{n'})(x_n - x_{n'})]\big|_{\mathrm{ret}}. \qquad (11.19)$$

Inserting here the dipole part of the current, one obtains

$$A^m(x^n) = \int w^{ms'} \delta^4_{,s'}[x^m - a^m(\tau)]G_{\mathrm{ret}}(x^n - x^{n'})\mathrm{d}\tau \, \mathrm{d}^4 x'$$
$$= -\int w^{ns'}[\partial G_{\mathrm{ret}}(x^n - x^{n'})/\partial x^{s'}]\big|_{x^{n'}=a^n} \, \mathrm{d}\tau \qquad (11.20)$$
$$= \int w^{ns}[\partial G_{\mathrm{ret}}(x^n - a^n)/\partial x^s]\mathrm{d}\tau.$$

Using, as in Section 10.4 and in Fig. 10.2, the notation

$$R^n = x^n - a^n(\tau) \to \mathrm{d}R^n/\mathrm{d}\tau = -u^n, \quad \mathrm{d}(R^n R_n)/\mathrm{d}\tau = -2R^n u_n, \quad (11.21)$$

and taking into account the definition

$$\delta_{\mathrm{ret}}(R^n R_n) = \delta(\tau_{\mathrm{ret}})/2\,|R^a u_a| \qquad (11.22)$$

of the retarded part of the δ-function (cp. the rules for $\delta[f(\tau)]$), we get

$$A^m(x^n) = \frac{1}{2\pi} \int \delta'_{\mathrm{ret}}(R^n R_n)2R_s w^{ms} \, \mathrm{d}\tau$$

$$= -\frac{1}{2\pi} \int [\mathrm{d}\delta'_{\mathrm{ret}}(R^n R_n)/\mathrm{d}\tau] \frac{R_s}{R^a u_a} w^{ms} \, \mathrm{d}\tau \qquad (11.23)$$

$$= \frac{1}{2\pi} \int \frac{\mathrm{d}}{\mathrm{d}\tau} \left(\frac{R_s}{R^a u_a} w^{ms} \right) \delta_{\mathrm{ret}}(R^n R_n) \mathrm{d}\tau,$$

and finally

$$A^m(x^n) = \frac{1}{4\pi\,|R^a u_a|} \frac{\mathrm{d}}{\mathrm{d}\tau} \left(\frac{R_s}{R^a u_a} w^{ms} \right) \Big|_{\mathrm{ret}}. \qquad (11.24)$$

The compactness of this formula is a little bit misleading: when really calculating the field tensor from this potential, one will get lengthy expressions, as in the case of the Liénard–Wiechert potentials in Section 10.4.

11.3 The force exerted on moving dipoles

As shown in Section 7.4, the density $\widehat{f}^a$ of the force exerted on any current by the surrounding electromagnetic field is given by

$$\widehat{f}^a = F^{an} j_n / c. \tag{11.25}$$

To get the (three-dimensional) force $\mathbf{F}$, we have to integrate this expression over the three-dimensional volume. For the dipole part of the current (11.16), we thus have

$$(\mathbf{F})^\alpha = \int \widehat{f}^a \, \mathrm{d}^3 x / c = \int w^{ns} \delta^4_{,s} [x^m - a^m(\tau)] F^\alpha{}_n \, \mathrm{d}\tau \, \mathrm{d}^3 x. \tag{11.26}$$

Because of $\mathrm{d}\tau = \mathrm{d}t \sqrt{1 - v^2/c^2}$, and the properties of the δ-function, we can perform the four-dimensional integration and obtain finally

$$(\mathbf{F})^\alpha(\tau) = -w^{ns}(\tau) F^\alpha{}_{n,s}(\tau) c^{-1} \sqrt{1 - v(\tau)^2/c^2}. \tag{11.27}$$

Exercises

11.1   Calculate the potential of a magnetic dipole at rest from the m^{ms}-part of the current, in analogy to equation (11.17).

11.2   Apply the formalism of Green's function to the monopole part of the current and show that the Liénard–Wiechert potentials result.

11.3   Use (11.27) to find the force on a moving electric dipole in an electric field.

12

Electrodynamics in media

12.1 Field equations and constitutive relations

For a medium, some of the fundamental properties of Maxwell fields are the same as in vacuo, and we can make use of results of Chapter 7.

The first set of Maxwell's equations, $\text{curl } \mathbf{E} + \partial \mathbf{B}/c\partial t = 0$, $\text{div } \mathbf{B} = 0$, guarantees the existence of a four-potential A^n from which the field tensor F^{ab} can be derived,

$$F_{mn} = A_{n,m} - A_{m,n}, \quad \widetilde{F}^{mn}{}_{,n} = 0. \tag{12.1}$$

Similarly, the second set of Maxwell's equations, $\text{curl } \mathbf{H} = (\mathbf{j} + \partial \mathbf{D}/\partial t)/c$, $\text{div } \mathbf{D} = \rho$, is equivalent to

$$H^{mn}{}_{,n} = \frac{1}{c}j^m, \quad H^{mn} = \begin{pmatrix} 0 & H_z & -H_y & -D_x \\ -H_z & 0 & H_x & -D_y \\ H_y & -H_x & 0 & -D_z \\ D_x & D_y & D_z & 0 \end{pmatrix}. \tag{12.2}$$

These equations have to be completed by a set of constitutive relations which relate F^{mn} to H^{mn} (and j^m to F^{mn}). In many cases a *linear* relation between H^{ab} and F^{ab} is taken. In general, this is of the form

$$H_{ab} = \lambda_{ab}{}^{mn} F_{mn}, \tag{12.3}$$

with the appropriate antisymmetry conditions on the tensor λ. But to mirror the simple relations $\mathbf{D} = \varepsilon \mathbf{E}$, $\mathbf{B} = \mu \mathbf{H}$, a less ambitious approach suffices, which we shall present now.

If $u^n = u^n(x^a)$ is the four-velocity of the medium (which need not be constant throughout the medium), then one can define the electric and magnetic fields by

$$E^a = F^{ab}\frac{u_b}{c}, \quad B^a = -\widetilde{F}^{ab}\frac{u_b}{c}, \quad H^a = -\widetilde{H}^{ab}\frac{u_b}{c}, \quad D^a = H^{ab}\frac{u_b}{c}. \tag{12.4}$$

In the local rest frame $u^n = (0,0,0,c)$, the spatial components of these four-vectors are exactly the usual fields; hence the constitutive equations are

$$H^{ab}u_b = \varepsilon F^{ab}u_b, \quad \widetilde{H}^{ab}u_b = \mu^{-1}\widetilde{F}^{ab}u_b. \tag{12.5}$$

In three-dimensional notation these equations read

$$\mathbf{D} + \frac{\mathbf{v}}{c} \times \mathbf{H} = \varepsilon \left(\mathbf{E} + \frac{\mathbf{v}}{c} \times \mathbf{B}\right), \quad \mathbf{H} - \frac{\mathbf{v}}{c} \times \mathbf{D} = \frac{1}{\mu}\left(\mathbf{B} - \frac{\mathbf{v}}{c} \times \mathbf{E}\right), \tag{12.6}$$

or, after decomposition into the parts parallel and perpendicular to $\mathbf{v}$ and some reshuffling,

$$\mathbf{D}_{\parallel} = \varepsilon \mathbf{E}_{\parallel}, \quad \mathbf{D}_{\perp} = \frac{1}{1 - v^2/c^2}\left[\mathbf{E}_{\perp}\left(\varepsilon - \frac{v^2}{\mu c^2}\right) + \mathbf{v} \times \mathbf{B}\left(\frac{\varepsilon}{c} - \frac{1}{\mu c}\right)\right], \tag{12.7}$$

$$\mathbf{H}_\parallel = \frac{1}{\mu}\mathbf{B}_\parallel, \mathbf{H}_\perp = \frac{1}{1 - v^2/c^2}\left[\mathbf{B}_\perp\left(\frac{1}{\mu} - \frac{\varepsilon v^2}{c^2}\right) - \mathbf{v}\times\mathbf{E}\left(\frac{1}{\mu c} - \frac{\varepsilon}{c}\right)\right].$$

We now have to translate Ohm's law. In general, the three-dimensional current consists of two parts: a *convection* current due to charges moving with the medium, and a *conduction* current within the medium. The convection current j^n_{conv} thus is proportional to the four-velocity,

$$j^n_{\text{conv}} = \rho_0 u^n. \tag{12.8}$$

It is a timelike vector whose spatial part vanishes in the local rest system of the medium, whereas the conduction current j^n_{cond} is spacelike, with $j^4_{\text{cond}=0}$ in the local rest system. So if only these two types of current are present (there are for example no electrons moving freely through the medium), the general current can be split as follows:

$$j^n = j^n_{\text{conv}} + j^n_{\text{cond}} = -u^n(u^a j_a)/c^2 + \left[j^n + u^n(u^a j_a)/c^2\right]. \tag{12.9}$$

Ohm's law then can be written as

$$j^n + u^n(u^a j_a)/c^2 = \sigma F^{na} u_a/c. \tag{12.10}$$

As can be seen from its right hand side, the driving force for the current is the Lorentz force.

Media can be characterized by their polarization $\mathbf{P}$ and their magnetization $\mathbf{M}$ defined by

$$\mathbf{D} = \mathbf{E} + \mathbf{P}, \quad \mathbf{H} = \mathbf{B} - \mathbf{M}, \tag{12.11}$$

which can be interpreted in terms of (densities of) magnetic and electric dipoles. The corresponding tensor is the polarization tensor

$$P^{mn} = F^{mn} - H^{mn} = \begin{pmatrix} 0 & M_z & -M_y & -P_x \\ -M_z & 0 & M_x & -P_y \\ M_y & -M_x & 0 & -P_z \\ P_x & P_y & P_z & 0 \end{pmatrix}. \tag{12.12}$$

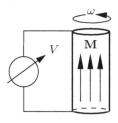

Fig. 12.1. Electric field of a rotating magnet.

A magnetized medium may produce an electric field when rotated, see Fig. 12.1. This effect may lead to enormous electric fields in a rotating neutron star.

12.2 Remarks on the matching conditions at moving surfaces

It is well known that if a boundary $f(x^n) = 0$ divides the medium into two parts with different physical properties, some care is needed to satisfy Maxwell's equations across that boundary. We shall now formulate the conditions appropriate for a moving boundary. For a boundary (locally) at rest, we have $f_{,4} = 0$. So in the general case the vector n_a normal to that boundary,

$$n_a = f_{,a}/(f_{,b}f^{,b})^{1/2}, \quad n_a n^a = 1, \quad f(x^n) = 0, \quad (12.13)$$

will be spacelike. We chose n^a to point into the interior of medium 1. If there are charges or currents in the surface $f(x^n) = 0$, characterized by j_B^a, we assume that they do not leave it,

$$j_B^a n_a = 0. \quad (12.14)$$

The boundary conditions then read

$$H^{ab} n_b \big|_1 - H^{ab} n_b \big|_2 = j_B^a/c, \quad \tilde{F}^{ab} n_b \big|_1 - \tilde{F}^{ab} n_b \big|_2 = 0, \quad (12.15)$$

where $|_A$ indicates on which side the value has to be taken.

To prove that these conditions are correct, at any point of the boundary one can take the local rest frame $n^a = (\mathbf{n}, 0)$ and show that the usual boundary conditions follow; one then gets for example $H^{4b} n_b = D^a n_a = D_{(n)}$, $j_B^4/c = \rho_B$ and thus $D_{(n)}|_1 - D_{(n)}|_2 = \rho_B$ (ρ_B is the surface density of the charge).

12.3 The energy-momentum tensor

Before we try to construct an energy-momentum tensor for electromagnetic fields in media, we shall collect the conditions an energy-momentum tensor should satisfy in general.

The energy-momentum tensor T_{ab} of a physical system generalizes the concept of the energy-momentum four-vector of mechanics, and therefore also that of energy and momentum in Classical Mechanics. For all such systems a balance equation for energy and momentum can be formulated. In particular if the system is closed, i.e. if there are no outside forces and all forces are due to the interaction of the parts within that system, energy and momentum are conserved. In terms of the energy-momentum tensor of that system this means

$$T^{mn}{}_{,n} = 0 \quad \text{for closed systems.} \tag{12.16}$$

For the Maxwell field (with sources) in vacuo this equation is not satisfied: as equation (7.36) shows, the system is not closed, energy and momentum can be transferred to the current (to the particles which carry the charge). The system is closed only for $j^n \equiv 0$.

In Classical Mechanics there is one more balance equation: that of angular momentum. It is a consequence of the equations of motion. To generalize the concept of angular momentum to fields, we remind the reader of the structure $\mathbf{r} \times \mathbf{p}$ of angular momentum: it is the antisymmetrized product of $\mathbf{r}$ with the momentum. If we write T^{mn} instead of momentum, we get

$$M^{mna} = T^{mn}x^a - T^{an}x^m \tag{12.17}$$

as the angular momentum tensor. To check whether it is conserved we determine its divergence for closed systems:

$$M^{mna}{}_{,m} = T^{mn}{}_{,n}x^a + T^{mn}\delta_n^a - T^{an}{}_{,n}x^m - T^{an}\delta_n^m = T^{ma} - T^{am}. \tag{12.18}$$

Angular momentum will be conserved only if the energy-momentum tensor is symmetric,

$$T^{ma} = T^{am} \quad \text{for closed systems.} \tag{12.19}$$

We now return to our task of finding the energy-momentum tensor for an electromagnetic in a medium. From what has been said above it is obvious that the system 'electromagnetic field' is not closed, due to the presence of the medium and its currents. But the situation is even worse: it is not clear which part of the total energy-momentum should be called 'electromagnetic'. For the system

$$\begin{array}{ccccc} \text{pure} & & \text{inter-} & & \text{elastic} \\ \text{Maxwell field} & + & \text{action} & + & \text{medium} \end{array}$$

there is no clear way of determining which of the two sides the interaction should be added to, and how it might possibly be divided. This question was much debated in the beginning of the twentieth century, when all this was not yet clear.

In any case, as can be seen from the expression $w = \mathbf{ED}/2$ for the electrostatic energy inside a medium, the energy-momentum tensor should be bilinear in the components of F^{ab} and H^{ab}. The most preferred choice

$$T^{mn} = F^{am}H_a{}^n + \tfrac{1}{4}\eta^{mn}F^{ab}H_{ab} \qquad (12.20)$$

is due to Minkowski. This tensor is not symmetric, but its trace vanishes: $T^m{}_m = 0$. To understand the physical meaning of its components we observe that if the momentum balance is included in $T^{mn}{}_{,n}$ it should be in the form $\partial(\text{momentum})/\partial t + \cdots$, which means that the $T^{\mu 4}$-components are the components of the momentum. On the other hand, by calculating the components in a local rest system we find that

$$T^{mn} = \begin{pmatrix} -\sigma^{\mu\nu} & cg^{\mu} \\ & \\ S^{\nu}/c & w \end{pmatrix}, \quad \text{with} \quad \begin{array}{l} \sigma^{\mu\nu} : \text{stress tensor}, \\ cg^{\mu} = D \times E : \text{momentum} \\ S^{\nu} : \text{Poynting vector}, \\ w : \text{energy density}. \end{array}$$

Exercises

12.1   Show that F^{ab} can be written in terms of E^a and B^a.

12.2   Give the tensor λ_{abmn} for the case of the relations (12.5). Hint: use u_n, η_{mn} and ε_{abnm}.

13
Perfect fluids and other physical theories

13.1 Perfect fluids

For the *non-relativistic* description of perfect fluids one needs the density $\mu(\mathbf{r}, t)$, the pressure $p(\mathbf{r}, t)$ and the velocity field $\mathbf{v}(\mathbf{r}, t)$. The fluid obeys Euler's equation (momentum balance)

$$\mu\frac{d\mathbf{v}}{dt} = -\,\mathbf{grad}\,p + \mathbf{f}, \qquad \frac{d}{dt} = \frac{\partial}{\partial t} + v^{\alpha}\frac{\partial}{\partial x^{\alpha}}, \qquad \alpha = 1, 2, 3, \qquad (13.1)$$

where $\mathbf{f}$ is the density of the (exterior) forces. In addition to this the mass is conserved, which is expressed in the continuity equation

$$\partial\mu/\partial t + \text{div}(\mu\mathbf{v}) = 0. \qquad (13.2)$$

The system of the two equations has to be supplemented by an equation of state, for example by

$$f(p, \mu) = 0 \quad \leftrightarrow \quad p = p(\mu). \tag{13.3}$$

In a *relativistic* theory, the four-velocity field

$$u^a(\mathbf{r}, t) = u^a(x^n) = \left(\frac{v}{\sqrt{1 - v^2/c^2}}, \frac{c}{\sqrt{1 - v^2/c^2}} \right) \tag{13.4}$$

replaces the velocity field $v(\mathbf{r}, t)$. For an arbitrary (but fixed) point P_0 it can be transformed to

$$u^a(P_0) = (0, 0, 0, c) \quad \text{(local rest system)}. \tag{13.5}$$

To find the relativistic generalization for the dynamical equations, we will not try to generalize Euler's equation. Instead we shall construct the energy-momentum tensor of the fluid and get the dynamical equations from it.

A perfect fluid is characterized by the fact that in the local rest system of a fluid element there is only an isotropic pressure; other stresses, a heat current across the border of that element I, etc. do not occur. That is to say, the energy-momentum tensor has the form

$$\overset{0}{T}{}^{mn} = \begin{pmatrix} -\text{ stress} & c \times \text{momen-} \\ \text{tensor} & \text{tum} \\ c \times \text{energy} & \text{energy} \\ \text{current} & \end{pmatrix} = \begin{pmatrix} p & & \\ & p & \\ & & p \\ & & & \mu c^2 \end{pmatrix}, \tag{13.6}$$

with the energy density proportional to the rest mass density μ. This tensor can be easily be written in a covariant form as

$$T^{mn} = p\eta^{mn} + (\mu + p/c^2)u^a u^b. \tag{13.7}$$

The special case $p = 0$ is called *dust*, or *incoherent matter*. It will later be used to model galaxies in the universe.

To understand the physical meaning of this energy-momentum tensor, we shall now inspect some of its components. Momentum density $\mathbf{g}$ and energy density w are given by

$$\begin{aligned} g^\alpha &= T^{\alpha 4}/c = (\mu + p/c^2)u^\alpha u^4/c = v^\alpha(\mu + p/c^2)/(1 - v^2/c^2), \\ w &= T^{44} = (\mu + p/c^2)u^4 u^4 - p = (\mu c^2 + pv^2/c^2)/(1 - v^2/c^2). \end{aligned} \tag{13.8}$$

Combining the two equations, we get

$$g^\alpha = v^\alpha(w + p)/c^2. \tag{13.9}$$

Not only the mass (w/c^2), but also the pressure contributes to the momentum of a fluid element.

As we shall show now, the dynamical equations of the fluid are contained in the balance equations of its energy and momentum,

$$T^{mn}{}_{,n} = f^m, \tag{13.10}$$

where f^m is the four-vector of the external force density.

The timelike component of these equations is the energy balance

$$T^{4n}{}_{,n} = T^{4\alpha}{}_{,\alpha} + T^{44}{}_{,4} = cg^\alpha_{,\alpha} + \partial w/c\partial t = f^4, \tag{13.11}$$

which can be written as

$$\text{div}[(w + p)\mathbf{v}] + \partial w/\partial t = cf^4. \tag{13.12}$$

In the non-relativistic limit $w = \mu c^2$, $p \ll \mu c^2$, $f^4/c \ll 1$, this equation yields

$$\text{div}(\mu\mathbf{v}) + \partial\mu/\partial t = 0. \tag{13.13}$$

One sees that the continuity equation (13.2) (mass conservation) is the non-relativistic limit of the energy balance, the energy density is dominated by μc^2, and mass is not changed under the action of external forces as it is in the fully relativistic description.

The spacelike components of (13.10) read

$$T_\alpha{}^n{}_{,n} = p_{,\alpha} + [(\mu + p/c^2)u_\alpha u^\beta]_{,\beta} + [(\mu + p/c^2)u_\alpha u^4]_{,4} = f_\alpha \tag{13.14}$$

or

$$p_{,\alpha} + (g_\alpha v^\beta)_{,\beta} + \partial g_\alpha/\partial t = f_\alpha. \tag{13.15}$$

Substituting the expression derived above for g_α, and making use of the energy balance equation (13.12), one obtains

$$\frac{w + p}{c^2}\frac{d\mathbf{v}}{dt} + \mathbf{grad}\,p + (f_4 + p_{,t}/c^2)\mathbf{v} = \mathbf{f}. \tag{13.16}$$

One easily recognizes that this generalizes Euler's equation which is contained here for $w = \mu c^2$, $p \ll \mu c^2$, $\mathbf{v}f_4 \ll \mathbf{f}$.

One learns from this approach to relativistic fluid mechanics that it is advantageous to start with the energy-momentum tensor if one is asking for the correct form of the dynamical equations – this is true also for other theories.

We close this section with a few remarks on the balance equation for the mass. We have seen that the continuity equation (13.2) no longer holds. If one tries to give that equation a relativistic form, one may be tempted to write it as $(\mu u^n)_{,n} = 0$. But that is not true; instead one gets, with $\mu u^n = -T^{nm}{}_{,m}/c^2$ and $u^n_{,a}u^a u_n = \dot{u}^n u_n = 0$,

$$(\mu u^n)_{,n} = -f^n u_n/c^2 - p u^n_{,n}/c^2; \qquad (13.17)$$

in general, mass will be generated (or annihilated). Note that in particle dynamics we had $p = 0 = f^n u_n$, and mass was conserved there.

If not mass, then other properties of matter may be conserved, for example the baryon number n_0 (number of baryons per volume). This can be expressed by

$$(n_0 u^n)_{,n} = 0. \qquad (13.18)$$

13.2 Other physical theories – an outlook

At the end of the chapters on Special Relativity we add a few very short remarks on its incorporation into other physical theories.

The typical notions and theorems of *Thermodynamics* are centred around temperature, heat, work and energy (first law), entropy (second law), and their foundation in Statistical Mechanics. Since we know how to deal with mechanics, Statistical Mechanics offers the easiest way for the understanding of the transformation properties of thermodynamical quantities. The entropy S is defined there by the probability of a configuration. Probability is based on counting numbers, and since numbers do not change under Lorentz transformations, the entropy S is an invariant,

$$S = \text{inv.} \qquad (13.19)$$

A second typical expression in Statistical Mechanics, governing the distribution of states, is $e^{-E/kT}$, with k being Boltzmann's constant, E the energy, and T the temperature. Not said, but tacitly assumed, is that thermodynamics usually is done in the rest frame of the medium. The function in the exponent must be an invariant, and since E is the fourth component of a four-vector, we are forced to assume that E/kT is the product of two four-vectors,

$$-E/T = p^n \Theta_n, \qquad (13.20)$$

where $p^n = (\mathbf{p}, E/c)$ is the four-momentum – which is $p^n = (0,0,0,E/c)$ in the rest frame – and Θ is the *temperature four-vector*

$$\Theta^n = u^n/T. \qquad (13.21)$$

One can now take the invariant $T = c\sqrt{-\Theta^n \Theta_n}$ as the definition of temperature, or alternatively define the temperature from the fourth component Θ^4 by $T' = \Theta^4/c$, in which case one has the transformation law $T' = T\sqrt{1 - v^2/c^2}$.

For a detailed discussion of relativistic thermodynamics we refer the reader to specialized textbooks, see for example Neugebauer (1980).

All classical and quantum field theories, and elementary particle physics, use Special Relativity as an indisputable ingredient – with one important exception: the theory of the gravitational field. If one tries to generalize Newton's theory of gravitation, one is faced with the problem of generalizing the Poisson equation

$$\Delta U = \frac{\partial^2 U}{\partial x^2} + \frac{\partial^2 U}{\partial y^2} + \frac{\partial^2 U}{\partial z^2} = U^\alpha{}_{,\alpha} = 4\pi f \mu, \quad \alpha = 1, 2, 3, \quad (13.22)$$

which tells how the Newtonian potential U (defined by its action on a mass, $m\,\mathrm{d}^2\mathbf{r}/\mathrm{d}t^2 = -m\,\mathbf{grad}\,U$) is generated by the mass distribution $\mu = \mu(\mathbf{r})$, f being the Newtonian constant of gravitation. It is easy to write this Poisson equation Lorentz invariantly, namely as an inhomogeneous wave equation

$$\frac{\partial^2 U}{\partial x^2} + \frac{\partial^2 U}{\partial y^2} + \frac{\partial^2 U}{\partial z^2} - \frac{\partial^2 U}{c^2 \partial t^2} = U^n{}_{,n} = 4\pi f \mu, \quad n = 1, 2, 3, 4. \quad (13.23)$$

But this equation not only violates the spirit of relativity in that only the energy appears on the right hand side, and not the momentum four-vector, nor the energy-momentum tensor. It also does not explain why inertial and gravitational masses are equal, and – even worse – it turns out to be experimentally wrong. It needs more than a cheap invariance trick to obtain the correct theory of the gravitational field. We shall deal with that problem in the following chapters.

II. Riemannian geometry

14

Introduction: the force-free motion of particles in Newtonian mechanics

14.1 Coordinate systems

In theoretical mechanics one usually meets only a few simple coordinate systems for describing the motion of a particle. For the purposes of mechanics one can characterize the coordinate system best via the specification of the connection between the infinitesimal separation $\mathrm{d}s$ of two points and the difference of their coordinates. In describing the motion in three-dimensional space one chooses Cartesian coordinates x, y, z with

$$\mathrm{d}s^2 = \mathrm{d}x^2 + \mathrm{d}y^2 + \mathrm{d}z^2, \tag{14.1}$$

cylindrical coordinates ρ, φ, z with

$$\mathrm{d}s^2 = \mathrm{d}\rho^2 + \rho^2 \, \mathrm{d}\varphi^2 + \mathrm{d}z^2, \tag{14.2}$$

or spherical coordinates r, ϑ, φ with

$$\mathrm{d}s^2 = \mathrm{d}r^2 + r^2 \, \mathrm{d}\vartheta^2 + r^2 \sin^2 \vartheta \, \mathrm{d}\varphi^2. \tag{14.3}$$

If the motion is restricted to a surface which does not change with time, for example, a sphere, then one would use the corresponding two-dimensional section $(\mathrm{d}r = 0)$ of spherical coordinates

$$\mathrm{d}s^2 = r^2 \, \mathrm{d}\vartheta^2 + r^2 \sin^2 \vartheta \, \mathrm{d}\varphi^2. \tag{14.4}$$

For other arbitrary coordinate systems $\mathrm{d}s^2$ is also a quadratic function of the coordinate differentials:

$$\mathrm{d}s^2 = g_{\alpha\beta}(x^\nu) \, \mathrm{d}x^\alpha \, \mathrm{d}x^\beta; \quad \alpha, \beta, \nu = 1, 2, 3. \tag{14.5}$$

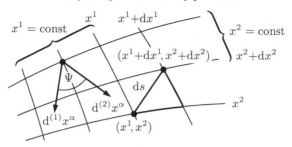

Fig. 14.1. Measurement of lengths and angles by the use of the metric tensor.

Here and in all following formulae indices occurring twice are to be summed, from one to three for a particle in three-dimensional space and from one to two for a particle in a plane.

The form (14.5) is called the *fundamental metric form*; the position-dependent coefficients $g_{\alpha\beta}$ form the components of the *metric tensor*. It is symmetric: $g_{\alpha\beta} = g_{\beta\alpha}$. The name 'metric tensor' refers to the fact that by its use the quantities length and angle which are fundamental to geometrical measurement can be defined and calculated. The displacement $\mathrm{d}s$ of two points with coordinates (x^1, x^2) and $(x^1 + \mathrm{d}x^1,\, x^2 + \mathrm{d}x^2)$ is given by (14.5), and the angle ψ between two infinitesimal vectors $\mathrm{d}^{(1)}x^\alpha$ and $\mathrm{d}^{(2)}x^\alpha$ diverging from a point can be calculated as

$$\cos\psi = \frac{g_{\alpha\beta}\,\mathrm{d}^{(1)}x^\alpha\,\mathrm{d}^{(2)}x^\beta}{\sqrt{g_{\rho\sigma}\,\mathrm{d}^{(1)}x^\rho\,\mathrm{d}^{(1)}x^\sigma}\,\sqrt{g_{\mu\nu}\,\mathrm{d}^{(2)}x^\mu\,\mathrm{d}^{(2)}x^\nu}}. \qquad (14.6)$$

Formula (14.6) is nothing other than the familiar vector relation $\boldsymbol{ab} = |\boldsymbol{a}||\boldsymbol{b}|\cos(\boldsymbol{a}, \boldsymbol{b})$ applied to infinitesimal vectors.

If the matrix of the metric tensor is diagonal, that is to say, $g_{\alpha\beta}$ differs from zero only when $\alpha = \beta$, then one calls the coordinate system *orthogonal*. As (14.6) shows, the coordinate lines $x^\alpha = \text{const.}$ are then mutually perpendicular.

If the determinant of $g_{\alpha\beta}$ is non-zero, the matrix possesses an inverse matrix $g^{\beta\mu}$ which satisfies

$$g_{\alpha\beta}g^{\beta\mu} = \delta_\alpha^\mu = g_\alpha^\mu. \qquad (14.7)$$

The immediate significance of the fundamental metric form (14.5) for mechanics rests on its simple connection with the square of the velocity v of the particle,

$$v^2 = \left(\frac{\mathrm{d}s}{\mathrm{d}t}\right)^2 = g_{\alpha\beta}\frac{\mathrm{d}x^\alpha}{\mathrm{d}t}\frac{\mathrm{d}x^\beta}{\mathrm{d}t}, \qquad (14.8)$$

which we need for the construction of the kinetic energy as one part of the Lagrangian.

14.2 Equations of motion

We can obtain the equations of motion most quickly from the Lagrangian L, which for force-free motion is identical with the kinetic energy of the particle

$$L = \frac{m}{2}v^2 = \frac{m}{2}g_{\alpha\beta}\frac{\mathrm{d}x^\alpha}{\mathrm{d}t}\frac{\mathrm{d}x^\beta}{\mathrm{d}t} = \frac{m}{2}g_{\alpha\beta}\dot{x}^\alpha\dot{x}^\beta. \tag{14.9}$$

The corresponding Lagrange equations (of the second kind)

$$\frac{\mathrm{d}}{\mathrm{d}t}\frac{\partial L}{\partial \dot{x}^\nu} - \frac{\partial L}{\partial x^\nu} = 0 \tag{14.10}$$

are easily set up. We have

$$\partial L/\partial \dot{x}^\nu = mg_{\alpha\nu}\dot{x}^\alpha, \quad \partial L/\partial x^\nu = L_{,\nu} = \tfrac{1}{2}mg_{\alpha\beta,\nu}\dot{x}^\alpha\dot{x}^\beta \tag{14.11}$$

(as done in Minkowski space, we use the comma followed by an index as an abbreviation for a partial derivative), and from (14.10) it follows immediately that

$$g_{\alpha\nu}\ddot{x}^\alpha + g_{\alpha\nu,\beta}\dot{x}^\alpha\dot{x}^\beta - \tfrac{1}{2}g_{\alpha\beta,\nu}\dot{x}^\alpha\dot{x}^\beta = 0. \tag{14.12}$$

If we first write the second term in this equation in the form

$$g_{\alpha\nu,\beta}\dot{x}^\alpha\dot{x}^\beta = \tfrac{1}{2}(g_{\alpha\nu,\beta} + g_{\beta\nu,\alpha})\dot{x}^\alpha\dot{x}^\beta, \tag{14.13}$$

then multiply (14.12) by $g^{\mu\nu}$ and sum over ν, then because of (14.7) we obtain

$$\ddot{x}^\mu + \Gamma^\mu_{\alpha\beta}\dot{x}^\alpha\dot{x}^\beta = 0, \tag{14.14}$$

where the abbreviation

$$\Gamma^\mu_{\alpha\beta} = \tfrac{1}{2}g^{\mu\nu}(g_{\alpha\nu,\beta} + g_{\beta\nu,\alpha} - g_{\alpha\beta,\nu}) \tag{14.15}$$

has been used.

Equations (14.14) are the required equations of motion of a particle. In the course of their derivation we have also come across the *Christoffel symbols* $\Gamma^\mu_{\alpha\beta}$, defined by (14.15), which play a great rôle in differential geometry. As is evident from (14.15), they possess the symmetry

$$\Gamma^\mu_{\alpha\beta} = \Gamma^\mu_{\beta\alpha}, \tag{14.16}$$

and hence there are eighteen distinct Christoffel symbols in three-dimensional space, and six for two-dimensional surfaces.

On contemplating (14.14) and (14.15), one might suppose that the Christoffel symbols lead to a particularly simple way of constructing the equations of motion. This supposition is, however, false; on the contrary, one uses the very equations of motion in order to construct the Christoffel symbols. We shall illustrate this method by means of an example. In spherical coordinates (14.3), $x^1 = r$, $x^2 = \vartheta$, $x^3 = \varphi$, the Lagrangian

$$L = \tfrac{1}{2}m(\dot{r}^2 + r^2\dot{\vartheta}^2 + r^2 \sin^2 \vartheta\, \dot{\varphi}^2) \tag{14.17}$$

implies the following Lagrange equations of the second kind:

$$\ddot{r} - r\dot{\vartheta}^2 - r\sin^2\vartheta\,\dot{\varphi}^2 = 0, \quad r\ddot{\varphi} + 2\dot{r}\dot{\varphi} + 2r\cot\vartheta\,\dot{\varphi}\dot{\vartheta} = 0,$$
$$r\ddot{\vartheta} + 2\dot{r}\dot{\vartheta} - r\sin\vartheta\,\cos\vartheta\,\dot{\varphi}^2 = 0. \tag{14.18}$$

Comparison with (14.14) shows that (noticing that, because of the symmetry relation (14.16), mixed terms in the speeds $\dot{r}$, $\dot{\vartheta}$, $\dot{\varphi}$ always occur twice) only the following Christoffel symbols are different from zero:

$$\Gamma^1_{22} = -r, \qquad \Gamma^2_{12} = \Gamma^2_{21} = r^{-1}. \qquad \Gamma^3_{13} = \Gamma^3_{31} = r^{-1},$$
$$\Gamma^1_{33} = -r\sin^2\vartheta, \quad \Gamma^3_{23} = \Gamma^3_{32} = \cot\vartheta, \quad \Gamma^2_{33} = -\sin\vartheta\,\cos\vartheta. \tag{14.19}$$

In the case of free motion of a particle in three-dimensional space the physical content of the equations of motion is naturally rather scanty; it is merely a complicated way of writing the law of inertia – we know beforehand that the particle moves in a straight line in the absence of forces. In the two-dimensional case, for motion on an arbitrary surface, the path of the particle can of course be rather complicated. As we shall show in the following section, however, a simple geometrical interpretation of the equations of motion (14.14) is then possible.

14.3 The geodesic equation

In three-dimensional space the path of a force-free particle, the straight line, has the property of being the shortest curve between any two points lying on it. We are going to generalize this relation, and therefore ask for the shortest curve connecting two points in a three-dimensional or two-dimensional space; that is, for that curve whose arclength s is a minimum for given initial-point and end-point:

$$s = \int_{P_{\mathrm{I}}}^{P_{\mathrm{E}}} ds = \text{extremum}. \tag{14.20}$$

In order to describe this curve we need an initially arbitrary parameter λ, which for all curves under comparison has the same value at the endpoints P_E and P_I; if for the differential arclength ds we substitute the expression (14.5), then (14.20) implies

$$s = \int_{\lambda_I}^{\lambda_E} \frac{ds}{d\lambda}\, d\lambda = \int_{\lambda_I}^{\lambda_E} \sqrt{g_{\alpha\beta} \frac{dx^\alpha}{d\lambda} \frac{dx^\beta}{d\lambda}}\, d\lambda = \text{extremum}, \qquad (14.21)$$

from which we shall determine the required shortest connecting curve, the geodesic, in the form $x^\alpha(\lambda)$.

The variational problem (14.21) has precisely the mathematical form of Hamilton's principle with the Lagrangian

$$L = \sqrt{g_{\alpha\beta} x'^\alpha x'^\beta} = \sqrt{F}, \qquad x'^\alpha \equiv \frac{dx^\alpha}{d\lambda}, \qquad (14.22)$$

and the parameter λ instead of the time t. Thus the geodesic must obey the associated Lagrange equations of the second kind

$$\frac{d}{d\lambda} \frac{\partial L}{\partial x'^\nu} - \frac{\partial L}{\partial x^\nu} = \frac{d}{d\lambda}\left(\frac{g_{\alpha\nu} x'^\alpha}{\sqrt{F}}\right) - \frac{1}{2\sqrt{F}} g_{\alpha\beta,\nu} x'^\alpha x'^\beta$$

$$= \frac{1}{2F\sqrt{F}}\left[-\frac{dF}{d\lambda} g_{\alpha\nu} x'^\alpha + 2F \frac{d}{d\lambda}(g_{\alpha\nu} x'^\alpha) - F g_{\alpha\beta,\nu} x'^\alpha x'^\beta\right] = 0.$$

$$(14.23)$$

We can simplify this differential equation for the geodesic by choosing the parameter λ appropriately (only for this extremal curve, not for the comparison curves); we demand that λ be proportional to the arclength s. From (14.21) and (14.22) it follows that $F = $ constant, and from (14.23) we get the differential equation of the geodesic

$$\frac{d^2 x^\mu}{d\lambda^2} + \Gamma^\mu_{\alpha\beta} \frac{dx^\alpha}{d\lambda} \frac{dx^\beta}{d\lambda} = 0. \qquad (14.24)$$

This differential equation not only has the same form as the equation of motion (14.14), it is also completely equivalent to it, since, of course, for a force-free motion the magnitude $v = ds/dt$ of the speed is constant because of the law of conservation of energy, and consequently the time t is one of the allowable possibilities in (14.22) for the parameter λ which is proportional to the arclength s.

If we choose as parameter λ the arclength s itself, then we can recapitulate our result in the following law:

A force-free particle moves on a geodesic

$$\frac{d^2 x^\mu}{ds^2} + \Gamma^\mu_{\alpha\beta} \frac{dx^\alpha}{ds} \frac{dx^\beta}{ds} = 0, \qquad (14.25)$$

of the three-dimensional space or of the surface to which it is constrained.

Its path is therefore always the shortest curve between any two points lying on it; for example, on the spherical surface the paths are great circles.

In the General Theory of Relativity we shall meet the problem of how to set up the equation of motion of a point mass in an arbitrary gravitational field. It will turn out that the formulation of the equation of motion for force-free motion just derived is a good starting point for the solution of this problem.

14.4 Geodesic deviation

In this section we shall turn to a question whose answer requires the help of Riemannian geometry, which we shall indeed use. The reader is therefore asked for indulgence if some of the formalism appears rather vague and the calculations inadequately motivated. He is recommended to read this section again after mastering Chapter 5.

If the surface to which the particle is constrained is a plane, or a surface which is due to the deformation of a plane (e.g. cylinder, cone) then the geodesics are straight lines of this plane, and the equations of motion of the point mass are very simple to integrate. With the use of unsuitable coordinates, however, the geodesic equation (14.25) can be rather complicated. In such a case how can one tell from the equation of motion, that is, from the Christoffel symbols $\Gamma^\mu_{\alpha\beta}$, that motion on a plane is being described?

To answer this question we examine a family $x^\alpha(s,p)$ of geodesics on a surface, see Fig. 14.2. Here the parameter p labels the different geodesics and the arclength s is the parameter along the curves fixing the different points of the same geodesic.

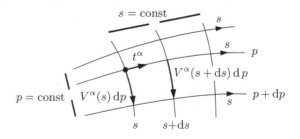

Fig. 14.2. The family of geodesics $x^\alpha(s,p)$.

A family of straight lines in the plane is now distinguished by the displacement of two neighbouring geodesics, as measured between points with the same value of the parameter s, being a *linear* function of arclength s. This is a hint that also in the general case of geodesics on an arbitrary surface we should examine the behaviour of the separation of neighbouring geodesics and from this draw conclusions about the properties of the surface.

We first form the partial derivatives

$$\frac{\partial x^\alpha}{\partial s} = t^\alpha, \qquad \frac{\partial x^\alpha}{\partial p} = V^\alpha, \qquad \frac{\partial t^\alpha}{\partial p} \equiv \frac{\partial V^\alpha}{\partial s}. \tag{14.26}$$

The unit tangent vector t^α points in the direction of the velocity, and $V^\alpha\, dp$ is just the displacement vector of two neighbouring geodesics. In order to see whether we are dealing with a plane or not, it is, however, insufficient to simply form $\partial^2 V^\alpha/\partial s^2$. Indeed, even for a straight line in the plane, the fact that the tangent vector $t^\mu = dx^\mu/ds$ is constant, that is, independent of s, is not expressed in an arbitrary coordinate system by $dt^\mu/ds = 0$, but, as a glance at the geodesic equation (14.25) shows, by

$$\frac{D}{Ds}t^\mu \equiv \frac{dt^\mu}{ds} + \Gamma^\mu_{\alpha\beta}t^\alpha \frac{dx^\beta}{ds} = 0. \tag{14.27}$$

We interpret (14.27) as the defining equation for the operator D/Ds, valid for every parameter s and applicable to every vector t^μ. If according to this prescription we form the expressions

$$\frac{D}{Dp}t^\alpha = \frac{\partial^2 x^\alpha}{\partial s\, \partial p} + \Gamma^\alpha_{\mu\nu}t^\mu V^\nu, \qquad \frac{D}{Ds}V^\alpha = \frac{\partial^2 x^\alpha}{\partial s\, \partial p} + \Gamma^\alpha_{\mu\nu}V^\mu t^\nu, \tag{14.28}$$

then we can at once read off the relation analogous to (14.26),

$$\frac{D}{Dp}t^\alpha = \frac{D}{Ds}V^\alpha. \tag{14.29}$$

We shall now calculate the quantity

$$\frac{D^2 V^\alpha}{Ds^2} = \frac{D}{Ds}\left(\frac{D}{Dp}t^\alpha\right) \tag{14.30}$$

in order to discuss with its help the behaviour of the separation of two neighbouring geodesics. Our first goal is to express the right-hand side in terms of the Christoffel symbols. Substitution of the defining equation (14.27) gives us immediately

$$\frac{\mathrm{D}^2 V^\alpha}{\mathrm{D}s^2} = \frac{\mathrm{D}}{\mathrm{D}s}\left(\frac{\partial t^\alpha}{\partial p} + \Gamma^\alpha_{\mu\nu} t^\mu V^\nu\right)$$

$$= \frac{\partial^2 t^\alpha}{\partial s\,\partial p} + \Gamma^\alpha_{\mu\nu,\beta} t^\beta t^\mu V^\nu + \Gamma^\alpha_{\mu\nu}\left(\frac{\partial t^\mu}{\partial s} V^\nu + t^\mu \frac{\partial V^\nu}{\partial s}\right) \quad (14.31)$$

$$+ \Gamma^\alpha_{\rho\tau}\left(\frac{\partial t^\rho}{\partial p} + \Gamma^\rho_{\mu\nu} t^\mu V^\nu\right) t^\tau.$$

We can simplify this equation by invoking the relation

$$0 = \frac{\mathrm{D}}{\mathrm{D}p}\frac{\mathrm{D}t^\alpha}{\mathrm{D}s} = \frac{\mathrm{D}}{\mathrm{D}p}\left(\frac{\partial t^\alpha}{\partial s} + \Gamma^\alpha_{\mu\nu} t^\mu t^\nu\right), \quad (14.32)$$

which follows from the geodesic equation (14.27). This leads to

$$\frac{\partial^2 t^\alpha}{\partial s\,\partial p} = -\Gamma^\alpha_{\mu\nu,\beta} V^\beta t^\mu t^\nu - \Gamma^\alpha_{\mu\nu}\left(\frac{\partial t^\mu}{\partial p} t^\nu + \frac{\partial t^\nu}{\partial p} t^\mu\right)$$

$$- \Gamma^\alpha_{\rho\tau}\left(\frac{\partial t^\rho}{\partial s} + \Gamma^\rho_{\mu\nu} t^\mu t^\nu\right) V^\tau, \quad (14.33)$$

which we can substitute into (14.31). Bearing in mind (14.26) and (14.16), we then find that

$$\frac{\mathrm{D}^2 V^\alpha}{\mathrm{D}s^2} = t^\beta t^\mu V^\nu (\Gamma^\alpha_{\mu\nu,\beta} - \Gamma^\alpha_{\mu\beta,\nu} + \Gamma^\alpha_{\rho\beta}\Gamma^\rho_{\mu\nu} - \Gamma^\alpha_{\rho\nu}\Gamma^\rho_{\mu\beta}). \quad (14.34)$$

The right-hand side of this equation gives us a measure of the change in separation of neighbouring geodesics, or, in the language of mechanics, of the relative acceleration of two particles moving towards one another on neighbouring paths ($V^\alpha\,\mathrm{d}p$ is their separation, $\mathrm{d}s$ is proportional to $\mathrm{d}t$ for force-free motion.) It is also called the *geodesic deviation*.

When the geodesics are straight lines in a plane, the right-hand side should vanish; it is therefore – geometrically speaking – also a measure of the curvature of the surface, of the deviation of the surface from a plane. This intuitive basis also makes understandable the name 'curvature tensor' for the expression $R^\alpha{}_{\mu\beta\nu}$, defined by

$$R^\alpha{}_{\mu\beta\nu} = \Gamma^\alpha_{\mu\nu,\beta} - \Gamma^\alpha_{\mu\beta,\nu} + \Gamma^\alpha_{\rho\beta}\Gamma^\rho_{\mu\nu} - \Gamma^\alpha_{\rho\nu}\Gamma^\rho_{\mu\beta}. \quad (14.35)$$

It can be determined from the Christoffel symbols by calculation or by measurement of the change in separation of neighbouring paths. If it vanishes, the surface is a plane and the paths are straight lines.

As an illustration we calculate the curvature tensor of the spherical surface

$$\mathrm{d}s^2 = K^2\left(\mathrm{d}\vartheta^2 + \sin^2\vartheta\,\mathrm{d}\varphi^2\right)$$

$$= K^2\left[(\mathrm{d}x^1)^2 + \sin^2 x^1\,(\mathrm{d}x^2)^2\right] = g_{\alpha\beta}\,\mathrm{d}x^\alpha\,\mathrm{d}x^\beta. \quad (14.36)$$

The only non-vanishing Christoffel symbols are

$$\Gamma^1_{22} = -\sin\vartheta\,\cos\vartheta, \qquad \Gamma^2_{12} = \Gamma^2_{21} = \cot\vartheta, \qquad (14.37)$$

and from them we find after a simple calculation the components of the curvature tensor

$$R^1{}_{221} = -R^1{}_{212} = -\sin^2\vartheta = -K^{-2}g_{22},$$
$$R^2{}_{121} = -R^2{}_{112} = 1 = K^{-2}g_{11}, \qquad R^\alpha{}_{\mu\beta\nu} = 0 \quad \text{otherwise.} \qquad (14.38)$$

This result can be summarized by the formula

$$R^\alpha{}_{\mu\beta\nu} = K^{-2}\left(\delta^\alpha_\beta g_{\mu\nu} - \delta^\alpha_\nu g_{\mu\beta}\right). \qquad (14.39)$$

It expresses the fact that, apart from the coordinate-dependent metric, the curvature tensor depends only upon the radius of the sphere K. For $K \to \infty$ the curvature tensor vanishes.

Exercise

14.1 Show that the curvature tensor of an arbitrary two-dimensional surface always has the form (14.39), with of course a position-dependent K.

15
Why Riemannian geometry?

It is one of the most important results of Special Relativity that basic physical laws are most simply expressed when they are formulated not in three-dimensional space but in four-dimensional space-time

$$ds^2 = \eta_{ab}\,dx^a\,dx^b = dx^2 + dy^2 + dz^2 - c^2\,dt^2,$$
$$\eta_{ab} = \eta^{ab} = \operatorname{diag}(1,1,1,-1), \qquad a,b = 1,\ldots,4. \qquad (15.1)$$

We shall now show that it is worthwhile replacing this Minkowski space by a yet more complicated mathematical space-time structure.

If we were to examine a circular disk which is at rest in an inertial system, from the standpoint of a coordinate system rotating around the axis of the disc, and try to measure the geometrical properties of the disc with the help of rulers, then the following result would be plausible: rulers laid out in the radial direction are not influenced by the rotation of the disc, and the radius of the circle is unchanged; rulers laid out along the periphery of the disc are shortened by the Lorentz contraction, the circumference of the circle being thereby decreased. The rotating observer thus establishes that the ratio of the circumference of the circle to its diameter is less than π; he finds geometrical relations similar to those on the curved surface of the sphere. Naturally the application of the Lorentz transformation to rotating systems, and the definition of simultaneity hidden in this measuring procedure, are questionable. But that would only support the result of our 'gedanken' investigation, that in going over to observers (coordinate systems) in arbitrary motion real changes in the space-time structure (the behaviour of rulers and clocks) can arise.

Physically even more significant is the indication to be deduced from the investigation of the equation of motion of a particle moving in a gravitational field $\mathbf{g}(\mathbf{r}, t)$. If we write this equation in a Cartesian coordinate system, whose origin moves with acceleration $\mathbf{a}$ with respect to an inertial system, and which rotates with angular velocity $\boldsymbol{\omega}$, then we get the familiar equation

$$m\ddot{\mathbf{r}} = m\mathbf{g} - m\mathbf{a} - 2m\boldsymbol{\omega} \times \dot{\mathbf{r}} - m\boldsymbol{\omega} \times (\boldsymbol{\omega} \times \mathbf{r}) - m\dot{\boldsymbol{\omega}} \times \mathbf{r}. \qquad (15.2)$$

All the terms of this equation of motion have the mass m as a factor. From the standpoint of Newtonian mechanics this factor possesses two distinct physical meanings: the force $m\mathbf{g}$, which acts upon a body in the gravitational field, is proportional to the gravitational mass m_{G}, whilst all other terms in (15.2) are an expression of the inertial behaviour of the body (which is the same for all kinds of forces) and consequently contain the inertial mass m_{I}. It was one of the most important discoveries of mechanics that for all bodies these two parameters are the same: all bodies fall equally fast, and hence

$$m_{\mathrm{G}} = m_{\mathrm{I}} = m. \qquad (15.3)$$

When testing this relation, one has to exclude that the ratio $m_{\mathrm{G}}/m_{\mathrm{I}}$ depends on the composition of the body. The current values of the Eötvös parameter $\eta_{\mathrm{E}}(1,2) = (m_{\mathrm{G}}/m_{\mathrm{I}})_1 - (m_{\mathrm{G}}/m_{\mathrm{I}})_2$ are $\eta_{\mathrm{E}}(1,2) \leq 10^{-12}$, see Will (1993) and Schäfer (2000).

The numerical identity of inertial and gravitational mass also points to a more essential identity. In the language of (15.2) gravitation is perhaps just as much an apparent force as the Coriolis force or the centrifugal force. One could therefore suppose that the particle moves weightlessly in reality, and that also the gravitational force can be eliminated by a suitable choice of coordinate system.

As an exact consideration shows, the gravitational force can really only be transformed away locally, that is, over a spatial region within which the gravitational field can be regarded as homogeneous: inside an Earth satellite or a falling box bodies move force-free for the co-moving observer. Globally, however, this is not attainable through a simple coordinate transformation (by changing to a moving observer): there is no Cartesian coordinate system in which two distantly separated satellites simultaneously move force-free.

If we therefore wish to adhere to the view that in spite of the existing gravitational field the particle moves force-free, and in the sense of Chapter 14 translate 'force-free' by 'along a geodesic', then we must alter the geometry of the space. Just as the geodesics on a surface fail to be straight lines only if the surface is curved and the curvature tensor defined by (14.34) and (14.35) does not vanish, so the planetary orbits are only geodesics of the space if this space is curved.

In fact this idea of Einstein's, to regard the gravitational force as a property of the space and thereby to geometrize it, turns out to be extraordinarily fruitful. In the following chapters we shall therefore describe in detail the properties of such curved spaces.

16
Riemannian space

16.1 The metric

The geometrical background to the Special Theory of Relativity is the pseudo-Euclidean space (15.1) with one timelike and three spacelike coordinates. In the generalization which we will develop now we also start with a four-dimensional manifold; that is, we shall assume that every point (within a small finite neighbourhood) can be fixed uniquely by the specification of four coordinates x^n. It can of course occur that it is not possible to cover the whole space-time with a single coordinate system. In order to be able to study physics in this manifold, we must be able to measure the spatial and temporal separations of neighbouring points. As the generalization of (15.1) and (14.5) we therefore introduce the metric

$$ds^2 = g_{mn}(x^i)\,dx^n\,dx^m \tag{16.1}$$

(summation occurs from 1 to 4 over Latin indices appearing twice). This *fundamental metric form* indicates how one measures on the small scale (in the infinitesimal neighbourhood of a point) the interval ds between the points (x^n) and $(x^n + dx^n)$ and the angle between two directions dx^n and $d\bar{x}^n$

$$\cos(dx^n, d\bar{x}^m) = \frac{g_{nm}\,dx^n\,d\bar{x}^m}{\sqrt{ds^2\,d\bar{s}^2}} \tag{16.2}$$

(see also Fig. 14.1). The *metric tensor* (the metric) g_{mn} characterizes the space completely (locally). It is symmetric, its determinant g is in general different from zero, and it possesses therefore an inverse g^{an}:

$$g_{mn} = g_{nm}, \qquad |g_{nm}| = g \neq 0, \qquad g^{an}g_{nm} = \delta^a_m = g^a_m. \tag{16.3}$$

A space with the properties (16.1) and (16.3) is accordingly a generalization as much of the two-dimensional surfaces as also of the four-dimensional uncurved (flat) Minkowski space. If ds^2 is positive definite, that is, zero only for $dx^i = 0$ and positive otherwise (and if the parallel transport of a vector is defined as in Chapter 18), then we are dealing with a *Riemannian space* in the narrower sense. But, as we know from the Special Theory of Relativity, the physical space-time must have an extra structure: we can distinguish between timelike and spacelike intervals, between clocks and rulers, and there is a light cone with $ds = 0$. Our space is therefore a *pseudo-Riemannian space*, ds^2 can be positive (spacelike), negative (timelike) or null (lightlike); it is a Lorentzian metric. Nevertheless we shall usually use the term Riemannian space (in the broader sense) for it.

In Section 16.4 we shall describe how one takes into account the requirement that there exist one timelike and three spacelike directions.

16.2 Geodesics and Christoffel symbols

On a two-dimensional surface we could define geodesics by making them the shortest curve between two points:

$$\int ds = \text{extremum}. \tag{16.4}$$

In a pseudo-Riemannian space, in which ds^2 can also be zero or negative, we encounter difficulties in the application of (16.4), especially for curves with $ds = 0$ (null lines). We therefore start here from the variational principle

$$\int L \, d\lambda = \int \left(\frac{ds}{d\lambda}\right)^2 d\lambda = \int g_{mn} \frac{dx^n}{d\lambda} \frac{dx^m}{d\lambda} \, d\lambda = \text{extremum}, \qquad (16.5)$$

which, as we have shown in Section 14.3, is equivalent to (16.4) for $ds \neq 0$. The Lagrange equations of the second kind for the Lagrangian $L = (ds/d\lambda)^2$ give (see Section 14.3)

$$\frac{d^2 x^m}{d\lambda^2} + \Gamma^m_{ab} \frac{dx^a}{d\lambda} \frac{dx^b}{d\lambda} = 0 \qquad (16.6)$$

as differential equations of the geodesics. These are four second-order differential equations for the four functions $x^m(\lambda)$, and accordingly geodesics are locally uniquely determined if the initial-point and the initial direction or the initial-point and the end-point are given. When later we speak briefly of the separation of two points, we always mean the arclength of the connecting geodesics. The Christoffel symbols occurring in (16.6) are of course defined as in (14.15) by

$$\Gamma^m_{ab} = \tfrac{1}{2} g^{mn} (g_{an,b} + g_{bn,a} - g_{ab,n}). \qquad (16.7)$$

The Lagrangian $L = g_{mn} \, dx^m \, dx^n/d\lambda^2$ is a homogeneous function of second degree in the 'velocities' $\dot{x}^n = dx^n/d\lambda$

$$\dot{x}^n \partial L/\partial \dot{x}^n = 2L. \qquad (16.8)$$

Because of (16.8) and the Lagrange equation of the second kind we have

$$\frac{dL}{d\lambda} = \frac{\partial L}{\partial x^n} \dot{x}^n + \frac{\partial L}{\partial \dot{x}^n} \ddot{x}^n = \left(\frac{d}{d\lambda} \frac{\partial L}{\partial \dot{x}^n}\right) \dot{x}^n + \frac{\partial L}{\partial \dot{x}^n} \ddot{x}^n$$

$$= \frac{d}{d\lambda} \left(\frac{\partial L}{\partial \dot{x}^n} \dot{x}^n\right) = 2 \frac{dL}{d\lambda}, \qquad (16.9)$$

and therefore

$$\frac{dL}{d\lambda} = 0, \quad L = g_{nm} \frac{dx^m}{d\lambda} \frac{dx^n}{d\lambda} = \left(\frac{ds}{d\lambda}\right)^2 = C = \text{const.} \qquad (16.10)$$

The constant C can be positive, negative or zero, and we distinguish correspondingly spacelike, timelike and null geodesics. We shall meet timelike geodesics again as paths for particles, and null geodesics as light rays. Because of (16.10) the (affine) parameter λ along a geodesic is clearly determined uniquely up to a linear transformation $\lambda' = a\lambda + b$; for timelike curves we shall identify λ with the proper time τ.

Christoffel symbols are important quantities in Riemannian geometry. We therefore want to investigate more closely their relations to the partial derivatives of the metric tensor given by (16.7). Because of the symmetry $g_{mn} = g_{nm}$ of the metric tensor, the Christoffel symbols too are symmetric in the lower indices:

$$\Gamma_{ab}^m = \Gamma_{ba}^m. \qquad (16.11)$$

In four dimensions there are $\binom{5}{2} = 10$ different components of the metric tensor, and therefore because of the additional freedom provided by the upper indices, $4 \times 10 = 40$ distinct Christoffel symbols. But this number is the same as the number of partial derivatives $g_{mn,a}$ of the metric tensor, and it should therefore be possible to express the partial derivatives through the Christoffel symbols, thus solving (16.7). In fact because of (16.3) we have

$$
\begin{aligned}
g_{mi}\Gamma_{ab}^m &= \tfrac{1}{2}(g_{ai,b} + g_{bi,a} - g_{ab,i}), \\
g_{ma}\Gamma_{ib}^m &= \tfrac{1}{2}(g_{ia,b} + g_{ab,i} - g_{ib,a}),
\end{aligned}
\qquad (16.12)
$$

and adding the two equations we get

$$g_{ia,b} = g_{mi}\Gamma_{ab}^m + g_{ma}\Gamma_{ib}^m. \qquad (16.13)$$

The partial derivatives of the determinant g of the metric tensor can also be calculated in a simple manner from the Christoffel symbols. The chain rule implies that

$$\frac{\partial g}{\partial x^b} = \frac{\partial g}{\partial g_{ia}} g_{ia,b}. \qquad (16.14)$$

If one now introduces the expansion of the determinant $g = |g_{mn}|$ along the ith row by

$$g = \sum_a g_{ia} G_{ia} \quad \text{(no summation over } i!) \qquad (16.15)$$

and uses the fact that the elements g^{ia} of the inverse matrix can be expressed through the co-factors G_{ia} according to

$$g g^{ia} = G_{ia}, \qquad (16.16)$$

then one finds that

$$\partial g / \partial g_{ia} = g g^{ia}. \qquad (16.17)$$

If one also takes into account (16.13), then it follows that

$$\partial g / \partial x^b = g g^{ia} g_{ia,b} = g(\Gamma_{ab}^a + \Gamma_{ib}^i) = 2g\Gamma_{ab}^a, \qquad (16.18)$$

and from this finally that

$$\frac{\partial \ln \sqrt{-g}}{\partial x^b} = \frac{1}{2g}\frac{\partial g}{\partial x^b} = \Gamma^a_{ab}. \tag{16.19}$$

In writing the formula thus we have already assumed that g is negative (see Section 17.2).

16.3 Coordinate transformations

Naturally the physical structure of our space-time manifold is not allowed to depend upon the choice of the coordinates with which we describe it. We now investigate which properties of the metric tensor and of the Christoffel symbols are derivable from this requirement, that is to say, how these quantities behave under a coordinate transformation.

All coordinate transformations of the old coordinates x^n into new coordinates $x^{n'}$ are permitted which guarantee a one-to-one relationship of the form

$$x^{n'} = x^{n'}(x^n), \qquad \left|\frac{\partial x^{n'}}{\partial x^n}\right| \neq 0. \tag{16.20}$$

We have in (16.20) made use of the convention of distinguishing the new coordinates from the old by a prime on the index, as explained in Section 1.3. With the abbreviation

$$A^{n'}_n = \frac{\partial x^{n'}}{\partial x^n}, \tag{16.21}$$

we obtain from (16.20) the transformation law for the coordinate differentials

$$dx^{n'} = A^{n'}_n dx^n. \tag{16.22}$$

The inverse transformation to (16.20),

$$x^n = x^n(x^{n'}), \tag{16.23}$$

implies analogously

$$dx^n = A^n_{n'} dx^{n'}, \qquad A^n_{n'} = \frac{\partial x^n}{\partial x^{n'}}, \tag{16.24}$$

and, from (16.21) and (16.24),

$$A^{n'}_n A^n_{m'} = \delta^{n'}_{m'}, \qquad A^n_{n'} A^{n'}_m = \delta^n_m. \tag{16.25}$$

We obtain the prescription for the transformation of the components of the metric tensor from the requirement that lengths and angles should not change under a coordinate transformation; that is, ds^2 is an invariant:

$$ds'^2 \equiv g_{n'm'} dx^{n'} dx^{m'} = ds^2 = g_{nm} dx^n dx^m = g_{nm} A^n_{n'} A^m_{m'} dx^{n'} dx^{m'}. \tag{16.26}$$

Since this equation must hold for arbitrary choice of the $\mathrm{d}x^{n'}$, it follows that

$$g_{n'm'} = g_{nm}A_{n'}^n A_{m'}^m. \tag{16.27}$$

The behaviour of the Christoffel symbols under transformations is most easily calculated using the geodesic equation (16.6). Since the variational principle (16.5) was formulated with the help of the invariant quantities $\mathrm{d}s$ and $\mathrm{d}\lambda$, the geodesic equation must have the form (16.6) in the new coordinates $x^{n'}$ as well:

$$\frac{\mathrm{d}^2 x^{m'}}{\mathrm{d}\lambda^2} + \Gamma_{a'b'}^{m'} \frac{\mathrm{d}x^{a'}}{\mathrm{d}\lambda}\frac{\mathrm{d}x^{b'}}{\mathrm{d}\lambda} = 0 \tag{16.28}$$

(the property of a curve, to be the shortest connection between two points, is independent of the choice of coordinates). If we substitute into equation (16.28) the equation

$$
\begin{aligned}
\frac{\mathrm{d}^2 x^{m'}}{\mathrm{d}\lambda^2} &= A_m^{m'}\frac{\mathrm{d}^2 x^m}{\mathrm{d}\lambda^2} + A_{m,b}^{m'}\frac{\mathrm{d}x^b}{\mathrm{d}\lambda}\frac{\mathrm{d}x^m}{\mathrm{d}\lambda} \\
&= -A_m^{m'}\Gamma_{ab}^m \frac{\mathrm{d}x^a}{\mathrm{d}\lambda}\frac{\mathrm{d}x^b}{\mathrm{d}\lambda} + A_{a,b}^{m'}\frac{\mathrm{d}x^b}{\mathrm{d}\lambda}\frac{\mathrm{d}x^a}{\mathrm{d}\lambda},
\end{aligned}
\tag{16.29}
$$

which follows from (16.22) and

$$\frac{\mathrm{d}x^{m'}}{\mathrm{d}\lambda} = A_m^{m'}\frac{\mathrm{d}x^m}{\mathrm{d}\lambda}, \tag{16.30}$$

and transform everything to dashed coordinates, we obtain finally

$$\Gamma_{a'b'}^{m'} = A_m^{m'}A_{a'}^a A_{b'}^b \Gamma_{ab}^m - A_{a,b}^{m'}A_{a'}^a A_{b'}^b. \tag{16.31}$$

In this transformation formula it should be noted that the new Christoffel symbols are not homogeneous linear functions of the old. It is therefore quite possible that in a Riemannian space the Christoffel symbols are non-zero in one coordinate system, whilst in another coordinate system they vanish identically. Thus in the usual three-dimensional space the Christoffel symbols are identically zero in Cartesian coordinates, whereas in spherical coordinates they have the values given in (14.19). We shall answer in Section 19.2 the question of whether the Christoffel symbols can always be made to vanish.

16.4 Special coordinate systems

For many calculations and considerations it is convenient to use a special coordinate system. But one must examine in each individual case whether a coordinate system with the desired properties really does exist; that is, whether it is possible for a given metric $g_{a'b'}$ to define the four functions $x^n(x^{n'})$ so that the transformed metric g_{ab} fulfils the chosen requirements.

Orthogonal coordinates If the matrix g_{ab} has only diagonal elements,

$$\mathrm{d}s^2 = g_{11}(\mathrm{d}x^1)^2 + g_{22}(\mathrm{d}x^2)^2 + g_{33}(\mathrm{d}x^3)^2 + g_{44}(\mathrm{d}x^4)^2, \qquad (16.32)$$

then we are dealing with orthogonal coordinates, and the coordinate lines (lines along which only one coordinate varies at any given time) form right-angles with one another. In the three-dimensional Euclidean space one uses and prefers such coordinates; for example, spherical coordinates or cylindrical coordinates. As a more exact analysis shows, such orthogonal coordinate systems do not in general exist in a four-dimensional Riemannian space, since for arbitrarily given functions $g_{a'b'}$ the system of differential equations

$$g_{a'b'} \frac{\partial x^{a'}}{\partial x^a} \frac{\partial x^{b'}}{\partial x^b} = 0 \quad \text{for } a \neq b \qquad (16.33)$$

has no solutions $x^{a'}(x^a)$ which satisfy the conditions (16.20). This result is plausible, since (16.33) is a system of six differential equations for four functions.

Time-orthogonal coordinates We shall customarily choose time as the fourth coordinate: $x^4 = ct$; time-orthogonal coordinates exist when $g_{4\alpha} = 0$. If, moreover, g_{44} has the value ± 1, then we are dealing with *Gaussian coordinates*, often also called *synchronous* coordinates. Since it follows from $g_{4\alpha} = 0$ that also $g^{4\alpha} = 0$ (and vice versa), in going over to time-orthogonal coordinates we have to satisfy the system

$$g^{4\alpha} = A_{a'}^{\alpha} A_{b'}^{4} g^{a'b'} = \frac{\partial x^{\alpha}}{\partial x^{a'}} \frac{\partial x^{4}}{\partial x^{b'}} g^{a'b'} = 0, \quad \alpha = 1, 2, 3. \qquad (16.34)$$

One can see that it is still possible to specify arbitrarily the function $x^4(x^{b'})$, and then for every one of the functions $x^{\alpha}(x^{a'})$ a partial differential equation has to be solved, the existence of the solution being guaranteed by general laws. Time-orthogonal coordinates,

$$\mathrm{d}s^2 = g_{\alpha\beta}\, \mathrm{d}x^{\alpha}\, \mathrm{d}x^{\beta} + g_{44}(\mathrm{d}x^4)^2, \qquad (16.35)$$

can therefore always be introduced (the fact that x^4 has the name 'time' plays no rôle at all), and also it is still possible to satisfy the additional condition $|g_{44}| = 1$ by choice of the function $x^4(x^{b'})$.

Comoving coordinates Later applications in Riemannian spaces often deal with a velocity field $u^n = dx^n/d\lambda$ (a flux of bodies, or of observers). Since λ is a coordinate-independent parameter, the components of this velocity transform like the coordinate differential

$$u^n = A^n_{n'} u^{n'} = \frac{\partial x^n}{\partial x^{n'}} u^{n'}. \qquad (16.36)$$

By means of a coordinate transformation it is always possible to make the three spatial components u^α of the velocity zero, since the differential equations

$$u^1 = \frac{\partial x^1}{\partial x^{1'}} u^{1'} + \frac{\partial x^1}{\partial x^{2'}} u^{2'} + \frac{\partial x^1}{\partial x^{3'}} u^{3'} + \frac{\partial x^1}{\partial x^{4'}} u^{4'} = 0, \ \ldots \qquad (16.37)$$

always have a solution $x^\alpha(x^{n'})$. In the resulting coordinate system, in which the velocity has the form $u^n = (0,0,0,u^4)$, the particles do not change their position; the coordinates move with the particles (one can visualize the coordinate values attached to the particles as names). Although the coordinate difference of two particles never alters, their separation can vary because of the time-dependence of the metric.

Local Minkowski system At an arbitrarily given point, which in the following we shall identify with the origin O of the coordinates, let the coordinate lines form right-angles with one another. That this is possible is intuitively obvious, and mathematically provable since, with the help of suitable transformation matrices $A^{a'}_a$, one can transform the constant matrix $g_{a'b'}(O)$ to principal axes. Then the metric at the point O,

$$ds'^2 = g_{11}(dx^{1'})^2 + g_{22}(dx^{2'})^2 + g_{33}(dx^{3'})^2 + g_{44}(dx^{4'})^2, \qquad (16.38)$$

can be further simplified to

$$ds'^2 = \pm(dx^1)^2 \pm (dx^2)^2 + \pm(dx^3)^2 \pm (dx^4)^2 \qquad (16.39)$$

by a stretching of the coordinates

$$x^1 = \sqrt{|g_{11}|} x^{1'}, \ \ldots. \qquad (16.40)$$

In the general case of an arbitrary metric no statement can be made about the signs occurring in (16.39). In order to make sure of the connection to the structure of Minkowski space we *demand* that the spaces used in the General Theory of Relativity have signature $(+2)$; that is, at every point under transformation of the metric to the form (16.39) three positive signs and one negative occur. We call such spaces *normal hyperbolic pseudo-Riemannian spaces*. One can show that the signature is an invariant, that is to say, it is independent of the choice of the initial coordinate system and of the (not uniquely determined) coordinate transformations, which lead to (16.39) (law of inertia of the quadratic forms).

In physically important spaces (e.g. gravitational fields), there can be singular points, however, at which the metric cannot be brought to the normal form (16.39). Obviously, at these points the structure of the space really does depart from that with which we are familiar.

Locally flat (geodesic) system After the introduction of a local Minkowski system the situation at a point is as in a flat four-dimensional space. One can also extend such a system into a (differential) neighbourhood of the point.

This can be illustrated by the example of an arbitrarily curved two-dimensional surface (see Fig. 16.1). Suppose one sets up the tangent plane to the surface at the point O under consideration and projects the Cartesian coordinates of the plane onto the surface. Since surface and tangent plane touch, the resulting coordinate lines on the surface differ from the straight lines of the plane only in second order. Applying the same consideration to a four-dimensional space-time, we would project the quasi-Cartesian coordinates of the tangential Minkowski space onto the Riemannian space and expect a metric of the form

$$g_{mn}(x^a) = \eta_{mn} + \tfrac{1}{2}d_{mnab}(O)x^a x^b + \cdots. \qquad (16.41)$$

We therefore call such a coordinate system locally flat.

In fact one can always locally transform an arbitrary metric

$$\bar{g}_{mn}(\bar{x}^a) = \bar{g}_{mn}(O) + \bar{g}_{mn,i}(O)\bar{x}^i + \cdots \qquad (16.42)$$

into the form (16.41). For if one introduces new coordinates x^a by

$$x^a = \bar{x}^a + \tfrac{1}{2}\bar{\Gamma}^a_{mn}(O)x^{\bar{m}}x^{\bar{n}} + \ldots \longleftrightarrow \bar{x}^a = x^a - \tfrac{1}{2}\bar{\Gamma}^a_{mn}(O)x^m x^n + \cdots,$$
$$(16.43)$$

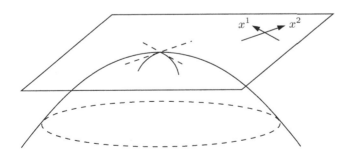

Fig. 16.1. Tangent plane and locally flat coordinate system.

then, because of

$$\partial \bar{x}^a / \partial x^n = \delta_n^a - \bar{\Gamma}^a_{mn}(O)x^m + \cdots, \tag{16.44}$$

the new metric tensor has the form (ignoring terms higher than linear in x^a)

$$g_{mn} = \bar{g}_{ab}\left(\delta_b^a - \bar{\Gamma}^a_{in}(O)x^i\right)\left(\delta_m^b - \bar{\Gamma}^b_{km}(O)x^k\right). \tag{16.45}$$

Its partial derivatives

$$\begin{aligned}
\frac{\partial g_{mn}}{\partial x^s}\bigg|_{x^i=0} &= \bar{g}_{ab}(O)\left(-\bar{\Gamma}^a_{sn}(O)\delta_m^b - \delta_n^a\bar{\Gamma}^b_{sm}(O)\right) + \bar{g}_{nm,s}(O) \\
&= -\bar{g}_{am}(O)\bar{\Gamma}^a_{sn}(O) - \bar{g}_{nb}(O)\bar{\Gamma}^b_{sm}(O) + \bar{g}_{nm,s}(O)
\end{aligned} \tag{16.46}$$

all vanish, however, since the last row is zero because of (16.13). We have therefore arrived at a metric

$$g_{mn}(x^i) = g_{mn}(O) + \tfrac{1}{2}d_{mnab}(O)x^a x^b, \tag{16.47}$$

which can be changed into (16.41) by transformation to principal axes and stretching of the axes.

Since in a locally flat coordinate system the partial derivatives of the metric vanish at the point $x^a = 0$, and with them the Christoffel symbols, the geodesic equation (16.6) simplifies locally to

$$\frac{\mathrm{d}^2 x^n}{\mathrm{d}\lambda^2} = 0, \tag{16.48}$$

that is, the coordinate lines (e.g. x^1 variable; x^2, x^3 and x^4 constant) are geodesics. One therefore also calls such a coordinate system locally geodesic (at $x^a = 0$).

A locally flat coordinate system offers the best approximation to a Minkowski space that is possible in Riemannian geometry. How good this substitution of a curved space by the tangent space is depends upon the magnitudes of the coefficients d_{mnab} in (16.41), from which we can therefore expect to obtain a measure for the curvature of the space.

16.5 The physical meaning and interpretation of coordinate systems

Coordinates are names which we give to events in the universe; they have in the first instance nothing to do with physical properties. For this reason all coordinate systems are also in principle equivalent, and the choice of a special system is purely a question of expediency. Just as in three-dimensional space for a problem with spherical symmetry one would use spherical coordinates, so, for example, for a static metric one will favour time-orthogonal coordinates (16.35). Because of the great mathematical difficulties in solving problems in the General Theory of Relativity, the finding of a coordinate system adapted to the problem is often the key to success.

In many applications one is interested in the outcome of measurements performed by a special observer (or a family of observers); then one will link the coordinate system with the observer and the objects which he studies (observer on the rotating Earth, in a satellite, ...). After having been thus fixed, the coordinate system naturally has a physical meaning, because it is tied to real objects.

In addition to the comoving coordinates the locally flat coordinate system possesses a particular significance. For an observer at the preferred origin of the coordinate system, particles whose paths are geodesics move force-free, because of (16.48). But geodesics are paths of particles in the gravitational field (as we have made plausible and shall later prove); that is, for the observers just mentioned there exists (locally) no gravitational field: the locally flat coordinate system is the system of a freely falling observer at the point in question of the space-time. This is the best approximation to the Minkowski world, that is, to an inertial system, that Riemannian geometry offers. It is determined in this manner only up to four-dimensional rotations (Lorentz transformations).

Even when one has decided on a particular coordinate system, one should always try to state results in an invariant form; that is, a form independent of the coordinate system. To this end it is clearly necessary to characterize the coordinate system itself invariantly. We shall later familiarize ourselves with the necessary means to do this.

Finally, a few remarks on the question of how one can determine the metric tensor g_{ab} when the coordinate system and auxiliary physical quantities have been specified. Specification of the coordinate system means physically that observers possessing rulers and clocks are distributed in the space. Locally, in the infinitesimal neighbourhood of a point, the question is very easy to answer. One takes a freely falling observer, who measures lengths and times in the manner familiar from Special Relativity, and one then knows the interval ds^2 of two points. One then transforms to the originally given coordinate system; that is, one expresses the result through the coordinates of the observers distributed in the space. Since ds^2 does not change, from $ds^2 = g_{ab}\,dx^a\,dx^b$ one can read off the g_{ab} for known ds^2 and dx^a.

In time measurement, which is especially important, one distinguishes between clocks which run (forwards) *arbitrarily* and thereby show coordinate times t (which therefore have no immediate physical significance), and standard clocks which show proper time τ, defined by $ds^2 = -c^2 \, d\tau^2$. For a clock at rest $(dx^\alpha = 0)$ the two times are related by

$$d\tau^2 = -g_{44} \, d\tau^2. \qquad (16.49)$$

Exercises

16.1   A matrix $A_i^{n'}$ has the structure (n': row, i: column)

$$A_i^{n'} = \begin{pmatrix} (x^1)^2 & & & \\ & 1 & & \\ & & (x^4)^2 & \\ & & & 1 \end{pmatrix}.$$

Can it represent an infinitesimal coordinate transformation $dx^{n'} = A_i^{n'} dx^i$?

16.2   Calculate the matrix $A_n^{n''}$ representing two successive coordinate transformations, and show that the matrices $A_n^{n'}$ form a group.

16.3   A given vector u^a can always be transformed into $u^{a'} = (0,0,0,1)$. Is that also true for u_a?

Further reading for Chapter 16

Eisenhart (1949), Schouten (1954).

17
Tensor algebra

In General Relativity physical quantities and laws are required to have a simple and well-defined behaviour under coordinate transformations

$$dx^{a'} = \frac{\partial x^{a'}}{\partial x^a} dx^a = A_a^{a'} dx^a, \qquad (17.1)$$

just as they do in Special Relativity. In contrast to Lorentz transformations

$$x^{a'} = L^{a'}_{\ a} x^a,\tag{17.2}$$

which are (special) linear transformations with position-independent coefficients $L^{a'}_{\ a}$, we shall now be dealing with linear transformations of coordinate differentials with position-dependent coefficients $A^{a'}_a$. But if we restrict ourselves to the investigation of physical quantities at a given point, without forming derivatives, then the differences from the rules used in calculating with Lorentz transformations will be trivial: they correspond to the difference between orthogonal and non-orthogonal Cartesian coordinates. In particular the formal rules for manipulating tensors are the same as those used in Sections 6.2–6.4 if we substitute $(L^{a'}_{\ a}, \eta_{ab}, \eta^{ab})$ by $(A^{a'}_a, g_{ab}, g^{ab})$, respectively. Referring to those sections, we shall give here only a rather concise review of tensor algebra, concentrating on the differences to Minkowski space and Lorentz transformations.

17.1 Scalars and vectors

Scalars (invariants) A scalar does not change under coordinate transformation,

$$\varphi' = \varphi,\tag{17.3}$$

its numerical value remains unchanged even if the coordinates it depends on are transformed.

Vectors The four quantities T^a are called the *contravariant components* of a vector if they transform like the coordinate differentials

$$T^{n'} = A^{n'}_n T^n.\tag{17.4}$$

This definition implies that the coordinates x^a themselves are *not* the components of a vector – in a Riemannian space there is no position vector.

Using the prescription

$$T_a = g_{an}T^n, \quad T^n = g^{na}T_a,\tag{17.5}$$

one can associate the *covariant components* T_a (index subscripted) with the contravariant components T^n (index superscripted). Because of the transformation laws (17.4) and (16.27) and the relation (16.25), the relations

$$T_{a'} = g_{a'n'}T^{n'} = g_{am}A^a_{a'}A^m_{n'}T^n A^{n'}_n = g_{an}T^n A^a_{a'}\tag{17.6}$$

hold, and therefore

$$T_{a'} = A^a_{a'}T_a.\tag{17.7}$$

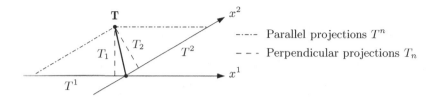

Fig. 17.1. Covariant and contravariant components of a vector **T**.

Covariant and contravariant components describe the same vector, the difference between them being typical of non-orthogonal coordinates.

Fig. 17.1 shows how in the x^1-x^2-plane one obtains the contravariant components by parallel projection onto, and the covariant components by dropping perpendiculars onto, the coordinate axes of a non-orthogonal Cartesian system (with $g_{11} = 1 = g_{22}$).

17.2 Tensors and other geometrical objects

'Geometrical object' is the collective name for all objects whose components Ω^k transform under a given coordinate transformation in such a way that the new components $\overline{\Omega}^k$ are unique functions of the old ones, of the transformation matrix $A_a^{a'}$, and of its derivatives:

$$\overline{\Omega}^k = \overline{\Omega}^k(\Omega^k; A_a^{a'}; A_{a,m}^{a'}; A_{a,mn}^{a'}; \ldots). \qquad (17.8)$$

In this section we shall encounter several geometrical objects which are especially important for physics.

Tensors The quantities $T^a{}_{bc\ldots}{}^{d\cdots}$ are the components of a tensor if, with respect to every upper (contravariant) index, they transform like the contravariant components of a vector, and, to every lower (covariant) index, like the covariant components of a vector:

$$T^{a'}{}_{b'c'\ldots}{}^{d'\cdots} = T^a{}_{bc\ldots}{}^{d\cdots} A_a^{a'} A_{b'}^b A_{c'}^c A_d^{d'} \ldots . \qquad (17.9)$$

The *rank* of a tensor is equal to the number of its indices.

Corresponding to the rule (17.5) for vectors, we can also transform between covariant and contravariant indices. For example, from (17.9) we can form the covariant tensor (tensor written out in purely covariant components)

$$T_{nbcm} = g_{na} g_{md} T^a{}_{bc}{}^d. \qquad (17.10)$$

Evidently the g_{ab} are the covariant components of a second-rank tensor, whose 'mixed' components coincide with the Kronecker symbol,

$$g_n^i = \delta_n^i, \tag{17.11}$$

cp. equations (16.3) and (16.27).

Tensor densities If we transform the determinant $g = |g_{ab}|$ of the metric tensor to another coordinate system, then we obtain

$$g' = |g_{a'b'}| = |g_{ab} A_{a'}^a A_{b'}^b| = |g_{ab}| \cdot |A_{a'}^a| \cdot |A_{b'}^b|, \tag{17.12}$$

that is,

$$g' = |A_{a'}^a|^2 g = \left| \frac{\partial x^a}{\partial x^{a'}} \right|^2 g. \tag{17.13}$$

The square of the functional determinant (Jacobian) $\left| \partial x^a / \partial x^{a'} \right|$ occurs in the transformation law for g; we are dealing with a scalar density of weight 2.

In general we speak of a tensor density of weight W, whenever

$$T^{a'}{}_{b'\ldots}{}^{d'\ldots} = |A_{n'}^n|^W T^a{}_{b\ldots}{}^{d\ldots} A_a^{a'} A_{b'}^b A_d^{d'} \ldots. \tag{17.14}$$

We can draw an important conclusion from equation (17.13); since we admit only those Riemannian spaces which at every point allow the introduction of a local Minkowski system with $g' = |\eta_{mn}| = -1$, then from (17.13) the sign of g does not change under an arbitrary coordinate transformation, and so g is always negative.

Pseudotensors In the transformation law of a pseudotensor (compared with that of a tensor) there occurs also the sign of one of the elements of the transformation matrix $A_{a'}^a$ or of a combination of its elements. A simple example is a pseudovector,

$$T^{n'} = \text{sgn} |A_{a'}^a| A_n^{n'} T^n, \tag{17.15}$$

in whose transformation law the sign of the functional determinant of the coordinate transformation occurs. Under coordinate transformations possessing locally the character of a rotation it behaves like a vector; under reflections it also changes its sign.

The ε-pseudotensor The most important pseudotensor occurring in General Relativity is the ε-pseudotensor, often also short: ε-tensor. We have learned in Section 6.1 that in a Minkowski space it is defined so that under interchange of two arbitrary indices its sign changes (it is completely antisymmetric), and that

$$\overset{M}{\varepsilon}{}^{1'2'3'4'} = 1 \qquad (17.16)$$

holds (we have added the superscript M to emphasize that this is valid only in a Minkowski system, and used coordinates $x^{n'}$). To define the ε-pseudotensor in a Riemannian space, we can use this equation and demand that (17.16) be valid in any local Minkowski system. If we transform from this local Minkowski system $x^{n'}$ to an arbitrary coordinate system x^n, then we have

$$\varepsilon^{abcd} = \pm \overset{M}{\varepsilon}{}^{a'b'c'd'} A^a_{a'} A^b_{b'} A^c_{c'} A^d_{d'}. \qquad (17.17)$$

The apparent ambiguity of sign appearing in (17.17) has its origin in the pseudotensorial property. For example, (17.16) must again hold after a pure reflection $x^{1'} = -x^1$, that is, for $A^1_{1'} = -1$, $A^2_{2'} = A^3_{3'} = A^4_{4'} = 1$, $A^a_{a'} = 0$ otherwise, which once again leads to a Minkowski system; the components of the ε-tensor must not change.

One can, however, bring (17.17) into a more easily manageable form. If one fixes the indices $abcd$ as 1234, then the right hand side of the formula

$$\varepsilon^{1234} = \pm \overset{M}{\varepsilon}{}^{a'b'c'd'} A^1_{a'} A^2_{b'} A^3_{c'} A^4_{d'} \qquad (17.18)$$

is (up to the $\pm$) precisely the determinant $|A^a_{a'}|$. Since in every Minkowski system $g' = -1$, then because of (17.13) this determinant has the value

$$|A^a_{a'}| = \pm 1/\sqrt{-g}, \qquad (17.19)$$

so that from (17.17) one obtains

$$\varepsilon^{abcd} = \pm \overset{M}{\varepsilon}{}^{abcd} |A^n_{n'}| = \overset{M}{\varepsilon}{}^{abcd}/\sqrt{-g}, \qquad (17.20)$$

and thus the relation

$$\varepsilon^{1234} = 1/\sqrt{-g}, \qquad (17.21)$$

which replaces (17.16). For the covariant components of the ε-tensor one gets by lowering the indices according to (17.13)

$$\begin{aligned}
\varepsilon_{abcd} &= g_{am}g_{bn}g_{cp}g_{dq}\varepsilon^{mnpq} \\
&= g_{am}g_{bn}g_{cp}g_{dq}\overset{M}{\varepsilon}{}^{mnpq}/\sqrt{-g} = -g\,\overset{M}{\varepsilon}{}_{abcd}/\sqrt{-g},
\end{aligned} \qquad (17.22)$$

so that

$$\varepsilon_{1234} = -\sqrt{-g}. \qquad (17.23)$$

Equation (17.21), together with the property of being completely antisymmetric, uniquely defines the ε-pseudotensor.

Two-point tensors Two-point tensors are not geometrical objects in the strict sense. They appear in the description of physical processes in which a cause at a point P brings about an effect at the point $\overline{P}$. Their indices refer to the points $\overline{P}$ and P, and are written respectively with and without a bar over the index. Accordingly the transformation law for a two-point tensor reads, for example,

$$T_{\overline{a}'n'}(\overline{P}, P) = T_{\overline{a}n}(\overline{P}, P) A_{\overline{a}'}^{\overline{a}}(\overline{P}) A_{n'}^{n}(P). \qquad (17.24)$$

An example of a two-point scalar is the arclength of a geodesic which connects the points P and $\overline{P}$. We shall meet an example of a second-rank two-point tensor in Section 18.3 (the parallel propagator).

17.3 Algebraic operations with tensors

The rules for addition, multiplication, contraction and inner product of tensors are the same as in Minkowski space, see Section 6.2, the properties (6.17)–(6.21) of products of ε-tensors remain unchanged, and also symmetries of tensors are defined the same way; only on some occasions one has to replace η_{mn} by g_{mn}, or to add an additional $\sqrt{-g}$.

So the decomposition (6.31) of an arbitrary second rank tensor now reads

$$T_{ab} = T_{[ab]} + T_{(ab)} = T_{[ab]} + \{T_{(ab)} - \tfrac{1}{4}T^n_n g_{ab}\} + \tfrac{1}{4}T^n_n g_{ab}, \qquad (17.25)$$

the eigenvector equation (6.33) for a symmetric tensor T_{ab} is to be replaced by

$$(T_{ab} - \lambda g_{ab}) w^b = 0, \qquad (17.26)$$

and the coefficients α_1 and α_4 occurring in (6.35) are now

$$\alpha_1 = T^n_n, \quad \alpha_4 = -\|T_{ab}\| \, g^{-1}. \qquad (17.27)$$

Similarly, the secular equation for the antisymmetric tensors now reads

$$\|F_{ab} - \lambda g_{ab}\| = 0. \qquad (17.28)$$

As in Minkowski space, every antisymmetric tensor F_{ab} can be *dualized* with the aid of the ε-tensor by defining

$$\widetilde{F}^{ab} = \tfrac{1}{2}\varepsilon^{abcd} F_{cd}. \qquad (17.29)$$

Because of the property (6.19) of the ε-tensor, a double application of the duality operation yields the original tensor, apart from a sign:

$$\widetilde{\widetilde{F}}_{nm} = \tfrac{1}{2}\varepsilon_{nmab}\widetilde{F}^{ab} = \tfrac{1}{4}\varepsilon_{nmab}\varepsilon^{abcd} F_{cd} = -F_{nm}. \qquad (17.30)$$

17.4 Tetrad and spinor components of tensors

Tetrads At every point of space one can introduce systems of four linearly independent vectors $h_a^{(r)}$, which are known as tetrads. The index in brackets is the tetrad index; it numbers the vectors from one to four. These four vectors can have arbitrary lengths and form arbitrary angles with one another (as long as they remain linearly independent). The matrix

$$g^{(r)(s)} = h_a^{(r)} h_b^{(s)} g^{ab} \tag{17.31}$$

is an arbitrary symmetric matrix with negative-definite determinant. Its inverse $g_{(s)(t)}$, which is defined by

$$g_{(s)(t)} g^{(t)(r)} = \delta_{(s)}^{(r)} = g_{(s)}^{(r)}, \tag{17.32}$$

can be used to define tetrad vectors with tetrad indices subscripted

$$h_{(r)a} = g_{(r)(s)} h_a^{(s)}, \tag{17.33}$$

and to solve (17.31) for g_{ab}:

$$g_{ab} = g_{(r)(s)} h_a^{(r)} h_b^{(s)}. \tag{17.34}$$

Tetrad components of tensors Just as one can write any arbitrary vector as a linear combination of the four tetrad vectors, so one can use them to describe any tensor

$$T^{ab\ldots}{}_{nm\ldots} = T^{(r)(s)\ldots}{}_{(p)(q)\ldots} h_{(r)}^a h_{(s)}^b h_n^{(p)} h_m^{(q)} \ldots . \tag{17.35}$$

The quantities $T^{(r)(s)\ldots}{}_{(p)(q)\ldots}$ are called the tetrad components of the tensor. They are calculated according to

$$T^{(r)(s)\ldots}{}_{(p)(q)\ldots} = T^{ab\ldots}{}_{nm\ldots} h_a^{(r)} h_b^{(s)} h_{(p)}^n h_{(q)}^m \ldots , \tag{17.36}$$

which is consistent with (17.31) and (17.33). Tetrad indices are raised and lowered by $g^{(r)(s)}$ and $g_{(r)(s)}$, respectively.

Coordinate and tetrad transformations The advantages offered in many cases by the use of the tetrad components, which at first look very complicated, become clear when one examines their transformation properties and when one introduces tetrads which are appropriate to the particular problem being investigated.

As one can see from a glance at the defining equation (17.36), the tetrad components behave like scalars under coordinate transformations; clearly the labelling of the tetrad vectors, that is, their tetrad indices, does not change under a coordinate transformation. One has therefore a good way of investigating the algebraic properties of tensors and can simplify tensor components (that is, tetrad components) in a coordinate-independent fashion by the choice of the tetrads.

Besides the coordinate transformations – and completely independently of them – one can introduce a new tetrad system through a linear (position-dependent) transformation of the tetrad vectors $h_a^{(r)}$ at every point in the space:

$$h_a^{(r)'} = A_{(r)}^{(r)'} h_a^{(r)}, \quad h_{(r)'a} = A_{(r)'}^{(r)} h_{(r)a}, \quad A_{(r)}^{(r)'} A_{(s)'}^{(r)} = \delta_{(s)'}^{(r)'}. \quad (17.37)$$

Under such transformations, of course, the tetrad components of tensors alter; indeed they will be transformed with the matrices $A_{(r)}^{(r)'}$ and $A_{(r)'}^{(r)}$, respectively, for example,

$$g_{(s)'(t)'} = g_{(s)(t)} A_{(s)'}^{(s)} A_{(t)'}^{(t)}. \quad (17.38)$$

Special tetrad systems We can choose the tetrads in such a way that the four vectors at each point are in the directions of the coordinate axes; that is, parallel to the four coordinate differentials dx^a:

$$h_{(r)}^a = \delta_{(r)}^a, \quad g_{rs} = g_{(r)(s)}. \quad (17.39)$$

This choice has the consequence that tetrad and tensor components coincide. But on the other hand, given an arbitrary tetrad system in the space, it is not always possible to transform the coordinates so that the tetrads become tangent vectors to the coordinate lines.

A second important possibility is the identification of the tetrad vectors with the base vectors of a Cartesian coordinate system in the local Minkowski system of the point concerned:

$$g_{(r)(s)} = h_{(r)}^a h_{(s)}^b g_{ab} = \eta_{(r)(s)} = \begin{pmatrix} 1 & & & \\ & 1 & & \\ & & 1 & \\ & & & -1 \end{pmatrix}. \quad (17.40)$$

The four tetrad vectors, which we shall call z_a, w_a, v_a and u_a/c, form an orthonormal system of one timelike and three spacelike vectors. From (17.34) and (17.40), it follows that the metric tensor can be written as

$$g_{ab} = z_a z_b + w_a w_b + v_a v_b - u_a u_b/c^2. \quad (17.41)$$

A third special case is the use of null vectors as tetrad vectors, a possibility we already exploited in Section 9.1. As explained there, we take two real null vectors

$$k_a = \tfrac{1}{\sqrt{2}}(u_a/c + v_a), \quad l_a = \tfrac{1}{\sqrt{2}}(u_a/c - v_a), \quad (17.42)$$

and the two complex null vectors

$$m_a = \tfrac{1}{\sqrt{2}}(z_a - \mathrm{i}\, w_a), \quad \overline{m}_a = \tfrac{1}{\sqrt{2}}(z_a + \mathrm{i}\, w_a). \tag{17.43}$$

The system $(k_a, l_a, m_a, \overline{m}_a)$ of four null vectors is called a *null tetrad*, or a *Sachs tetrad*, or a *Newman–Penrose* tetrad (the reader should be aware of different sign conventions in the literature). Only two of the products of these null vectors are non-zero,

$$k^a l_a = -1, \quad m^a \overline{m}_a = 1, \quad \text{all other products zero.} \tag{17.44}$$

We thus have

$$g_{(r)(s)} = \begin{pmatrix} 0 & 1 & & \\ 1 & 0 & & \\ & & 0 & -1 \\ & & -1 & 0 \end{pmatrix}. \tag{17.45}$$

and

$$g_{ab} = m_a \overline{m}_b + \overline{m}_a m_b - k_a l_b - l_a k_b. \tag{17.46}$$

Using this system, complex tetrad components can arise, although we have allowed only real coordinates and tensors.

Spinors First-rank spinors are elements of a two-dimensional, complex vector-space, in which an alternating scalar product

$$[\varphi, \psi] = -[\psi, \varphi] \tag{17.47}$$

is defined. A spinor φ can be represented either by its contravariant components φ^A or by its covariant components φ_A. The scalar product of two spinors can be formed from these components with the help of the *metric spinor* ϵ_{AB},

$$[\varphi, \psi] = \epsilon_{AB}\varphi^A \psi^B = -\epsilon_{AB}\psi^A \varphi^B, \quad A, B = 1, 2. \tag{17.48}$$

The metric spinor is antisymmetric:

$$\epsilon_{AB} = -\epsilon_{BA}. \tag{17.49}$$

Together with its inverse, defined by

$$\epsilon_{AB}\epsilon^{CB} = \delta_A^C, \tag{17.50}$$

it can be used to shift indices:

$$\varphi^A = \epsilon^{AB}\varphi_B, \qquad \varphi_B = \varphi^A \epsilon_{AB}. \tag{17.51}$$

The scalar products (17.47) and (17.48) do not change if one carries out a unimodular transformation

$$\varphi^{A'} = \Lambda_A^{A'}\varphi^A, \quad \varphi_{A'} = \Lambda_{A'}^A \varphi_A, \quad |\Lambda_A^{A'}| = 1, \quad \Lambda_A^{A'}\Lambda_{B'}^A = \delta_{B'}^{A'}. \tag{17.52}$$

The connection between the group of the unimodular transformations and the group of Lorentz transformations isomorphic to it plays a great rôle in special-relativistic field theory.

We denote quantities which transform with the complex matrix $\overline{(\Lambda_A^{\dot{A}'})}$ $= \Lambda_{\dot{A}}^{\dot{A}'}$ by a dot over the index $\varphi_{\dot{A}}$, $\psi^{\dot{B}}$, ... (in the literature, a prime on the index is also customary). They obey

$$\varphi^{\dot{A}'} = \Lambda_{\dot{A}}^{\dot{A}'} \varphi^{\dot{A}}. \tag{17.53}$$

Scalar products $\varphi_{\dot{A}} \psi^{\dot{A}} = \epsilon_{\dot{A}\dot{B}} \varphi^{\dot{A}} \psi^{\dot{B}}$ remain invariant under such transformations. According to this convention one forms the complex conjugate of a spinor by dotting the index (with $\ddot{A} \equiv A$, naturally):

$$\overline{(\varphi_A)} = \varphi_{\dot{A}}. \tag{17.54}$$

Spinors $\chi_{AB...}^{\ \ \ M\dot{N}...}$ of higher rank are structures which behave with respect to unimodular transformations of each index like the corresponding first-rank spinor. The rules for handling these spinors follow from the properties of first-rank spinors sketched above. Notice that upon multiplication and contraction, only summation over a contravariant and a covariant index of the same type (that is, dotted *or* undotted) yields a spinor again.

A spinor is *Hermitian* if it obeys the condition

$$\varphi_{\dot{A}B} = \varphi_{B\dot{A}}. \tag{17.55}$$

Spinor components and tensors With the aid of the metric spin-tensors $\sigma_m^{\ A\dot{B}} = \sigma_m^{\ \dot{B}A}$, which are generalizations of the Pauli spin-matrices, one can map the four complex components $\varphi_{A\dot{B}}$ of an arbitrary second-rank spinor onto the four (now also complex in general) components of a four-vector:

$$T^a = \sigma^a_{\ A\dot{B}} \varphi^{A\dot{B}} / \sqrt{2} \quad \leftrightarrow \quad \varphi_{A\dot{B}} = -\sigma_{aA\dot{B}} T^a / \sqrt{2},$$
$$T^a T_a = -\varphi_{A\dot{B}} \varphi^{A\dot{B}}. \tag{17.56}$$

Here the four 2×2 matrices $\sigma_a^{\ A\dot{B}}$ satisfy the equations

$$\sigma^a_{\ A\dot{B}} \sigma_{aC\dot{D}} = -2\epsilon_{AC} \epsilon_{\dot{B}\dot{D}}, \quad \sigma_m^{\ A\dot{B}} \sigma_{nA\dot{B}} = -2g_{mn}. \tag{17.57}$$

In analogous fashion one can map every nth-rank tensor to a spinor of rank $2n$.

From two basis spinors χ_A and μ_A, which satisfy the relations

$$\chi_A \mu^A = -\mu_A \chi^A = 1, \quad \chi_A \chi^A = \mu_A \mu^A = 0, \tag{17.58}$$

one can form four second-rank spinors

$$m_{A\dot{B}} = \chi_A \mu_{\dot{B}}, \quad \overline{m}_{A\dot{B}} = \mu_A \chi_{\dot{B}}, \quad k_{A\dot{B}} = \chi_A \chi_{\dot{B}}, \quad l_{A\dot{B}} = \mu_A \mu_{\dot{B}}. \quad (17.59)$$

The vectors m_a, $\bar{m}_a$, k_a and l_a associated with them according to the prescription (17.56) satisfy the relations (17.45) and (17.44) of the null-tetrad system. Thus there exists a close relation between the representation of a tensor by its spinor components and its representation by components related to a null tetrad.

Exercises

17.1 A vector **A** is given in a two-dimensional Cartesian coordinate system. Perform a coordinate transformation $x'^1 = x^1 - bx^2$, $x'^2 = x^2$, and draw the contravariant and the covariant components of that vector in both systems.

17.2 Show that the symmetry property $T_{ab} = \frac{1}{2}(T_{ab} + T_{ba})$ is invariant under coordinate transformations.

Further reading for Section 17.4

Eisenhart (1949), Penrose and Rindler (1984, 1986).

18

The covariant derivative and parallel transport

18.1 Partial and covariant derivatives

Physical laws are usually written down in mathematical form as differential equations. In order to guarantee that the laws are independent of the coordinate system, they should moreover have the form of tensor equations. We must therefore examine whether tensors can be differentiated in such a way that the result is again a tensor, and if so, how this can be done.

The partial derivative We denote the usual partial derivative of a position-dependent tensor by a comma:

$$\frac{\partial T^{ab\ldots}_{\quad c\ldots}}{\partial x^i} = T^{ab\ldots}_{\quad c\ldots,i}. \tag{18.1}$$

However, the components $T^{ab\ldots}_{\quad c\ldots,i}$ are not the components of a tensor, as we can show from the example of the derivative of a vector. For we have

$$(T^n_{,i})' = \left(\frac{\partial T^n}{\partial x^i}\right)' = \frac{\partial}{\partial x^{i'}}(A^{n'}_n T^n) = \frac{\partial x^i}{\partial x^{i'}}\frac{\partial}{\partial x^i}(A^{n'}_n T^n)$$
$$= A^i_{i'} A^{n'}_n T^n_{,i} + A^i_{i'} A^{n'}_{n,i} T^n. \tag{18.2}$$

That is, the $T^n_{,i}$ transform like the components of a tensor if, and only if, the transformation matrices $A^{n'}_n$ are independent of position (this is true, for example, for Lorentz transformations of Minkowski space).

The only exception is the generalized gradient $\varphi_{,a} = \partial\varphi/\partial x^a$ of a scalar φ; its components are those of a covariant vector. From $\varphi' = \varphi$, and hence $\mathrm{d}\varphi' = \mathrm{d}\varphi$, we have

$$\varphi_{,a}\,\mathrm{d}x^a = (\varphi_{,a})'\,\mathrm{d}x^{a'}, \tag{18.3}$$

and the quotient law ensures the vector property of $\varphi_{,a}$.

One can see why the partial derivatives of a tensor do not form a tensor if one describes a constant vector field in the plane by polar coordinates (Fig. 18.1). The vector components of this constant vector field become position-dependent, because the directives of the coordinate lines change from point to point; the partial derivative of the vector components is a measure of the actual position-dependence of the vector only in Cartesian coordinate systems.

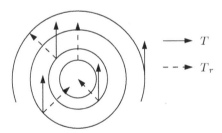

Fig. 18.1. Components of a constant vector field in polar coordinates.

Definition of the covariant derivative The above considerations suggest that a covariant derivative (which produces tensors from tensors) can be constructed from the partial derivative by making use of locally geodesic coordinates and defining: *The covariant derivative $T^{ab\ldots}_{\quad mn\ldots;i}$ of a tensor $T^{ab\ldots}_{\quad mn\ldots}$ is again a tensor, which coincides with the partial derivative in* *the locally geodesic coordinate system:*

$$T^{ab...}{}_{mn...;i} = T^{ab...}{}_{mn...,i} \quad \text{for } g_{mn} = \eta_{mn}, \ \Gamma^a_{bc} = 0. \tag{18.4}$$

This definition of the covariant derivative is unique. It ensures the tensor property and facilitates the translation of physical laws to the Riemannian space, if these laws are already known in the Minkowski space (using partial derivatives).

Properties of, and rules for handling, the covariant derivatives Since the partial derivatives of η_{mn} and $\overset{M}{\varepsilon}{}^{abcd}$ are zero, we have the equations

$$g_{ab;m} = 0, \qquad \varepsilon^{abcd}{}_{;n}. \tag{18.5}$$

The metric tensor and the ε-tensor are thus covariantly constant.

Because the product rule and the sum rule hold for partial derivatives, these rules also hold for covariant derivatives:

$$\begin{aligned}
(T^{ab} + S^{ab})_{;n} &= T^{ab}{}_{;n} + S^{ab}{}_{;n}, \\
(T^{ab}S_c)_{;n} &= T^{ab}{}_{;n}S_c + T^{ab}S_{c;n}.
\end{aligned} \tag{18.6}$$

Contraction, raising and lowering of indices, and taking the dual depend upon multiplication with the metric or the ε-tensor. These operations therefore commute with covariant differentiation. For example,

$$(T^a{}_a)_{;n} = T^{ab}{}_{;n}g_{ab} = (T^a{}_a)_{,n}, \tag{18.7}$$

in agreement with the fact that the covariant derivative of a scalar is equal to its partial derivative.

For practical calculations we naturally need also a formula for determining the covariant derivative in a given coordinate system – we certainly do not want to transform every time first to the local geodesic system, calculate the partial derivatives, and then transform back. This formula is given by the following prescriptions.

The covariant derivatives of the contravariant and covariant components of a vector are calculated according to the formulae

$$\begin{aligned}
T^a{}_{;n} &= T^a{}_{,n} + \Gamma^a_{nm}T^m, \\
T_{a;n} &= T_{a,n} - \Gamma^m_{an}T_m,
\end{aligned} \tag{18.8}$$

respectively. The covariant derivative of an arbitrary tensor is calculated by applying the prescription (18.8) to every contravariant and covariant index, for example,

$$T^a{}_{bc;d} = T^a{}_{bc,d} + \Gamma^a_{dm}T^m{}_{bc} - \Gamma^m_{bd}T^a{}_{mc} - \Gamma^m_{cd}T^a{}_{bm}. \tag{18.9}$$

To show that these prescriptions meet the definition of the covariant derivative given above, we first observe that in a locally geodesic system all Christoffel symbols disappear, so that, there, covariant and partial derivatives coincide. The formulae (18.8) really do produce a tensor, although the two terms on the right-hand sides are not separately tensors; we leave the proof of this to the reader.

Although the covariant derivative always produces a covariant index, one also writes it as a contravariant index; for example $T^{ab;n}$ is an abbreviation for

$$T^{ab;n} = g^{ni}T^{ab}{}_{;i}. \tag{18.10}$$

18.2 The covariant differential and local parallelism

There is an obvious geometric meaning to the covariant derivative which we shall describe in the following. One can visualize the covariant derivative – like the partial derivative – as the limiting value of a difference quotient. In this context one does not, however, simply form the difference in the value of the tensor components at the points x^i and $x^i + dx^i$:

$$dT^a = T^a(x^i + dx^i) - T^a(x^i) = T^a{}_{,i}\, dx^i \tag{18.11}$$

(this would correspond to the partial derivative), but rather uses

$$DT^a = dT^a + \Gamma^a{}_{nm}T^n\, dx^m = (T^a{}_{,m} + \Gamma^a{}_{nm}T^n)\, dx^m = T^a{}_{;m}\, dx^m. \tag{18.12}$$

The deeper reason for this more complicated formula lies in the fact that tensors at two different points x^i and $x^i + dx^i$ obey different transformation laws, and hence their difference is not a tensor. Before forming the difference, the tensor at the point $x^i + dx^i$ must therefore be transported in a suitable manner (preserving the tensor property) to the point x^i, without of course changing it during the process. In our usual three-dimensional space we would translate 'without changing it' as 'keeping it parallel to itself'. We shall take over this way of speaking about the problem, but we must keep clearly in mind that the meaning of 'parallelism at different points' and 'parallel transport' is not at all self-evident in a non-Euclidean space.

Three simple examples may illustrate this. Referring to Fig. 18.2, we ask the following questions. (*a*) Are two vectors in a plane section still parallel after bending of the plane? (*b*) Are the two vectors, which are parallel in three-dimensional space, also parallel in the curved surface? (Obviously not, for vectors *in* the surface can have only two components,

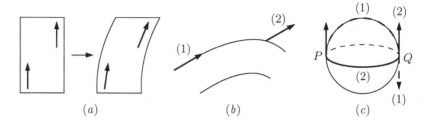

Fig. 18.2. Parallelism of vectors on surfaces.

whereas vector (2) juts out of the surface and has three components –
but should one perhaps take the projection onto the surface?) (*c*) Which
of the two vectors at the point Q of the sphere of Fig. 18.2*c* is parallel to
that at the point P? Clearly *both* were parallelly transported, the one
along the equator always perpendicularly, and the other over the poles,
always parallel to the curve joining P and Q!

What we should realize from these examples is that in a curved space
one must *define* what one means by parallelism and parallel displacement. The definition used in the construction of the covariant derivative
obviously reads: two vectors at infinitesimally close points are parallel
if, and only if, we have

$$DT^a = dT^a + \Gamma^a_{nm} T^n \, dx^m = T^a_{\;;m} \, dx^m = 0; \qquad (18.13)$$

that is to say, their covariant differential disappears. A vector field is
parallel in the (infinitesimal) neighbourhood of a point if its covariant
derivative is zero there:

$$T^a_{\;;n} = T^a_{\;,n} + \Gamma^a_{nm} T^m = 0. \qquad (18.14)$$

If in a general affine (hence possibly even non-Riemannian) curved
space one were also to use this definition, then the Γ^a_{mn} would in that
case be arbitrary functions. A Riemannian space is distinguished by
the fact that Γ^a_{mn} are precisely the Christoffel symbols formed from
the metric tensor. The definitions (18.13) and (18.14) are of course so
constructed that in the local geodesic system they lead to the usual
parallel displacement in Minkowski space.

18.3 Parallel displacement along a curve and the parallel propagator

Let an arbitrary curve in our Riemannian space be given parametrically by $x^n = x^n(\lambda)$. It is then always possible to construct a parallel vector field along this curve from the requirement that the covariant differential of a vector along the curve vanishes; that is, from

$$\frac{DT^a}{D\lambda} \equiv \frac{T^a{}_{;n}\,\mathrm{d}x^n}{\mathrm{d}\lambda} = \frac{\mathrm{d}T^a}{\mathrm{d}\lambda} + \Gamma^a_{nm}T^m\frac{\mathrm{d}x^n}{\mathrm{d}\lambda} = 0. \tag{18.15}$$

One can in fact specify arbitrarily the value of the vector components T^a at some initial point $\lambda = \bar{\lambda}$ and uniquely determine the vector at some other arbitrary point λ of the curve from the system of differential equations (18.15).

The geodesic equation

$$\frac{\mathrm{d}^2x^a}{\mathrm{d}\lambda^2} + \Gamma^a_{nm}\frac{\mathrm{d}x^n}{\mathrm{d}\lambda}\frac{\mathrm{d}x^m}{\mathrm{d}\lambda} = \frac{D}{D\lambda}\frac{\mathrm{d}x^a}{\mathrm{d}\lambda} = 0 \tag{18.16}$$

is obviously an example of such an equation which expresses the parallel transport of a vector. It says that the tangent vector $t^a = \mathrm{d}x^a/\mathrm{d}\lambda$ of a geodesic remains parallel to itself. The geodesic is thus not only the shortest curve between two points, but also the straightest. The straight line in Euclidean space also has these two properties.

There is precisely one geodesic between two points if one excludes the occurrence of conjugate points. (Such points of intersection of geodesics which originate in one point occur, for example, on a sphere: all great circles originating at the north pole intersect one another at the south pole.) The result of parallelly transporting a vector (or a tensor) from the point $\bar{P}$ to the point P *along a geodesic* is therefore uniquely determined, while in general it certainly depends upon the choice of route (see Section 19.2). Since the differential equation (18.15) to be integrated is linear in the components of the vector T^a to be transported, the vector components at the point P are linear functions of the components at the point $\bar{P}$:

$$T_a(P) = g_{a\bar{b}}(P,\bar{P})T^{\bar{b}}(\bar{P}). \tag{18.17}$$

For tensor components we have analogously

$$T_a{}^b{}_{n...} = g_{a\bar{a}}g^{b\bar{b}}g_{n\bar{n}}\dots T^{\bar{a}}{}_{\bar{b}}{}^{\bar{n}...}. \tag{18.18}$$

The quantities $g_{a\bar{b}}$ are the components of the *parallel propagator*. It is a two-point tensor of the type (17.24); the indices (barred or not) of such a tensor also indicate the coordinates of which points it depends upon. For more details, see Synge (1960).

18.4 Fermi–Walker transport

The parallel displacement of a vector appears to be the most natural way of comparing vectors at two different points of the space with one another or of transporting one to the other point. There are, however, physically important cases in which another kind of transport is more useful for the formulation of physical laws.

An observer who moves along an arbitrary timelike curve $x^n(\tau)$ under the action of forces will regard as natural, and use, a (local) coordinate system in which he himself is at rest and his spatial axes do not rotate. He will therefore carry along with him a tetrad system whose timelike vector is always parallel to the tangent vector $t^n = \mathrm{d}x^n/\mathrm{d}\tau$ of his path, for only then does the four-velocity of the observer possess no spatial components (he really is at rest), and he will regard as constant a vector whose components do not change with respect to this coordinate system.

The fact that the tangent vector to his own path does not change for the observer cannot, however, be expressed by saying that it is parallelly transported along the path. Indeed the observer does not in general move along a geodesic, and therefore under parallel transport a vector v^n pointing initially in the direction of the motion will later make an angle with the world line (see Fig. 18.3).

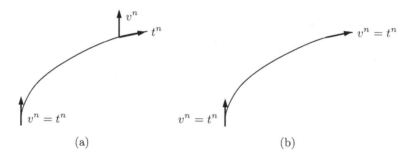

Fig. 18.3. Parallel transport (a) and Fermi–Walker transport (b) of a vector v^n.

If, however, for every vector T^n one uses *Fermi–Walker transport*, defined by the vanishing of the *Fermi derivative*, that is, by

$$\frac{\mathrm{D}T^n}{\mathrm{D}\tau} - T_a \frac{1}{c^2}\left(\frac{\mathrm{d}x^n}{\mathrm{d}\tau}\frac{\mathrm{D}^2 x^a}{\mathrm{D}\tau^2} - \frac{\mathrm{d}x^a}{\mathrm{d}\tau}\frac{\mathrm{D}^2 x^n}{\mathrm{D}\tau^2}\right) = 0, \tag{18.19}$$

then one can establish that the tangent vector t^n to an arbitrary timelike curve in the space is indeed Fermi–Walker-transported, since for $T^n = t^n$ (18.19) is satisfied identically as a consequence of the relation

$$\frac{D}{D\tau}\left(\frac{dx_n}{d\tau}\frac{dx^n}{d\tau}\right) = 0 = 2\frac{dx_n}{d\tau}\frac{D^2x^n}{D\tau^2}, \tag{18.20}$$

which follows from $dx^n\, dx_n = -c^2\, d\tau^2$. If the observer moves on a geodesic $D^2x^n/D\tau^2 = 0$, then parallel transport and Fermi–Walker transport coincide.

For a given curve $x^n(\tau)$ through the space equation (18.19) provides a definition of how the change of a vector T^n under advance along the curve is to be calculated from the initial values of the vector. The reader may confirm that the scalar product of vectors does not change under this type of transport, and therefore that lengths and angles remain constant.

A Fermi–Walker-transport tetrad-system is the best approximation to the coordinate system of an observer who employs locally a non-rotating inertial system in the sense of Newtonian mechanics (cp. Section 21.2).

18.5 The Lie derivative

If in a space a family of world lines (curves) is available which covers the space smoothly and continuously, one speaks of a congruence of world lines. Such curves can be the world lines of particles of a fluid, for example. With every such congruence is associated a vector field $a^n(x^i)$, which at any given time has the direction of the tangent to the curve going through the point in question.

Let a vector field $T^n(x^i)$ also be given. One can now ask the question, how can the change of the vector T^n under motion of the observer in the direction of the vector field a^n be defined in an invariant (coordinate-system-independent) manner? Of course one will immediately think of the components $T^n_{\;;i}a^i$ of the covariant derivative of T^n in the direction of a^i. There is, however, yet a second kind of directional derivative, independent of the covariant derivative, namely the Lie derivative.

This derivative corresponds to the change determined by an observer who goes from the point P (coordinates x^i) in the direction of a^i to the infinitesimally neighbouring point $\bar{P}$ (coordinates $\bar{x}^i = x^i + \varepsilon a^i(x^n)$) and takes his coordinate system with him (see Fig. 18.4).

If, however, at the point $\bar{P}$ he uses the coordinate system appropriate for P, then this corresponds to a coordinate transformation which associates with the point $\bar{P}$ the coordinate values of point P; that is, the transformation

$$x^{n'} = x^n - \varepsilon a^n(x^i), \quad A^{n'}_i = \delta^n_i - \varepsilon a^n_{\;,i}. \tag{18.21}$$

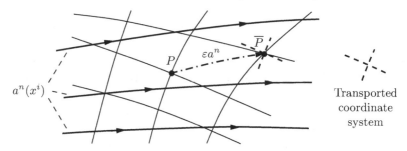

Fig. 18.4. How the Lie derivative is defined.

He will therefore regard as components of the vector T^n at the point $\bar{P}$ the quantities

$$
\begin{aligned}
T^{n'}(\bar{P}) &= A_i^{n'} T^i(x^k + \varepsilon a^k) = (\delta_i^n - \varepsilon a^n{}_{,i})\left[T^i(P) + \varepsilon T^i{}_{,k}(P)a^k\right] \\
&= T^n(P) + \varepsilon T^n{}_{,k}(P)a^k - \varepsilon a^n{}_{,k}T^k(P),
\end{aligned} \tag{18.22}
$$

(ignoring terms in ε^2) and compare them with $T^n(P)$.

This consideration leads us to define the Lie derivative in the direction of the vector field a^n as the limiting value

$$
\mathcal{L}_{\mathbf{a}}T^n = \lim_{\varepsilon \to 0} \frac{1}{\varepsilon}\left[T^{n'}(\bar{P}) - T^n(P)\right], \tag{18.23}
$$

or the expression, which is equivalent because of (18.22),

$$
\mathcal{L}_{\mathbf{a}}T^n = T^n{}_{,k}a^k - T^k a^n{}_{,k}. \tag{18.24}
$$

The Lie derivative of the covariant components T_n follows analogously as

$$
\mathcal{L}_{\mathbf{a}}T_n = T_{n,i}a^i + T_i a^i{}_{,n}. \tag{18.25}
$$

One forms the Lie derivative of a tensor of higher rank by carrying over (18.24) or (18.25) to every contravariant or covariant index, respectively; thus for example

$$
\mathcal{L}_{\mathbf{a}}g_{mn} = g_{mn,i}a^i + g_{in}a^i{}_{,m} + g_{mi}a^i{}_{,n}. \tag{18.26}
$$

Obviously the Lie derivative and the usual directional derivative coincide if the partial derivatives $a^i{}_{,n}$ are zero, for example, in a comoving coordinate system $a^i = (0,0,0,1)$.

The Christoffel symbols are not used in the calculation of the Lie derivative. One can, however, in (18.24)–(18.26) replace the partial derivatives by covariant derivatives, according to (18.8), obtaining

$$\mathcal{L}_{\mathbf{a}}T^n = T^n{}_{;i}a^i - T^i a^n{}_{;i},$$

$$\mathcal{L}_{\mathbf{a}}T_n = T_{n;i}a^i + T_i a^i{}_{;n}, \qquad (18.27)$$

$$\mathcal{L}_{\mathbf{a}}g_{mn} = a_{m;n} + a_{n;m}.$$

This result shows explicitly that the Lie derivative of a tensor is again a tensor, although only partial derivatives were used in its definition.

The Lie derivative of tensors has the following properties, which we list here without proof.

(a) It satisfies the Leibniz product rule.
(b) It commutes with the operation of contraction (although the Lie derivative of the metric tensor does not vanish).
(c) It can be applied to arbitrary, linear geometrical objects, to Christoffel symbols, for example.
(d) It commutes with the partial derivative.

The Lie derivative plays an important rôle in the investigation of symmetries of Riemannian spaces, see Chapter 33.

Exercises

18.1 Use equations (18.8), (16.21) and (16.31) to show that $(T^a{}_{;n})'$ $= \partial T^{a'}/\partial x^{n'} + \Gamma^{a'}_{n'm'}T^{m'}$ transforms like a tensor, i.e. that $(T^a{}_{;n})' = A^{a'}_a A^n_{n'} T^a{}_{;n}$ holds.

18.2 Apply $h_{(a)i} = g_{i\bar{n}}h^{\bar{n}}_{(a)}$ to a tetrad system (17.40) with (17.32) to show that $g_{i\bar{n}} = g_{\bar{n}i}$ holds.

18.3 Show that the scalar product of any two vectors does not change under Fermi–Walker transport (18.19).

18.4 Show that the Lie derivative really has properties (a), (b), and (d).

18.5 In a space with a given metric g_{ab}, a covariant derivative is defined by $T^a{}_{\|n} = T^a{}_{,n} + D^a_{nm}T^m$. Calculate $f_{,nm} - f_{,m\|n}$ and show that $S^a_{nm} = D^a_{nm} - D^a_{mn}$ is a tensor! Can S^a_{nm} be determined by demanding $g_{ab\|n} = 0$?

18.6 To any vector $\mathbf{a}$, with components a^n, an operator $\mathbf{a} = a^n\partial/\partial x^n$ can be assigned. Use this notation to give the Lie derivative of the vector T^n a simple form.

Further reading for Chapter 18
Eisenhart (1949), Schouten (1954), Yano (1955).

19

The curvature tensor

19.1 Intrinsic geometry and curvature

In the previous chapters of this book we have frequently used the concept 'Riemannian space' or 'curved space'. Except in Section 14.4 on the geodesic deviation, it has not yet played any rôle whether we were dealing only with a Minkowski space with complicated curvilinear coordinates or with a genuine curved space. We shall now turn to the question of how to obtain a measure for the deviation of the space from a Minkowski space.

If one uses the word 'curvature' for this deviation, one most often has in mind the picture of a two-dimensional surface in a three-dimensional space; that is, one judges the properties of a two-dimensional space (the surface) from the standpoint of a flat space of higher dimensionality. This way of looking at things is certainly possible mathematically for a four-dimensional Riemannian space as well – one could regard it as a hypersurface in a ten-dimensional flat space. But this higher-dimensional space has no physical meaning and is no more easy to grasp or comprehend than the four-dimensional Riemannian space. Rather, we shall describe the properties of our space-time by four-dimensional concepts alone – we shall study 'intrinsic geometry'. In the picture of the two-dimensional surface we must therefore behave like two-dimensional beings, for whom the third dimension is inaccessible both practically and theoretically, and who can base assertions about the geometry of their surface through measurements on the surface alone.

The surface of, for example, a cylinder or a cone, which in fact can be constructed from a plane section without distortion, could not be distinguished locally from a plane by such beings (that is, without their going right around the cylinder or the cone and returning to their starting point). But they would be able to establish the difference between a plane and a sphere, because on the surface of the sphere:

(a) The parallel displacement of a vector depends upon the route (along route 1 in Fig. 18.2c the vector is a tangent vector of a geodesic; along route 2 it is always perpendicular to the tangent vector).

(b) The sum of the angles of triangles bounded by 'straight lines' (geo-

desics) deviates from 180°; it can amount to 270°, for example.
(c) The circumference of a circle (produced by drawing out geodesics from a point and marking off a constant distance on them as radius) deviates from π multiplied by the diameter.
(d) The separation between neighbouring great circles is not proportional to the distance covered (cp. geodesic deviation).

As a detailed mathematical analysis shows, these four possibilities carried over to a four-dimensional space all lead to the concept of curvature and to that of the curvature tensor. In the following sections we shall become more familiar with this tensor, beginning with an investigation of the parallel transport of vectors.

19.2 The curvature tensor and global parallelism of vectors

The covariant derivative enables us to give, through (18.15),

$$T_{a;n}\frac{\mathrm{d}x^n}{\mathrm{d}\lambda} = \frac{\mathrm{d}T_a}{\mathrm{d}\lambda} - \Gamma^m_{an}T_m\frac{\mathrm{d}x^n}{\mathrm{d}\lambda} = 0, \qquad (19.1)$$

a unique formula for the parallel displacement of a vector along a fixed curve. When the result of the parallel displacement between two points is *independent* of the choice of the curve, one speaks of global parallelism. A necessary condition for its existence is evidently that the parallel displacement should be independent of the route already for infinitesimal displacements.

Upon applying (19.1) to the parallel displacement of the vector a_m along the sides of the infinitesimal parallelogram of Fig. 19.1 we obtain:

from P_1 to P': $\qquad \mathrm{d}_1 a_m = \Gamma^i_{mn}a_i\,\mathrm{d}_1 x^n,$ $\qquad (19.2)$

from P_1 to P'': $\qquad \mathrm{d}_2 a_m = \Gamma^i_{mn}a_i\,\mathrm{d}_2 x^n,$ $\qquad (19.3)$

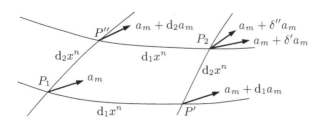

Fig. 19.1. Parallel displacement of a vector.

from P_1 over P' to P_2 (Christoffel symbols are to be taken at P', therefore $\Gamma^r_{mq} + \Gamma^r_{mq,s} \, d_1 x^s$):

$$
\begin{aligned}
\delta' a_m &= (\Gamma^r_{mq} + \Gamma^r_{mq,s} \, d_1 x^s)(a_r + d_1 a_r) \, d_2 x^q + d_1 a_m \\
&\approx \Gamma^r_{mq} \Gamma^i_{rn} \, d_2 x^q \, d_1 x^n a_i + \Gamma^r_{mq,s} a_r \, d_1 x^s \, d_2 x^q \\
&\quad + \Gamma^i_{mn} a_i (d_1 x^n + d_2 x^n),
\end{aligned}
\tag{19.4}
$$

from P_1 over P'' to P_2 (Christoffel symbols are to be taken at P''):

$$
\begin{aligned}
\delta'' a_m &= (\Gamma^r_{mq} + \Gamma^r_{mq,s} \, d_2 x^s)(a_r + d_2 a_r) \, d_1 x^q + d_2 a_m \\
&\approx \Gamma^r_{mq} \Gamma^i_{rn} \, d_1 x^q \, d_2 x^n a_i + \Gamma^r_{mq,s} a_r \, d_2 x^s \, d_1 x^q \\
&\quad + \Gamma^i_{mn} a_i (d_1 x^n + d_2 x^n).
\end{aligned}
\tag{19.5}
$$

The vectors transported to P_2 by different routes thus differ by

$$
\delta'' a_m - \delta' a_m = (-\Gamma^r_{mq,s} + \Gamma^r_{ms,q} + \Gamma^r_{nq}\Gamma^n_{ms} - \Gamma^r_{ns}\Gamma^n_{mq}) a_r \, d_1 x^s \, d_2 x^q. \tag{19.6}
$$

The parallel transport is therefore independent of the route for all vectors a_r and all possible infinitesimal parallelograms ($d_1 x^n$ and $d_2 x^n$ arbitrary) if and only if the *Riemann curvature tensor (Riemann–Christoffel tensor)*, defined by

$$
R^r{}_{msq} = \Gamma^r_{mq,s} - \Gamma^r_{ms,q} + \Gamma^r_{ns}\Gamma^n_{mq} - \Gamma^r_{nq}\Gamma^n_{ms}, \tag{19.7}
$$

vanishes. If this condition is satisfied, then one can also define global parallelism for *finite* displacements; the parallel transport will be independent of path (as one can show by decomposing the surface enclosed by a curve into infinitesimal parallelograms).

Path independence of the parallel displacement is the pictorial interpretation of the commutation of the second covariant derivatives of a vector; in fact for every arbitrary vector a_m we have

$$
\begin{aligned}
a_{m;s;q} &= a_{m;s,q} - \Gamma^r_{mq} a_{r;s} - \Gamma^r_{qs} a_{m;r} \\
&= a_{m,s,q} - \Gamma^r_{ms,q} a_r - \Gamma^r_{ms} a_{r,q} - \Gamma^r_{qm} a_{r,s} \\
&\quad + \Gamma^n_{mq} \Gamma^r_{ns} a_r - \Gamma^r_{qs} a_{m,r} + \Gamma^n_{qs} \Gamma^r_{mn} a_r,
\end{aligned}
\tag{19.8}
$$

and, after interchange of q and s bearing in mind (19.7), we obtain

$$
a_{m;s;q} - a_{m;q;s} = R^r{}_{msq} a_r. \tag{19.9}
$$

Covariant derivatives commute if, and only if, the curvature tensor vanishes. One can also take (19.9) as the definition of the curvature tensor.

We can see the justification for the word *curvature* tensor in the fact that it disappears if, and only if, the space is flat, that is, when a Cartesian coordinate system can be introduced in the whole space. In Cartesian coordinates all the Christoffel symbols do indeed vanish, and with them the curvature tensor (19.7). Conversely, if it does disappear, then one can create a Cartesian coordinate system throughout the space by (unique) parallel displacement of four vectors which are orthogonal at one point. That $R^r{}_{msq}$ really is a *tensor* can be most quickly realized from (19.9).

To summarize, we can thus make the following completely equivalent statements. The curvature tensor defined by (19.7) and (19.9) vanishes if, and only if (a) the space is flat, that is, Cartesian coordinates with $g_{ab} = \eta_{ab}$ and $\Gamma^a_{bc} = 0$ can be introduced throughout the space; or (b) the parallel transport of vectors is independent of path; or (c) covariant derivatives commute; or (d) the geodesic deviation (the relative acceleration) of two arbitrary particles moving force-free vanishes (cp. Section 14.4).

19.3 The curvature tensor and second derivatives of the metric tensor

The curvature tensor (19.7) contains Christoffel symbols and their derivatives, and hence the metric tensor and its first and second derivatives. We shall now examine more precisely the connection between the metric and the components of the curvature tensor.

To this end we carry out in a locally geodesic coordinate system,

$$\bar{g}_{ab} = \eta_{ab}, \quad \bar{\Gamma}^a_{bc} = 0 \quad \text{for } \bar{x}^n = 0, \tag{19.10}$$

a coordinate transformation

$$x^n = \bar{x}^n + \tfrac{1}{6} D^n{}_{pqr} \bar{x}^p \bar{x}^q \bar{x}^r, \quad A^n_i = \partial x^n / \partial \bar{x}^i = \delta^n_i + \tfrac{1}{2} D^n{}_{pqi} \bar{x}^p \bar{x}^q, \tag{19.11}$$

the constants $D^n{}_{pqi}$ being initially arbitrary, but symmetric in the lower indices. This transformation does not change the metric or the Christoffel symbols at the point $x^n = \bar{x}^n = 0$, but it can serve to simplify the derivatives of the Christoffel symbols. Because of the general transformation formula (16.31) we have

$$\Gamma^m_{ab,n} = \left(\bar{\Gamma}^r_{ik} A^m_r A^i_a A^k_b - A^m_{i,k} A^i_a A^k_b \right)_{,n} \tag{19.12}$$

from which because of (19.11) follows that at the point $x^n = 0$

$$\Gamma^m_{ab,n} = \bar{\Gamma}^m_{ab,n} - D^m{}_{abn}. \tag{19.13}$$

Since the coefficients $D^m{}_{abn}$ of formula (19.11) are symmetric in the three lower indices, whereas the derivatives $\bar{\Gamma}^m_{ab,n}$ of the Christoffel symbols, which are to be regarded as specified, do not possess this symmetry property, not all the derivatives $\Gamma^m_{ab,n}$ can be made to vanish. Through the choice

$$D^m{}_{abn} = \tfrac{1}{3}\left(\bar{\Gamma}^m_{ab,n} + \bar{\Gamma}^m_{na,b} + \bar{\Gamma}^m_{bn,a}\right), \tag{19.14}$$

however, one can always ensure that

$$\Gamma^m_{ab,n} + \Gamma^m_{na,b} + \Gamma^m_{bn,a} = 0. \tag{19.15}$$

If (19.15) and (19.10) are satisfied at a point, one speaks of *canonical coordinates*.

In such a canonical coordinate system it follows from (19.7) that the components of the curvature tensor satisfy

$$R^r{}_{msq} = \Gamma^r_{mq,s} - \Gamma^r_{ms,q}, \tag{19.16}$$

and therefore, using also (19.15), that

$$R^r{}_{msq} + R^r{}_{smq} = -3\Gamma^r_{ms,q}. \tag{19.17}$$

From the definition of the Christoffel symbols (16.13), on the other hand, it follows that

$$g_{ia,bn} = g_{mi}\Gamma^m_{ab,n} + g_{ma}\Gamma^m_{ib,n}, \tag{19.18}$$

and (19.17) and (19.18) together yield finally

$$g_{ia,bn} = -\tfrac{1}{3}\left(R_{iabn} + R_{iban} + R_{aibn} + R_{abin}\right). \tag{19.19}$$

Using in advance the symmetry relations (19.24), this is equivalent to

$$g_{ia,bn} = -\tfrac{1}{3}\left(R_{iban} + R_{abin}\right). \tag{19.20}$$

The equations (19.19) and (19.20) lead to an important conclusion. At first sight they merely state that in canonical coordinates the second derivatives of the metric tensor can be constructed from the components of the curvature tensor. But because of (19.10) and (19.20), in canonical coordinates *all* tensors which can be formed out of the metric and its first and second derivatives can be expressed in terms of the curvature tensor and the metric tensor itself. This relation between tensors must be coordinate-independent, and so any tensor containing only the metric and its first and second derivatives can be expressed in terms of the curvature tensor and the metric tensor.

If one wants to apply this law to pseudotensors, then one must also admit the ε-tensor as an additional building block.

Canonical coordinates permit a simple geometrical interpretation.

Their coordinate lines are pairwise orthogonal geodesics, and the coordinates of an arbitrary point are given by the product of the direction cosines of the geodesic to the point from the zero point with the displacement along this geodesic.

19.4 Properties of the curvature tensor

Symmetry properties The symmetry properties of the curvature tensor can, of course, immediately be picked out from the defining equation (19.7), or from

$$R_{amsq} = g_{ar}(\Gamma^r_{mq,s} - \Gamma^r_{ms,q} + \Gamma^r_{ns}\Gamma^n_{mq} - \Gamma^r_{nq}\Gamma^n_{ms}). \qquad (19.21)$$

But, in the geodesic coordinate system, in which the Christoffel symbols vanish, and in which it follows from (19.21) that

$$R_{amsq} = (g_{ar}\Gamma^r_{mq}),_s - (g_{ar}\Gamma^r_{ms}),_q, \qquad (19.22)$$

and hence finally

$$R_{amsq} = \tfrac{1}{2}(g_{aq,ms} + g_{ms,aq} - g_{as,mq} - g_{mq,as}), \qquad (19.23)$$

they are more quickly recognized. As one can immediately see from (19.23), the curvature tensor is antisymmetric under interchange of the first and second index, or of the third and fourth,

$$R_{amsq} = -R_{masq} = -R_{amqs} = R_{maqs}, \qquad (19.24)$$

but it does not alter under exchange of the first and last pairs of indices,

$$R_{amsq} = R_{sqam}, \qquad (19.25)$$

and, further, also satisfies the relation

$$3R_{a[msq]} = R_{amsq} + R_{asqm} + R_{aqms} = 0. \qquad (19.26)$$

The equations (19.24) imply that under the relabelling (12) → 1, (23) → 2, (34) → 3, (41) → 4, (13) → 5, (24) → 6 the independent components of the curvature tensor can be mapped onto a 6×6 matrix R_{AB}. Because of (19.25) this matrix is symmetric, and therefore has at most $\binom{7}{2} = 21$ different components. The cyclic relation (19.26) is independent of (19.24) and (19.25) (that is, not trivially satisfied) if, and only if, all four indices of the curvature tensor are different, and (19.26) hence supplies only *one* additional equation. The result of this count is thus that in a four-dimensional space the Riemann curvature tensor has a maximum of twenty algebraically independent components. One can show that in an N-dimensional space there are precisely $N^2(N^2-1)/12$ independent components.

Ricci tensor, curvature tensor and Weyl tensor Because of the symmetry properties of the curvature tensor there is (apart from a sign) only one tensor that can be constructed from it by contraction, namely, the *Ricci tensor*:

$$R_{mq} = R^a{}_{maq} = -R^a{}_{mqa}. \qquad (19.27)$$

It is symmetric, and has therefore ten different components. Its trace

$$R = R^m{}_m \qquad (19.28)$$

is called the *curvature scalar R*.

Just as (17.25) decomposes a symmetric tensor into a trace-free part and a term proportional to the metric tensor, the curvature tensor can be split into the *Weyl tensor* (or conformal curvature tensor) $C^{am}{}_{sq}$, and parts which involve only the Ricci tensor and the curvature scalar:

$$R^{am}{}_{sq} = C^{am}{}_{sq} + \tfrac{1}{2}(g_s^a R_q^m + g_q^m R_s^a - g_s^m R_q^a - g_q^a R_s^m) \\ - \tfrac{1}{6}(g_s^a g_q^m - g_q^a g_s^m)R. \qquad (19.29)$$

The Weyl tensor defined by (19.29) is 'trace-free',

$$C^{am}{}_{aq} = 0, \qquad (19.30)$$

and has all the symmetry properties of the full curvature tensor. The name 'conformal curvature tensor' or 'conformal tensor' relates to the fact that two different Riemannian spaces with the fundamental metric forms $d\widehat{s}^2$ and ds^2 which are conformally related

$$ds^2 = \Omega^2(x^i)d\widehat{s}^2 \qquad (19.31)$$

(all lengths are multiplied by the position-dependent conformal factor Ω^2, independent of direction) have the same conformal curvature tensor, although their Riemann curvature tensors are different.

In summary we can therefore make the following statement. At every point of a four-dimensional Riemannian space of the 100 possible different second derivatives of the metric tensor only twenty cannot be eliminated by coordinate transformations; they correspond to the twenty algebraically independent components of the curvature tensor. These twenty components can always be expressed by the ten components of the Ricci tensor and the ten of the Weyl tensor, as (19.29) shows.

In three-dimensional space the curvature tensor has only six independent components, exactly as many as the Ricci tensor, and the curvature tensor can be expressed in terms of the Ricci tensor:

$$R^{\alpha\mu}{}_{\sigma\tau} = (g_\sigma^\alpha R_\tau^\mu + g_\tau^\mu R_\sigma^\alpha - g_\sigma^\mu R_\tau^\alpha - g_\tau^\alpha R_\sigma^\mu) - \tfrac{1}{2}R(g_\sigma^\alpha g_\tau^\mu - g_\tau^\alpha g_\sigma^\mu). \quad (19.32)$$

On a two-dimensional surface (as for example on the sphere (14.36)) the curvature tensor has essentially only one component, the curvature scalar R:

$$R^{AM}{}_{ST} = R(g_S^A g_T^M - g_T^A g_S^M). \quad (19.33)$$

Bianchi identities Until now we have always thought of the metric as given, and derived the curvature tensor from it. Conversely, one can also ask the question whether the curvature tensor (with the correct symmetry properties, of course) can be specified as an arbitrary function of position, and the metric belonging to it determined. The answer to this apparently abstract mathematical question will reveal a further property of the Riemann tensor, which is particularly important for gravitation.

The determination of the metric from a specified curvature tensor amounts, because of (19.23), to the solution of a system of twenty second-order differential equations for the ten metric functions g_{ab}. In general such a system will possess no solutions; given a tensor with the algebraic properties of the curvature tensor there does not correspond a metric whose curvature tensor it is. Rather, additional integrability conditions must be satisfied. Although (19.20) holds only at one point, and therefore may not be differentiated, one can recognize the basis of the integrability condition in it; since the third partial derivatives of the metric commute, there must be some relations among the derivatives of components of the curvature tensor.

To set up these relations we write down the covariant derivative of the curvature tensor

$$R_{amsq;i} = g_{ar}(\Gamma^r_{mq,s} - \Gamma^r_{ms,q} + \Gamma^r_{ns}\Gamma^n_{mq} - \Gamma^r_{nq}\Gamma^n_{ms})_{;i} \quad (19.34)$$

in locally geodesic coordinates. Since the Christoffel symbols vanish in these coordinates one can replace the covariant derivative by the partial and drop the products of Christoffel symbols:

$$R_{amsq;i} = g_{ar}(\Gamma^r_{mq,si} - \Gamma^r_{ms,qi}). \quad (19.35)$$

If we add to this equation the two produced on permuting indices,

$$R_{amqi;s} = g_{ar}(\Gamma^r_{mi,sq} - \Gamma^r_{mq,is}),$$
$$R_{amis;q} = g_{ar}(\Gamma^r_{ms,iq} - \Gamma^r_{mi,sq}), \quad (19.36)$$

then we obtain the *Bianchi identities*:

$$3R_{am[is;q]} = R_{amis;q} + R_{amsq;i} + R_{amqi;s} = 0. \tag{19.37}$$

Every curvature tensor must satisfy these equations; if they hold, then one can determine the metric for a given curvature tensor, and conversely, if one expresses the curvature tensor through the metric, then they are satisfied identically. Because of the symmetry properties of the curvature tensor exhibited in (19.24)–(19.26), many of the Bianchi identities are trivially satisfied, for example, if not all the indices i, q, s are different. In four-dimensional space-time the system (19.37) contains only twenty non-trivial independent equations.

Upon contracting the Bianchi identities, we obtain identities for the Ricci tensor. We have

$$3g^{aq}R_{am[is;q]} = R^a{}_{mis;a} - R_{ms;i} + R_{mi;s},$$
$$3g^{ms}g^{aq}R_{am[is;q]} = R^a{}_{i;a} - R_{,i} + R^a{}_{i;a}, \tag{19.38}$$

and therefore

$$(R^{ai} - \tfrac{1}{2}g^{ai}R)_{;i} = 0. \tag{19.39}$$

Finally in this section on the properties of the curvature tensor we should point out that various sign conventions occur in the literature. With respect to our definition, the Riemann tensor can have the opposite sign, and the Ricci tensor can be formed by contraction over a different pair of indices and hence again change its sign. When reading a book or an article it is recommended that the convention used there be written out in order to avoid mistakes arising from comparison with this book or with other publications.

19.5 Spaces of constant curvature

An N-dimensional Riemannian space is of constant curvature if its curvature tensor obeys

$$R_{abcd} = \frac{R}{N(N-1)}(g_{ac}g_{bd} - g_{ad}g_{bc}), \quad R = \text{const.}, \tag{19.40}$$

where $R/N(N-1) = \varepsilon K^{-2}$ is called the Gaussian curvature. Those spaces will frequently occur as (sub-) spaces of physically interesting gravitational fields. They are the spaces with the greatest possible number of symmetries (see Section 33.4). We will list here some of their properties.

It is known since Riemann (1826 – 1866) and Christoffel (1829 – 1900) that locally a space of constant curvature, of any signature, can be written as

$$ds^2 = \frac{\overset{\circ}{\eta}_{PQ}\, dx^P dx^Q}{\left(1 + \frac{1}{4}\varepsilon K^{-2}\, \overset{\circ}{\eta}_{PQ}\, x^P x^Q\right)^2}, \qquad \overset{\circ}{\eta}_{PQ} = \mathrm{diag}\ (\epsilon_1, ..., \epsilon_N), \quad (19.41)$$

where $P, Q = 1, \ldots, N$, and $\epsilon_P = \pm 1$ as appropriate. A space of non-zero curvature, $R/N(N-1) = \varepsilon K^{-2}$, $K \neq 0$, can be considered as a hypersurface

$$\overset{\circ}{\eta}_{PQ}\, Z^P Z^Q + k(Z^{N+1})^2 = \varepsilon K^2, \qquad \varepsilon = \pm 1 \qquad (19.42)$$

in an $(N+1)$-dimensional pseudo-Euclidean space with the metric

$$ds^2 = \overset{\circ}{\eta}_{PQ}\, dZ^P dZ^Q + \varepsilon(dZ^{N+1})^2. \qquad (19.43)$$

Any suitable parametrization of (19.42), for example in terms of angular coordinates, will give rise to a special form of the metric of the space of constant curvature.

Two-dimensional spaces of constant curvature can be regarded (at least to some extent) as surfaces in flat three-dimensional space. Their metrics can be given in many different though locally equivalent forms.

Surfaces of positive curvature ($\varepsilon = +1$) are spheres of radius K,

$$ds^2 = K^2(d\vartheta^2 + \sin^2 \vartheta\, d\varphi^2), \quad 0 \leq \varphi \leq 2\pi, \quad 0 \leq \vartheta \leq \pi, \qquad (19.44)$$

cp. (14.36)–(14.39); they occur with spherically symmetric solutions.

Surfaces with negative curvature ($\varepsilon = -1$) are pseudospheres, and their metric can be written as

$$ds^2 = K^2[d\vartheta^2 + \exp(2\vartheta)\, d\varphi^2], \quad 0 \leq \varphi \leq 2\pi, \quad -\infty \leq \vartheta \leq \infty. \quad (19.45)$$

They can be realized by surfaces of revolution of the tractrix

$$z = K \ln \left|(K \pm \sqrt{k^2 - \rho^2})/\rho\right| \mp \sqrt{K^2 - \rho^2} \qquad (19.46)$$

about the z-axis ($\rho^2 = x^2 + y^2$). The name tractrix is due to the fact that precisely this curve results if a man runs along the z-axis pulling behind him an object on the end of a rope of length K, the object not lying on the z-axis initially, see Fig. 19.2.

It is easy to convince oneself that the surface (19.46) really does have the metric (19.45) by inserting the differential equation of the tractrix

$$\left(\frac{dz}{d\rho}\right)^2 = \frac{K^2 - \rho^2}{\rho^2} \qquad (19.47)$$

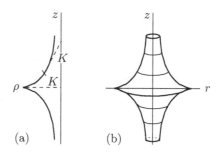

Fig. 19.2. Tractrix (a) and surface of constant negative curvature (b).

into the line element of flat space and then making the substitution $\rho = K \exp \vartheta$.

While the line element (19.45) of the space of constant negative curvature is regular for all values of φ and ϑ, the surface has a singular line for $z = 0$. One can prove quite generally that there exists no realization of that space by a surface in flat three-dimensional space which is regular everywhere. This two-dimensional Riemannian space is *not* globally embeddable in a flat three-dimensional space; such an embedding is only possible *locally*.

Finally, a surface of zero curvature is locally a plane,

$$ds^2 = d\vartheta^2 + d\varphi^2, \quad -\infty \leq \varphi \leq \infty, \quad -\infty \leq \vartheta \leq \infty, \qquad (19.48)$$

but can also realized (after bending the plane appropriately), for example, by the surface of a cone or a cylinder. If one identifies the points on the baseline of a cylindrical surface with those at height H, that is, of one allows the variables ϑ and φ to occupy only the intervals $0 \leq \varphi \leq 2\pi R$ and $0 \leq \vartheta \leq H$ by identification of the endpoints, then one has constructed a closed space of zero curvature.

The three types (19.44), (19.45) and (19.48) can be summarized in the form of (19.41), which here reads (after a rescaling of the coordinates)

$$ds^2 = K^2 \frac{dx^2 + dy^2}{\left(1 + \varepsilon r^2/4\right)^2}. \qquad (19.49)$$

Another frequently used version of the metric is

$$ds^2 = K^2 \left[\frac{d\bar{r}^2}{1 - \varepsilon \bar{r}^2} + \bar{r}^2 \, d\varphi^2 \right], \quad \bar{r} = \frac{r}{1 + \varepsilon r^2/4}. \qquad (19.50)$$

Three-dimensional spaces of constant curvature likewise split into three types if we restrict ourselves to positive definite metrics; those spaces occur in the interior Schwarzschild solution (Section 26.4), and in cosmology as part of the Robertson–Walker metrics.

A space of positive constant curvature ($\varepsilon = +1$) corresponds to a hypersphere

$$ds^2 = K^2[d\chi^2 + \sin^2 \chi(d\vartheta^2 + \sin^2 \vartheta \, d\varphi^2)],$$

$$0 \le \chi \le \pi, \quad 0 \le \vartheta \le \pi, \quad 0 \le \varphi \le 2\pi, \tag{19.51}$$

which can be embedded in a four-dimensional flat space according to

$$Z^1 = K \cos \chi, \qquad Z^3 = K \sin \chi \sin \vartheta \cos \varphi,$$

$$Z^2 = K \sin \chi \cos \vartheta, \qquad Z^4 = K \sin \chi \sin \vartheta \sin \varphi. \tag{19.52}$$

The volume of this hypersphere (the surface area, as regarded from the four-dimensional space) is

$$V = \int \sqrt{g} \, d\chi \, d\vartheta \, d\varphi = \int K^3 \sin^2 \chi \sin \vartheta \, d\chi \, d\vartheta \, d\varphi = 2\pi^2 K^3. \tag{19.53}$$

The 'radial' coordinate χ can take only the maximal value $\chi = \pi$; there is one point maximally distant from the null point $\chi = 0$, namely the antipodal point $\chi = \pi$.

A space of constant negative curvature ($\varepsilon = -1$) has the metric

$$ds^2 = K^2[d\chi^2 + \sinh^2 \chi(d\vartheta^2 + \sin^2 \vartheta \, d\varphi^2)]. \tag{19.54}$$

The 'radial' coordinate χ can vary arbitrarily, and the space can have infinite extent.

Finally, a space of zero curvature ($\varepsilon = 0$) is (locally) a flat space:

$$ds^2 = d\chi^2 + d\vartheta^2 + d\varphi^2. \tag{19.55}$$

The metric of all three types of space can be written in the form

$$ds^2 = K^2 \frac{dx^2 + dy^2 + dz^2}{(1 + \varepsilon r^2/4)^2} = K^2 \left[\frac{d\bar{r}^2}{1 - \varepsilon \bar{r}^2} + \bar{r}^2(d\vartheta^2 + \sin^2 \vartheta \, d\varphi^2) \right]. \tag{19.56}$$

Spaces with $\varepsilon = 1$ are called *closed*, since although they are of course unbounded, they contain a finite volume, and the separation of two points is bounded. Spaces with $\varepsilon = 0$ or $\varepsilon = -1$ are frequently designated *open*. Since, however, amongst the spaces of negative curvature and the flat spaces, closed models which result from a suitable identification of points can readily by found (see the example discussed in connection with (19.48)), this designation is rather misleading. The problem of finding all possible realizations of a space of constant curvature is called the Cayley–Klein space-structure problem.

Four-dimensional spaces of constant curvature are known, as cosmological models, under the name of de Sitter universes; sometimes one uses this name only for the space of constant positive curvature, whereas those of negative curvature are called anti-de Sitter universes. The de Sitter universes can all be represented by a hyperboloid

$$x^2 + y^2 + z^2 + \varepsilon w^2 - v^2 = K^2 \tag{19.57}$$

in a flat space of metric

$$\mathrm{d}s^2 = \mathrm{d}x^2 + \mathrm{d}y^2 + \mathrm{d}z^2 + \varepsilon\,\mathrm{d}w^2 - \mathrm{d}v^2. \tag{19.58}$$

For positive curvature the metric can be written as

$$\mathrm{d}s^2 = K^2[\cosh^2 ct\{\mathrm{d}\chi^2 + \sinh^2\chi(\mathrm{d}\vartheta^2 + \sin^2\vartheta\,\mathrm{d}\varphi^2)\} - c^2\,\mathrm{d}t^2], \tag{19.59}$$

the space of zero curvature is Minkowski space, and spaces of negative curvature are represented locally by

$$\mathrm{d}s^2 = K^2[\cos^2 ct\{\mathrm{d}\chi^2 + \sinh^2\chi(\mathrm{d}\vartheta^2 + \sin^2\vartheta\,\mathrm{d}\varphi^2)\} - c^2\,\mathrm{d}t^2]. \tag{19.60}$$

Amongst the spaces of negative curvature there are some with closed timelike curves.

De Sitter universes contain three-dimensional spaces of constant curvature and hence belong to the Robertson–Walker metrics (to be discussed in Chapter 40).

Exercises

19.1   Show (e.g. by using locally geodesic coordinates (16.41)) that the two metrics $\widehat{g}_{ab}$ and $g_{ab} = \Omega^2\,\widehat{g}_{ab}$ have the same Weyl tensor.

19.2   Show that the tractrix (19.47), with $\rho = K\exp\vartheta$, really gives the metric (19.45).

19.3   Find all spaces of constant curvature with metric $\mathrm{d}s^2 = K^2[\mathrm{d}\vartheta^2 + f^2(\vartheta)\mathrm{d}\varphi^2]$, $K = $ const.

19.4   Show that $R_{mabn} = F(g_{an}g_{mb} - g_{ab}g_{mn})$ and $R_{mabn;i} = R_{mibn;a}$ imply $F = $ const.

Further reading for Chapter 19

Eisenhart (1949), Schouten (1954).

20

Differential operators, integrals and integral laws

20.1 The problem

In the formulation of physical laws in three-dimensional flat space one often uses the vector operators div, grad, curl and $\Delta = \text{div grad}$, which in Cartesian coordinates can also be applied to tensor components. Because of the integral laws

$$\oint \text{div } \mathbf{A} \, dV = \oint \mathbf{A} \, d\mathcal{S} \quad \text{(Gauss)} \tag{20.1}$$

$$\oint \text{curl } \mathbf{A} \, d\mathcal{S} = \oint \mathbf{A} \, dr \quad \text{(Stokes)}, \tag{20.2}$$

they make an integral formulation of physical statements possible, for example in electrodynamics. The integral laws can also be applied to tensors of higher rank.

While the differential operators can be carried over relatively easily to a four-dimensional curved space, the generalization of integral laws leads to difficulties. One cause of the difficulties is that integrals can never be taken over tensor components, but only over scalars, if the result is to be a tensor. A second cause is the fact that the reverse of an integration is really a partial differentiation, whereas for tensor equations we have to choose the covariant derivative; for this reason we shall be especially interested in those differential operators which are covariant, and yet which can be expressed simply by partial derivatives.

The comprehensibility of the calculations is further obscured by the complicated way in which we write volume and surface elements in covariant form. The use of differential forms here can indeed produce some improvement, but for actual calculations the gain is small.

20.2 Some important differential operators

The covariant derivative is the generalized gradient; for a scalar, covariant and partial derivatives coincide:

$$\varphi_{;a} = \varphi_{,a}. \tag{20.3}$$

The generalized curl of a vector A_m is the antisymmetric part of the tensor $A_{n;m}$:

$$A_{n;m} - A_{m;n} = A_{n,m} - A_{m,n} - \Gamma^a_{nm} A_a + \Gamma^a_{mn} A_a. \qquad (20.4)$$

Because of the symmetry of the Christoffel symbols in the lower indices, one here can replace covariant derivatives by partial derivatives:

$$A_{n;m} - A_{m;n} = A_{n,m} - A_{m,n}. \qquad (20.5)$$

One obtains the generalized divergence by contraction over the index with respect to which the covariant derivative has been taken. Because of the relation (16.19), $\Gamma^a_{ab} = (\ln \sqrt{-g})_{,b}$, we have for a vector

$$B^n{}_{;n} = B^n{}_{,n} + \Gamma^n_{na} B^a = B^n{}_{,n} + (\ln \sqrt{-g})_{,a} B^a, \qquad (20.6)$$

and therefore

$$B^n{}_{;n} = \frac{1}{\sqrt{-g}} (B^n \sqrt{-g})_{,n}. \qquad (20.7)$$

For an antisymmetric tensor $F_{ab} = -F_{ba}$ we have, because of the symmetry property of the Christoffel symbol,

$$F^{ab}{}_{;b} = F^{ab}{}_{,b} + \Gamma^a_{bm} F^{mb} + \Gamma^b_{bm} F^{am} = F^{ab}{}_{,b} + (\ln \sqrt{-g})_{,m} F^{am}; \qquad (20.8)$$

thus, just as for a vector, its divergence can also be expressed as a partial derivative:

$$F^{ab}{}_{;b} = \frac{1}{\sqrt{-g}} (\sqrt{-g}\, F^{ab})_{,b}. \qquad (20.9)$$

Similarly, for every completely antisymmetric tensor we have

$$F^{[mn...ab]}{}_{;b} = \frac{1}{\sqrt{-g}} (\sqrt{-g}\, F^{[mn...ab]})_{,b}. \qquad (20.10)$$

For the divergence of a symmetric tensor there is no comparable simple formula.

The generalized Δ-operator is formed from div and grad; from (20.3) and (20.7) we have

$$\Delta\varphi = \varphi^{,n}{}_{;n} = \frac{1}{\sqrt{-g}} (\sqrt{-g}\, g^{na} \varphi_{,a})_{,n}. \qquad (20.11)$$

20.3 Volume, surface and line integrals

In an N-dimensional ($N \leq 4$) space, an s-dimensional hypersurface element ($s \leq N$) is spanned by s infinitesimal vectors $d_1 x^n$, $d_2 x^n, \ldots, d_s x^n$, which are linearly independent and do not necessarily have to point in the direction of the coordinate axes (see Fig. 20.1).

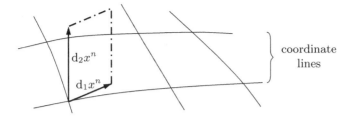

Fig. 20.1. A surface element.

We shall need the generalized Kronecker symbol $\delta^{n_1...n_s}_{m_1...m_s}$, which is antisymmetric both in all upper and all lower indices, and for $n_i = m_i$ takes the value 1 (when these numbers are all different), so that

$$\delta^{n_1...n_s}_{m_1...m_s} = \delta^{[n_1...n_s]}_{m_1...m_s} = \delta^{n_1...n_s}_{[m_1...m_s]}, \quad \delta^{n_1...n_s}_{m_1...m_s} = 1 \ \text{ for } n_i = m_i. \quad (20.12)$$

We next define the object

$$\mathrm{d}V^{n_1...n_s} = \delta^{n_1...n_s}_{m_1...m_s} \, \mathrm{d}_1 x^{m_1} \cdots \mathrm{d}_s x^{m_s} \quad (20.13)$$

as a hypersurface (volume) element. As one can see, and can verify from examples, this is a tensor which is antisymmetric in all indices. Its components become particularly simple when the $\mathrm{d}_i x^n$ point in the directions of the coordinate axes,

$$\mathrm{d}_1 x^n = (\mathrm{d}x^1, 0, 0, \ldots), \quad \mathrm{d}_2 x^n = (\mathrm{d}x^2, 0, 0, \ldots), \ldots. \quad (20.14)$$

For $s = 1$, (20.13) simply defines the line element

$$\mathrm{d}V^n = \mathrm{d}x^n. \quad (20.15)$$

For $s = 2$, since

$$\delta^{n_1 n_2}_{m_1 m_2} = \delta^{n_1}_{m_1} \delta^{n_2}_{m_2} - \delta^{n_1}_{m_2} \delta^{n_2}_{m_1}, \quad (20.16)$$

the hypersurface element is associated in a simple manner with the surface element $\mathrm{d}_1 \mathbf{r} \times \mathrm{d}_2 \mathbf{r}$

$$\mathrm{d}V^{n_1 n_2} = \mathrm{d}_1 x^{n_1} \, \mathrm{d}_2 x^{n_2} - \mathrm{d}_1 x^{n_2} \, \mathrm{d}_2 x^{n_1}, \quad (20.17)$$

and for $s = N$ it has, using the differentials (20.14), essentially one component

$$\mathrm{d}V^{12...N} = \mathrm{d}x^1 \, \mathrm{d}x^2 \cdots \mathrm{d}x^N. \quad (20.18)$$

Since we can in principle integrate only over scalars, if we demand that the integral be a tensor, then we must always contract the hypersurface element with a tensor of the same rank. Thus only integrals of the form

$$\int_{G_s} T_{n_1 \dots n_s} \, \mathrm{d}V^{n_1 \dots n_s} = I_s, \quad 1 \le s \le N, \tag{20.19}$$

are allowed.

In an N-dimensional space there are therefore precisely N different types of integral, each corresponding to the dimension s of the hypersurface being integrated over. We may suppose that the tensors $T_{n_1 \dots n_s}$ are completely antisymmetric, because in contraction with $\mathrm{d}V^{n_1 \dots n_s}$ all symmetric parts would drop out anyway. G_s denotes the region over which the s-dimensional integration is to be carried out.

In four-dimensional space there are thus four types of integral contained in (20.19). When $s = 1$ we have, for example, the simple line integral

$$I_1 = \int_{G_1} T_n \, \mathrm{d}x^n. \tag{20.20}$$

For $s = 3$, as we have shown in (17.4), one can map the tensor $T_{n_1 n_2 n_3}$ onto a vector, according to (6.29):

$$T_{n_1 n_2 n_3} = \varepsilon_{a n_1 n_2 n_3} T^a / 3! = \sqrt{-g} \overset{M}{\varepsilon}_{a n_1 n_2 n_3} T^a / 3!. \tag{20.21}$$

Here it is meaningful to introduce by

$$\mathrm{d}f_a = \varepsilon_{a n_1 n_2 n_3} \, \mathrm{d}V^{n_1 n_2 n_3} / 3! \tag{20.22}$$

the pseudovector $\mathrm{d}f_a$ which is perpendicular to the hypersurface element (that is, perpendicular to the vectors $\mathrm{d}_i x^n$), and whose length in a system (20.14) is just $\sqrt{-g} \, \mathrm{d}x^1 \, \mathrm{d}x^2 \, \mathrm{d}x^3$. Thus $\mathrm{d}f_a$ is the generalized surface element. In this way we obtain the simpler form

$$I_3 = \int_{G_3} T^a \, \mathrm{d}f_a \quad \left(= \int_{G_3} T^a \sqrt{-g} \, \mathrm{d}x^1 \, \mathrm{d}x^2 \, \mathrm{d}x^3 \, \delta_a^4 \right) \tag{20.23}$$

of the hypersurface integral (the expression in parentheses is valid only in the system in which G_3 is the surface $x_4 = \mathrm{const.}$ and in which (20.14) holds).

Finally, for $s = 4$, every antisymmetric tensor $T_{n_1 n_2 n_3 n_4}$ is proportional to the ε-tensor:

$$T_{n_1 n_2 n_3 n_4} = T \varepsilon_{n_1 n_2 n_3 n_4} / 4!. \tag{20.24}$$

Because

$$\delta^{n_1 n_2 n_3 n_4}_{m_1 m_2 m_3 m_4} = -\varepsilon^{n_1 n_2 n_3 n_4} \varepsilon_{m_1 m_2 m_3 m_4} \tag{20.25}$$

(cp. (6.17)), it is appropriate to introduce the volume element $\mathrm{d}V$ by

$$\mathrm{d}V = -\varepsilon_{m_1 m_2 m_3 m_4} \, \mathrm{d}_1 x^{m_1} \, \mathrm{d}_2 x^{m_2} \, \mathrm{d}_3 x^{m_3} \, \mathrm{d}_4 x^{m_4}, \tag{20.26}$$

which in the preferred system (20.14) has the form

$$dV = \sqrt{-g}\,dx^1\,dx^2\,dx^3\,dx^4. \tag{20.27}$$

Volume integrals thus always have the simple form

$$I_4 = \int_{G_4} T\,dV \quad \left(= \int_{G_4} T\sqrt{-g}\,dx^1\,dx^2\,dx^3\,dx^4\right) \tag{20.28}$$

(the expression in parentheses is the form when (20.14) is valid).

20.4 Integral laws

Integral laws which are valid in a Riemannian space all have the form
of a Stokes law, that is, they reduce the integral over a generalized curl

$$T_{[n_1 n_2 \ldots n_{s-1}; n_s]} = T_{[n_1 \ldots n_{s-1}, n_s]} \tag{20.29}$$

to an integral over the boundary G_{s-1} of the original (simply connected)
region of integration G_s:

$$\int_{G_s} T_{n_1 \ldots n_{s-1}, n_s}\,dV^{n_s n_1 \ldots n_{s-1}} = \int_{G_{s-1}} T_{n_1 \ldots n_{s-1}}\,dV^{n_1 \ldots n_{s-1}}. \tag{20.30}$$

(Because of the antisymmetry of the volume element we are able to drop
the antisymmetrizing brackets on the tensor field.) In spite of the partial
derivative, (20.30) is a tensor equation – one can in all cases replace the
partial by the covariant derivative.

We shall not go through the proof of this law here, but merely indicate
the idea on which it is based. Just as with the proof of the Stokes law for
a two-dimensional surface, one decomposes the region G_s into infinitesimal elements, demonstrates the validity of the law for these elements,
and sums up over all elements. In the summation the contributions from
the 'internal' boundary surfaces cancel out, because in every case they
are traversed twice, in opposite directions (see Fig. 20.2).

In three-dimensional flat space there are consequently three integral

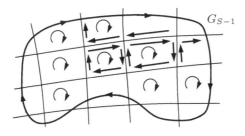

Fig. 20.2. The Stokes law for a surface G_s.

laws. For $s = 1$ we obtain from (20.30)

$$\int_{P_1}^{P_2} T_{,n}\, dx^n = T(P_2) - T(P_1) \qquad (20.31)$$

(the boundary of a curve is represented by the two end points P_1 and P_2). The Stokes law proper corresponds to $s = 2$, and $s = 3$ yields the Gauss law.

In four-dimensional space, too, the Gauss law is a special case of the general Stokes law (20.30). Because of (20.21) and (20.22), for $s = N = 4$ we obtain from (20.30)

$$\int_{G_4} \varepsilon_{a n_1 n_2 n_3} T^a{}_{;n_4}\, dV^{n_4 n_1 n_2 n_3}/3! = \int_{G_3} T^a\, df_a. \qquad (20.32)$$

We next substitute for $dV^{n_4 n_1 n_2 n_3}$ from (20.13), (20.25), (20.26):

$$-\int_{G_4} \varepsilon_{a n_1 n_2 n_3} T^a{}_{;n_4} \varepsilon^{n_4 n_1 n_2 n_3}\, dV/3! = \int_{G_3} T^a\, df_a, \qquad (20.33)$$

and, finally, taking into account the rule (6.20), we obtain the Gauss law

$$\int_{G_4} T^a{}_{;a}\, dV = \int_{G_3} T^a\, df_a. \qquad (20.34)$$

When making calculations with integrals and integral laws one has to make sure that the orientation of the hypersurface element is correctly chosen and remains preserved; under interchange of coordinates the sign of the hypersurface element $dV^{n_1 \cdots n_s}$ changes. Such a fixing of the orientation occurs also, of course, in the case of the usual Stokes law in three dimensions, where the sense in which the boundary curve is traversed is related to the orientation of the surface.

20.5 Integral conservation laws

We want to describe in detail a particularly important physical application of the Gauss integral law. From Special Relativity one already knows that a mathematical statement of the structure

$$T^{a \cdots c}{}_{,a} = 0 \qquad (20.35)$$

(the vanishing of the divergence of a tensor field) corresponds physically to a conservation law, establishing that some physical quantity does not change with time. In order to prove this connection, one uses the Gauss law, which there is also valid for tensor components.

In a Riemannian space the number of possible integral conservation laws is already restricted by the fact that the Gauss law (20.34) can only be applied to the divergence of a vector. To draw conclusions from

$$T^a{}_{;a} = 0, \qquad (20.36)$$

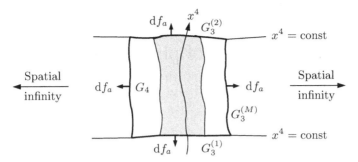

Fig. 20.3. The region of integration used in deriving the conservation law
(20.38).

let us imagine a vector field $T^a(x^i)$, which differs from zero only within a
finite spatial region, and apply (20.34) to a four-dimensional 'cylindrical'
region (hatched in Fig. 20.3) whose three-dimensional lateral surface
$G_3^{(M)}$ lies outside this region of space.

  Since the contributions from the lateral surface $G_3^{(M)}$ vanish, it follows
from (20.34) and (20.36) that

$$\int_{G_3^{(1)}} T^a \, df_a + \int_{G_3^{(2)}} T^a \, df_a = 0. \tag{20.37}$$

If we now let the lateral surfaces go to (spatial) infinity, then the regions
of integration $G_3^{(1)}$ and $G_3^{(2)}$ cover the whole space $x^4 = $ const. And if
we further notice the opposite orientations of df_a in the two regions,
then it follows from (20.37) and (20.22) that

$$\int_{x^4=\text{const.}} T^a \, df_a = \int_{x^4=\text{const.}} T^4 \sqrt{-g} \, dx^1 \, dx^2 \, dx^3 = \text{const.} \tag{20.38}$$

The integral (20.38) defines a quantity whose value does not depend
upon the (arbitrary) time coordinate x^4; it defines a conserved quantity.
We have derived this law under the supposition of a so-called isolated
vector field T^a, that is, one restricted to a finite region of space. It is,
however, also valid when there are no lateral surfaces $G_3^{(M)}$, that is, when
the space is closed (like a two-dimensional spherical surface), or when
the integral over the lateral surface tends to zero (T^a falls off sufficiently
quickly when the convex surface is pushed to spatial infinity).

Further reading for Chapter 20

Straumann (1984).

21

Fundamental laws of physics in Riemannian spaces

21.1 How does one find the fundamental physical laws?

Before turning in the next chapter to the laws governing the gravitational field, that is, to the question of how the matter existing in the universe determines the structure of the Riemannian space, we shall enquire into the physical laws which hold in a *given* Riemannian space; that is to say, how a given gravitational field influences other physical processes. How can one transcribe a basic physical equation, formulated in Minkowski space without regard to the gravitational force, into the Riemannian space, and thereby take account of the gravitational force?

In this formulation the word 'transcribe' somewhat conceals the fact that it is really a matter of searching for entirely new physical laws, which are very similar to the old laws only because of the especially simple way in which the gravitational field acts. It is clear that we shall not be forced to the new form of the laws by logical or mathematical considerations, but that we can attain the answer only by observation and experiment. In searching for a transcription principle we therefore want our experience to be summarized in the simplest possible formulae.

In the history of relativity theory the principle of covariance plays a large rôle in this connection. There is no clear and unique formulation of this principle; the opinions of different authors diverge here. Roughly speaking, the principle of covariance expresses the fact that physical laws are to be written covariantly by the use of tensors, to ensure the equivalence, in principle, of all coordinate systems. Many criticisms have been raised against this principle, their aim being to assert that neither is it a *physical* principle, nor does it guarantee the correctness of the equations thus obtained. An example from Special Relativity will illustrate this. The potential equation

$$\Delta V = \eta^{\alpha\beta}V_{,\alpha\beta} = 0, \quad \alpha,\beta = 1,2,3 \tag{21.1}$$

is certainly not Lorentz invariant. But we can make it so by introducing an auxiliary field u^n which in a special coordinate system (in which (21.1) holds) has the form $u^n = (0,0,0,c)$. The equation

$$\Delta V = (\eta^{ab} + u^a u^b/c^2)V_{,ab} = 0, \quad a,b = 1,\dots,4, \qquad (21.2)$$

thus obtained is certainly Lorentz invariant (covariant), but it is definitely false, because according to it effects always propagate with infinitely large velocity. Of course one has to criticize (21.2) on the grounds that a vector field u^a was introduced *ad hoc* which singles out the three-dimensional coordinates used in (21.1) and thereby favours the rest system of an 'aether'.

It is instructive to compare this example with the transition from the Lorentz-invariant wave equation,

$$\Box V = \eta^{ab}V_{,ab} = 0, \qquad (21.3)$$

to the generally covariant equation

$$\Box V = g^{ab}V_{;ab} = 0. \qquad (21.4)$$

In place of the auxiliary quantity u^n, the auxiliary quantity g^{ab} has entered, which also singles out special coordinate systems (for example, locally geodesic ones). How do we know whether (21.4) is correct? The fundamental difference between (21.2) and (21.4) consists of the fact that g^{ab}, in contrast to u^a, possesses a physical significance; the metric describes the influence of the gravitational field. One can therefore interpret the requirement that physical equations should be covariant, and that all the metric quantities being introduced to guarantee covariance should correspond to properties of the gravitational field, as the physical basis of the principle of covariance.

A much more meaningful transcription formula follows from the principle of equivalence. Consistent with experience, we can generalize the identity of inertial and gravitational mass. *All* kinds of interactions between the constituent parts of a body (nuclear forces in the nuclei, electromagnetic forces in atoms and molecules) contribute to its mass. The principle of equivalence says that locally (in a region of space-time not too large) one cannot *in principle* distinguish between the action of a gravitational field and an acceleration. In other words, a freely falling observer in a gravitational field cannot detect the gravitational field by physical experiments in his immediate neighbourhood; for him *all* events occur as in an inertial system.

We have already encountered coordinate systems, local geodesic coordinates, in which the orbits of freely moving particles are described by $d^2 x^a / d\tau^2 = 0$ as in an inertial system. Because of this coincidence we shall identify inertial systems and local geodesic coordinate systems. As we know, such a local geodesic system can only be introduced in the immediate neighbourhood of a point; it is only useful so long as derivatives of the Christoffel symbols, and hence the influence of the space curvature, can be ignored. Accordingly the freely falling observer too can establish the action of the gravitational force by examining larger regions of space-time; for him the planetary orbits are not straight lines, and upon bouncing on the Earth the freely falling box is no longer an inertial system.

The identification (by the freely falling observer) of inertial system and local geodesic coordinates and the definition (18.4) of the covariant derivative make plausible the following transcription principle: one formulates the physical laws in a Lorentz-invariant manner in an inertial system and substitutes covariant for partial derivatives. This prescription ensures simultaneously the covariance of the resulting equations and their validity upon using curvilinear coordinates in Minkowski space.

Two criticisms can at once be raised, pointing out that this prescription is neither unique nor logically provable. The first criticism concerns the order of higher derivatives. Partial derivatives commute, covariant ones do not. Practical examples nevertheless show that one can solve this problem simply in most cases. The second objection concerns the question of how we know that the curvature tensor and its derivatives do not also enter the basic physical laws. The resulting covariant equations would then *not* go over to the corresponding equations of Minkowski space in local-geodesic coordinates; the difference would certainly be small, however, and would be difficult to detect. Such a modification of our transcription formula cannot be excluded in principle. But up until now no experiments or other indications are known which make it necessary.

In the following sections we shall formulate the most important physical laws in Riemannian spaces, without referring every time to the transcription prescription 'partial → covariant derivative' we are using.

21.2 Particle mechanics

The momentum p^n of a particle is the product of the mass m_0 and the four-velocity u^n:

$$p^n = m_0 \frac{\mathrm{d}x^n}{\mathrm{d}\tau} = m_0 u^n, \tag{21.5}$$

in which τ is the proper time, defined by

$$\mathrm{d}s^2 = g_{nm} \, \mathrm{d}x^n \, \mathrm{d}x^m = -c^2 \, \mathrm{d}\tau^2. \tag{21.6}$$

For force-free motion in Minkowski space the momentum is constant. Accordingly, a particle upon which no force acts apart from the gravitational force moves on a geodesic of the Riemannian space,

$$\frac{Dp^n}{D\tau} = m_0 \frac{D^2 x^n}{D\tau^2} = m_0 u^n_{;i} u^i = m_0 \left(\frac{d^2 x^n}{d\tau^2} + \Gamma^n_{ab} \frac{dx^a}{d\tau} \frac{dx^b}{d\tau} \right) = 0. \quad (21.7)$$

External forces F^n cause a deviation from the geodesic equation:

$$\frac{Dp^n}{D\tau} = m_0 \frac{D^2 x^n}{D\tau^2} = F^n. \quad (21.8)$$

Since the magnitude $-c^2 = u_n u^n$ of the four-velocity is constant, we have

$$u_{n;i} u^n = 0; \quad (21.9)$$

that is, the four-velocity is perpendicular to the four-acceleration $\dot{u}^n = u^n_{;i} u^i$ and the force F^n,

$$u^n u_{n;i} u^i = u^n \dot{u}_n = 0 = F^n u_n. \quad (21.10)$$

The four equations of motion (21.8) are therefore not independent of one another (the energy law is a consequence of the momentum law).

In order to understand better the connection between Newtonian mechanics and mechanics in a Riemannian space we shall sketch how the guiding acceleration $\mathbf{a}$ and the Coriolis force $2\boldsymbol{\omega} \times \dot{\mathbf{r}}$, which an accelerated observer moving in a rotating coordinate system would experience, are contained in the geodesic equation (21.7) which is valid for all coordinate systems.

We therefore imagine an observer who is moving along an arbitrary (timelike) world line and carries with him an orthogonal triad of vectors, whose directions he identifies with the directions of his spatial coordinate axes. For the description of processes in his immediate neighbourhood he will therefore prefer a coordinate system with the following properties: the observer is permanently at the origin O of the spatial system; as time he uses his proper time; along his world line $x^\alpha = 0$ he always uses a Minkowski metric (see Fig. 21.1). Summarizing, this gives up to terms quadratic in the x^n

$$ds^2 = \eta_{ab} \, dx^a \, dx^b + g_{ab,\nu}(O) x^\nu \, dx^a \, dx^b, \quad g_{ab,4}(O) = 0. \quad (21.11)$$

As spatial coordinate lines he will take the lines which arise from 'straight' extension of his triad axes (and are thus geodesic), marking off as coordinates along them the arclength, and so completing this system that in his space $x^4 =$ const. *all* geodesics have locally the form of straight

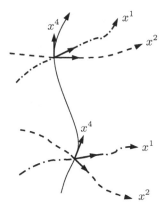

Fig. 21.1. Coordinate system of an arbitrarily moving observer.

lines $x^\alpha = s\lambda^\alpha$ (s is the arclength, and λ^α are the direction cosines). For these geodesics we have then for arbitrary constant λ^α

$$\frac{\mathrm{d}^2 x^a}{\mathrm{d}s^2} + \Gamma^a_{mn} \frac{\mathrm{d}x^m}{\mathrm{d}s} \frac{\mathrm{d}x^n}{\mathrm{d}s} = \Gamma^a_{\mu\nu}\lambda^\mu\lambda^\nu = 0; \qquad (21.12)$$

that is, all Christoffel symbols $\Gamma^a_{\mu\nu}$ ($a = 1, \ldots, 4$; $\mu, \nu = 1, 2, 3$) vanish. Because of (16.13) the derivatives of the metric (21.11) therefore satisfy the conditions

$$g_{\alpha\beta,\nu} = 0, \qquad g_{4\beta,\nu} = -g_{4\nu,\beta}. \qquad (21.13)$$

The equations (21.13) show that there are only three independent components of the derivatives $g_{4\beta,\nu}$; one can thus map these onto the components of a 'three-vector' ω^μ,

$$g_{4\beta,\nu} = -\varepsilon_{4\beta\nu\mu}\omega^\mu/c = -\varepsilon_{\beta\nu\mu}\omega^\mu/c. \qquad (21.14)$$

The derivatives $g_{44,\nu}$ not yet taken into account in (21.11), (21.13) and (21.14) can be expressed through the acceleration a_ν of the observer, for whose world line $x^\alpha = 0$, $x^4 = ct = c\tau$ we have

$$a^\nu \equiv \frac{\mathrm{d}^2 x^\nu}{\mathrm{d}t^2} + \Gamma^\nu_{ab} \frac{\mathrm{d}x^a}{\mathrm{d}t} \frac{\mathrm{d}x^b}{\mathrm{d}t} = \Gamma^\nu_{44}c^2, \qquad (21.15)$$

and hence

$$g_{44,\nu} = -2a^\nu/c^2 = -2g_{\nu b}a^b/c^2. \qquad (21.16)$$

To summarize, an observer, who carries with him his local Minkowski system and in whose position space all geodesics diverging from him are straight lines, uses in a neighbourhood of his world line $x^\alpha = 0$, $x^4 = c\tau$, the coordinate system

$$ds^2 = \eta_{\alpha\beta}\,dx^\alpha\,dx^\beta - 2\varepsilon_{\beta\nu\mu}x^\nu\omega^\mu\,dx^4\,dx^\beta/c - \left(1 + 2a_\nu x^\nu/c^2\right)(dx^4)^2.$$
$$(21.17)$$

For him the only non-vanishing Christoffel symbols are

$$\Gamma^\alpha_{4\nu} = \varepsilon_\nu{}^\alpha{}_\mu\omega^\mu/c, \qquad \Gamma^\alpha_{44} = a^\alpha/c^2, \qquad \Gamma^4_{4\nu} = a_\nu/c^2. \qquad (21.18)$$

If the observer moves on a geodesic, then a_ν vanishes. In the coordinates (21.17) the equation (18.19) which defines Fermi–Walker transport has the form

$$\frac{dT^\mu}{d\tau} + \varepsilon_\alpha{}^\mu{}_\nu\omega^\nu T^\alpha = 0, \qquad \frac{dT^4}{d\tau} = 0. \qquad (21.19)$$

Hence for an observer who subjects this triad, formed out of vectors which he regards as constant, to a Fermi–Walker transport, the vector ω^μ must vanish. If a^ν disappears as well as ω^μ, then the coordinate system (21.17) is an inertial system along the entire world line of the observer.

To describe the motion of a particle the observer will naturally use his coordinate system (21.17) and examine the acceleration d^2x^α/dt^2 of this particle in it. From the geodesic equation (21.12) we have for the three spatial components of the acceleration the relation

$$\frac{d^2x^\alpha}{dt^2} = 2\varepsilon^\alpha{}_{\nu\mu}\omega^\mu\frac{dx^\nu}{dt} - a^\alpha - \frac{d\lambda}{dt}\frac{d}{dt}\left(\frac{dt}{d\lambda}\right)\frac{dx^\alpha}{dt}. \qquad (21.20)$$

We can take the connection between λ and t from the time component of the geodesic equation

$$\frac{d^2t}{d\lambda^2} + 2\frac{a_\nu}{c^2}\frac{dt}{d\lambda}\frac{dx^\nu}{d\lambda} = 0. \qquad (21.21)$$

Substitution of (21.21) into (21.20) yields

$$\frac{d^2x^\alpha}{dt^2} = -a^\alpha + 2\varepsilon^\alpha{}_{\nu\mu}\omega^\mu\frac{dx^\nu}{dt} + \frac{2a_\nu}{c^2}\frac{dx^\nu}{dt}\frac{dx^\alpha}{dt}, \qquad (21.22)$$

or, in vector form,

$$\ddot{\mathbf{r}} = -\mathbf{a} - 2\boldsymbol{\omega}\times\dot{\mathbf{r}} + 2(\mathbf{a}\dot{\mathbf{r}})\dot{\mathbf{r}}/c^2. \qquad (21.23)$$

One recognizes the guiding acceleration $\mathbf{a}$ and its relativistic correction $2(\mathbf{a}\dot{\mathbf{r}})\dot{\mathbf{r}}/c^2$ (both vanish if the observer is moving freely on a geodesic), and also the Coriolis acceleration $2\boldsymbol{\omega}\times\dot{\mathbf{r}}$, caused by the rotation of the triad carried by the observer relative to a Fermi–Walker-transported triad. The vanishing of the Coriolis term in the Fermi–Walker system justifies the statement that for an observer who is not falling freely $(a^\nu \neq 0)$ a local coordinate system produced by Fermi–Walker transport of the spatial triad of vectors is the best possible realization of a non-rotating system.

21.3 Electrodynamics in vacuo

The field equations As in Minkowski space, the electromagnetic field is described by an antisymmetric field-tensor F_{mn}. Because it satisfies the equations

$$3F_{[mn;a]} = F_{mn;a} + F_{na;m} + F_{am;n} = F_{mn,a} + F_{na,m} + F_{am,n} = 0, \quad (21.24)$$

it can be represented as the curl of a four-potential A_n:

$$F_{mn} = A_{n;m} - A_{m;n} = A_{n,m} - A_{m,n}. \quad (21.25)$$

This potential is determined only up to a four-dimensional gradient. The field is produced by the four-current j^m:

$$F^{mn}{}_{;n} = \frac{1}{\sqrt{-g}} \left(\sqrt{-g}\, F^{mn}\right)_{,n} = \frac{1}{c} j^m. \quad (21.26)$$

Because of the antisymmetry of F_{mn}, (21.26) is only integrable (self-consistent) if the continuity equation

$$j^m{}_{;m} = \frac{1}{\sqrt{-g}} \left(\sqrt{-g}\, j^m\right)_{,m} = 0 \quad (21.27)$$

is satisfied. For an isolated charge distribution the conservation law for the total charge Q follows from it (cp. Section 20.5):

$$\int_{x^4 = \text{const.}} j^a \, \mathrm{d}f_a = \text{const.} = Q. \quad (21.28)$$

By substituting (21.25) into (21.26) one can derive the generalized inhomogeneous wave equation for the potential. Using the expressions written with covariant derivatives, one obtains

$$A^{n;m}{}_{;n} - A^{m;n}{}_{;n} = A^n{}_{;n}{}^{;m} + R_n{}^m A^n - A^{m;n}{}_{;n} = j^m/c. \quad (21.29)$$

If, on the other hand, one sets out directly from the special-relativistic equation,

$$(A^n{}_{,n})^{,m} - A^{m,n}{}_{,n} = j^m/c, \quad (21.30)$$

and in it replaces the partial by covariant derivatives, then one obtains (21.29) *without* the term in the Ricci tensor $R_n{}^m$ (which arises by interchange of covariant derivatives). One clearly sees here that the transcription formula 'partial $\rightarrow$ covariant derivative' is not unique when applied to the potential. Potentials, however, are not directly measurable, and for the physically important field strengths and their derivatives the prescription which we gave above is unique. Since (21.29) follows directly from this prescription it is considered to be the correct generalization of the inhomogeneous wave equation.

For practical calculations it is often convenient to use partial derivatives; (21.25) and (21.29) give

$$\left[\sqrt{-g}\,g^{ma}g^{nb}(A_{b,a}-A_{a,b})\right]_{,n}=\sqrt{-g}\,j^m/c. \tag{21.31}$$

Lagrangian and energy-momentum tensor   Maxwell's equations can be derived from the action principle

$$W=\int \mathcal{L}\,\mathrm{d}^4x=\int\left[j^a A_a/c-\tfrac14(A_{n,m}-A_{m,n})(A^{n,m}-A^{m,n})\right]\sqrt{-g}\,\mathrm{d}^4x$$

$$=\text{extremum}, \tag{21.32}$$

where the components of the potential are varied as the independent field quantities.

The symmetric energy-momentum tensor,

$$T^{mn}=F^{am}F_a{}^n-\tfrac14 g^{mn}F_{ab}F^{ab}, \tag{21.33}$$

is trace-free, $T^n{}_n=0$. Its divergence is, up to a sign, equal to the Lorentz force density,

$$T^{mn}{}_{;n}=-F^{mn}j_n/c. \tag{21.34}$$

Description of the solutions in terms of the sources   In a Riemannian space it is still possible to express the solution to the differential equation (21.29) in the form of an integral

$$A^m(x^i)=\int G^{m\bar n}(x^i,\bar x^i)j_{\bar n}(\bar x^i)\sqrt{-\bar g}\,\mathrm{d}^4\bar x. \tag{21.35}$$

The two-point tensor $G^{m\bar n}$, the generalization of the Green function, is now in general a very complicated function. We want to point out (without proof) a notable difference in the way in which effects (for example, light pulses) propagate in a Riemannian space in comparison with that in a Minkowski space. While in Minkowski space the propagation of effects *in vacuo* takes place exactly on the light cone, that is, a flash of light at the point $\bar P$ of Fig. 21.2(a) reaches the observer at precisely the point P, in Riemannian space the wave can also propagate inside the future light cone, a (weak) flash of light being noticeable also at points later that P (for example, P'). The reason for this deviation, which one can also interpret as a deviation from Huygens' principle, can be

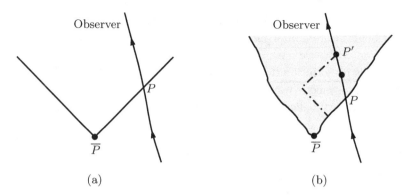

Fig. 21.2. Propagation of effects between source $\bar{P}$ and observer (a) Minkowski space: on the light cone, (b) Riemannian space: within the entire (shadowed) interior of the light cone. (Dashed line = possible light path ('dispersion').)

thought of as a kind of scattering of the light wave by the space curvature. In particularly simple Riemannian spaces this effect does not occur; for example, the Robertson–Walker metrics belong to this class (see Chapter 40).

Special properties of source-free fields Since one can convert (21.24) into the system

$$\tilde{F}^{ab}{}_{;b} = 0 \tag{21.36}$$

by use of the dual field tensor,

$$\tilde{F}_{ab} = \tfrac{1}{2}\varepsilon_{abmn}F^{mn}, \tag{21.37}$$

then for $j^m = 0$ Maxwell's equations are equivalent to the equations

$$\Phi^{ab}{}_{;b} = \left(F^{ab} + \mathrm{i}\tilde{F}^{ab}\right)_{;b} = 0 \tag{21.38}$$

for the complex field tensor Φ^{ab}. A solution Φ^{ab} remains a solution after multiplication by a complex number $\mathrm{e}^{\mathrm{i}\alpha}$ (a 'duality rotation'). The energy-momentum tensor

$$T^{ab} = \tfrac{1}{2}\Phi^{ac}\bar{\Phi}^b{}_c \tag{21.39}$$

does not change under such a duality rotation.

The source-free Maxwell's equations are 'conformally invariant'. A conformal transformation is a transformation between two spaces $\widehat{M}$ (with metric $\mathrm{d}\hat{s}^2$) and M (with metric $\mathrm{d}s^2$) such that

$$\mathrm{d}s^2 = \Omega^2\,\mathrm{d}\hat{s}^2 \quad\leftrightarrow\quad g_{ab} = \Omega^2\,\hat{g}_{ab}, \tag{21.40}$$

all distances are (locally) scaled by the same factor, independent of their directions.

Suppose now in the space M we have a Maxwell field satisfying

$$\sqrt{-g}F^{mn}{}_{;n} = \left(\sqrt{-g}F^{mn}\right)_{,n} = 0. \qquad (21.41)$$

Taking then in $\widehat{M}$ the same fields, $\widehat{A}_a = A_a$, $\widehat{F}_{ab} = F_{ab}$, we have with $g = \Omega^8\,\widehat{g}$ and $F^{mn} = \Omega^{-4}\widehat{F}^{mn}$ (here the dimension of the space enters!)

$$\left(\Omega^4\sqrt{-\widehat{g}}\,\Omega^{-4}\widehat{F}^{mn}\right)_{,n} = 0 = \left(\sqrt{-\widehat{g}}\,\widehat{F}^{mn}\right)_{,n}, \qquad (21.42)$$

Maxwell's equations are valid in $\widehat{M}$, too.

Null electromagnetic fields As in Minkowski space (cp. Section 6.4 and (7.44)) the electromagnetic field tensor possesses two invariants, namely

$$I_1 = F_{ab}F^{ab}, \qquad I_2 = F_{ab}\tilde{F}^{ab}. \qquad (21.43)$$

Null electromagnetic fields are fields for which both invariants vanish. They are therefore generalizations of plane waves in flat space. As shown in Section 9.3, the field F_{mn} and energy-momentum tensor T_{mn} have the form

$$F_{mn} = (p_m k_n - k_m p_n), \quad k^n k_n = 0 = p_n k^n, \quad T_{mn} = (p_n p^n)k_m k_n. \ (21.44)$$

21.4 Geometrical optics

The transition from wave solutions of the source-free Maxwell equations to geometrical optics can be accomplished by substituting into the field equations

$$(A^{n;m} - A^{m;n})_{;n} = \left[\sqrt{-g}\,g^{ma}g^{nb}(A_{b,a} - A_{a,b})\right]_{,n}\big/\sqrt{-g} = 0 \quad (21.45)$$

the ansatz

$$A_a = \widehat{A}_a(x^n)\,\mathrm{e}^{\mathrm{i}\omega S(x^n)} \quad (\widehat{A}_a \text{ complex, } S \text{ real}) \qquad (21.46)$$

and setting the coefficients of ω^2 and ω separately to zero. As in flat space, this splitting into amplitude $\widehat{A}_a$ and eikonal (phase) S is meaningful only in certain finite regions of space and represents a good approximation only for large ω.

Substitution of (21.46) into (21.45) gives, on taking into account only the terms in ω^2,

$$S^{,m}(\widehat{A}^n S_{,n}) - \widehat{A}^m(S_{,n}S^{,n}) = 0. \qquad (21.47)$$

Since the part of the field tensor proportional to ω is

$$F_{mn} = (\hat{A}_n S_{,m} - \hat{A}_m S_{,n}) \, \mathrm{i}\,\omega\, \mathrm{e}^{\mathrm{i}\omega S} = \mathrm{i}\,\omega\,(A_n S_{,m} - A_m S_{,n}), \qquad (21.48)$$

this part vanishes if $\hat{A}_m$ is parallel to $S_{,m}$. We are therefore interested only in the solution

$$S_{,n} S^{,n} = 0, \qquad A_n S^{,n} = 0 \qquad (21.49)$$

of (21.47). The gradient $S_{,n}$ of the surfaces of equal phase is therefore a null vector and the field tensor (21.48) has the structure (21.44) of the field tensor of a null field, with k_m proportional to $S_{,m}$ and p_n proportional to $\mathrm{Re}\,A_n$ (note that for comparison purposes one must take the real part of the complex field quantities used here). In this approximation the field consequently behaves locally like a plane wave.

Differentiating (21.49) gives

$$S^{,n} S_{,n;m} = 0. \qquad (21.50)$$

Since the curl of a gradient vanishes ($S_{,n;m} = S_{,m;n}$), this is equivalent to

$$S_{,m;n} S^{,n} = 0. \qquad (21.51)$$

This equation says that the curves $x^m(\lambda)$, whose tangent vector is $S^{,m}$,

$$\frac{\mathrm{d}x^m}{\mathrm{d}\lambda} = S^{,m}, \qquad \frac{\mathrm{D}^2 x^m}{\mathrm{D}\lambda^2} = \frac{\mathrm{D}S^{,m}}{\mathrm{D}\lambda} = S^{,m;n}\frac{\mathrm{d}x^m}{\mathrm{d}\lambda} = 0, \qquad (21.52)$$

are geodesics, and because $S_{,n} S^{,n} = 0$ they are null geodesics.

If we characterize the wave not by $\hat{A}_a$ and the surfaces of constant phase $S = \mathrm{const.}$, but by the curves $x^n(\lambda)$ orthogonal to them (which we call light rays), then we have accomplished the transition from wave optics to geometrical optics. In words, (21.51) then says that light rays are null geodesics.

We shall take the approximation one step further, investigating the terms in Maxwell's equations proportional to ω, and hence obtaining statements about how the intensity and polarization of the wave change along a light ray.

From (21.45) and (21.49) one obtains immediately

$$-\mathrm{i}\,\omega\left[2\hat{A}^m{}_{;n} S^{,n} + \hat{A}^m S^{,n}{}_{;n} - \hat{A}^n{}_{;n} S^{,m}\right] = 0. \qquad (21.53)$$

If one contracts this equation with the vector $\bar{\hat{A}}_m$, which is the complex conjugate of $\hat{A}_m$, and takes note of (21.49), then the result can be written in the form

$$\left(\bar{\hat{A}}_m \hat{A}^m\right)_{;n} S^{,n} + \left(\bar{\hat{A}}_m \hat{A}^m\right) S^{,n}{}_{;n} = 0, \qquad (21.54)$$

or in the equivalent form

$$\left(\bar{\hat{A}}_m \hat{A}^m S^{,n}\right)_{;n} = 0. \tag{21.55}$$

Because the intensity of the wave is proportional to $\bar{\hat{A}}_m \hat{A}^m$, (21.54) can be read as a statement about the change in intensity of the light ray in the direction $S^{,n}$. Even clearer, however, is the picture suggested by (21.55). If one interprets

$$J^n = \bar{\hat{A}}_m \hat{A}^m S^{,n} \tag{21.56}$$

as a photon current, then this current is source-free (conservation of photon number) and in the direction of the light rays.

We obtain a further physical consequence from (21.53) if we decompose the vector $\hat{A}_m$ into its magnitude a and the unit vector P_m:

$$\hat{A}_m = aP_m. \tag{21.57}$$

Then (21.54) is equivalent to

$$a_{,n}S^{,n} = -\tfrac{1}{2}aS^{,n}_{\ ;n}, \tag{21.58}$$

and from (21.53) we have

$$P^m_{\ ;b}S^{,b} = \tfrac{1}{2}\left(P^n a_{,n}/a + P^n_{\ ;n}\right)S^{,m}. \tag{21.59}$$

This means, however, that the tensor f_{mn} associated with the field tensor (21.48),

$$f_{mn} = P_n S_{,m} - P_m S_{,n}, \tag{21.60}$$

which contains the characteristic directions of the wave (direction of propagation $S^{,n}$ and polarization P^m), is parallelly transported along the rays; we have

$$f_{mn;i}S^{,i} = 0. \tag{21.61}$$

21.5 Thermodynamics

Thermodynamical systems can be extraordinarily complicated; for example, a great number of processes can be going on in a star simultaneously. We want to try to explain the basic general ideas, restricting ourselves to the simplest systems.

During thermodynamical processes certain elements of matter, with their properties, remain conserved, for example, in non-relativistic thermodynamics molecules or atoms and their masses. In the course of transformations in stars and during nuclear processes the baryons with their rest mass are conserved instead. We shall therefore relate all quantities to these baryons. If, for example, we choose a volume element of the system, then we shall take as four-velocity u^i of this element the average baryon velocity. The flow (motion) of the system will therefore be characterized by a four-velocity field

$$u^i = u^i(x^m), \qquad u^i u_i = -c^2. \tag{21.62}$$

To set up the basic thermodynamical equations one first goes to the local rest system,

$$u^i = (0, 0, 0, c), \tag{21.63}$$

of the volume element under consideration and regards this volume element as a system existing in equilibrium (of course it interacts with its surroundings, so that the whole system is not necessarily in equilibrium); that is, one introduces for this volume element the fundamental thermodynamic state variables, for example,

n baryon number density,	s entropy per baryon mass,
ρ baryon mass density,	p isotropic pressure,
T temperature,	$\bar{\mu}$ chemical potential,
u internal energy per unit mass,	f free energy per unit mass.

$$\tag{21.64}$$

'Density' here always means 'per three-dimensional volume in the local rest system'; the entropy density, for example, would be given by $s\rho$. There exist relationships between these state variables which in the simplest case express the fact that only two of them are really independent, and, from knowledge of the entropy as a function of the energy and the density, or of the specific volume $v = 1/\rho$,

$$s = s(u, v), \tag{21.65}$$

one can calculate the other quantities, for example,

$$\frac{\partial s}{\partial u} = \frac{1}{T}, \qquad \frac{\partial s}{\partial v} = \frac{p}{T}, \qquad f = u - Ts. \tag{21.66}$$

For the interaction of the volume element with its surroundings we have balance equations. These are the law of conservation of baryon number,

$$(\rho u^a)_{;a} = 0 \tag{21.67}$$

(generalized mass conservation), the balance equations for energy and momentum, formulated as the vanishing of the divergence of the energy-momentum tensor T^{ma},

$$T^{ma}{}_{;a} = 0 \tag{21.68}$$

(generalized first law), and the balance equation for the entropy,

$$s^a_{;a} = \sigma \geq 0, \tag{21.69}$$

which says that the density of entropy production σ is always positive or zero (generalized second law of thermodynamics). Of course these equations take on a physical meaning only if the entropy current density s^a and the energy-momentum tensor T^{ma} are tied up with one another and with the thermodynamic quantities (21.64).

This can be done as follows. One uses the projection tensor,

$$h_{ab} = g_{ab} + u_a u_b/c^2, \tag{21.70}$$

to decompose the energy-momentum tensor into components parallel and perpendicular to the four-velocity,

$$T_{ab} = \mu u_a u_b + ph_{ab} + (u_a q_b + q_a u_b)/c^2 + \pi_{ab},$$
$$q_a u^a = 0, \qquad \pi_{ab} u^a = 0, \qquad \pi^a_a = 0, \tag{21.71}$$

and links the quantities which then occur to the thermodynamic state variables and to the entropy current vector. The coefficient of h_{ab} is the isotropic pressure p, the internal energy per unit mass u is coupled to the mass density μ in the rest system of the matter by

$$\mu = \rho(1 + u/c^2), \tag{21.72}$$

and the heat current q^i (momentum current density in the rest system) goes into the entropy current density:

$$s^i = \rho s u^i + q^i/T. \tag{21.73}$$

Equation (21.73) says that the entropy flows in such a way that it is carried along convectively with the mass (first term) or transported by the flow of heat (generalization of $dS = dQ/T$).

We now want to obtain an explicit expression for the entropy production density σ. Upon using (21.67) and the equation

$$s_{,n} u^n = \frac{1}{T}\left(\frac{\mu c^2}{\rho}\right)_{,n} u^n + \frac{p}{T}\left(\frac{1}{\rho}\right)_{,n} u^n = \frac{1}{\rho T}\left[(p + \mu c^2)u^i_{;i} + \mu_{,n} u^n c^2\right], \tag{21.74}$$

which follows from (21.65), (21.66) and (21.72), we obtain

$$\sigma = s^n_{;n} = \frac{1}{T}\left[(p + \mu c^2)u^i_{;i} + \mu_{,n} u^n c^2\right] + \left(\frac{q^n}{T}\right)_{;n}. \tag{21.75}$$

Since the terms in square brackets can be written in the form

$$(p + \mu c^2)u^i_{;i} + \mu_{,n} u^n c^2 = -(\mu u_a u_b + ph_{ab})^{;b} u^a, \tag{21.76}$$

and the divergence of the energy-momentum tensor vanishes, (21.75) implies the relation

$$\sigma = \left[T^{mn} - \mu u^m u^n - ph^{mn}\right]_{;m} \frac{u_n}{T} + \left(\frac{q^n}{T}\right)_{;n}, \tag{21.77}$$

which, bearing in mind the definition (21.71) of q^n, and using (18.27), can be cast finally into the form

$$\begin{aligned}\sigma &= -(T^{mn} - \mu u^m u^n - ph^{mn})\left(\frac{u_m}{T}\right)_{;n} \\ &= -\tfrac{1}{2}(T^{mn} - \mu u^m u^n - ph^{mn})\mathcal{L}_{(\mathbf{u}/T)}g_{mn}.\end{aligned} \tag{21.78}$$

In irreversible thermodynamics one can satisfy the requirement that $\sigma \geq 0$ in many cases by writing the right-hand side of (21.78) as a positive-definite quadratic form, that is, by making an assumption of linear phenomenological equations. For the particular case $\pi^{mn} = 0$, when (21.78) reduces, because of (21.71), to

$$\sigma = -q_a \frac{1}{T^2}\left(\frac{T}{c^2}\dot{u}^a + T^{,a}\right), \quad \dot{u}^a \equiv u^a_{;n}u^n, \tag{21.79}$$

this ansatz means that

$$q_a = -\bar{\kappa}\left(T_{,n} + \dot{u}_n T/c^2\right)h_a{}^n, \tag{21.80}$$

which represents the relativistic generalization of the linear relation between heat current and temperature gradient.

In many cases one can ignore irreversible processes. If the system is determined by only two state quantities in the sense of (21.65), this means because of (21.78) that complete, exact reversibility ($\sigma = 0$) is possible either only for certain metrics (whose Lie derivatives vanish, and the system is then in thermodynamic equilibrium), or for especially simple media, whose energy-momentum tensor has the form

$$T^{mn} = \mu u^m u^n + ph^{mn} = \left(\mu + p/c^2\right)u^m u^n + pg^{mn}. \tag{21.81}$$

Such a medium is called a *perfect fluid* or, for $p = 0$, *dust*. In a local Minkowski system, the energy-momentum tensor has the form (13.6).

When superposing incoherent electromagnetic fields one has to add (and average) the energy-momentum tensors and not the field strengths (a field tensor can no longer be associated with this superposition). If the fields being superposed single out locally no spatial direction in the rest system $u^m = (0,0,0,c)$ of an observer, then the resulting energy-momentum tensor also has the perfect fluid form (21.81). Under the superposition the properties of vanishing trace, $T^n{}_n = 0$, and vanishing divergence, $T^{mn}{}_{;n} = 0$, are of course preserved. Consequently the radiation pressure p and the energy density μc^2 are related by

$$3p = \mu c^2. \tag{21.82}$$

Such a medium is called an *incoherent radiation field*.

21.6 Perfect fluids and dust

According to the definition given in Section 21.5, a perfect fluid is characterized by having an energy-momentum tensor of the form (21.81).

The equation of motion of this flow reads, using the notation $\dot\mu = \mu_{,n} u^n$, $\dot u^n = u^n{}_{;i} u^i$, etc.,

$$T^{mn}{}_{;n} = \left(\mu + p/c^2\right) u^n{}_{;n} u^m + \left(\mu + p/c^2\right)\dot{}\, u^m + \left(\mu + p/c^2\right)\dot u^m + p^{,m} = 0. \tag{21.83}$$

Contraction with u_m/c^2 gives the energy balance

$$\dot\mu + \left(\mu + p/c^2\right) u^n{}_{;n} = 0, \tag{21.84}$$

and contraction with the projection tensor, $h^i_m = g^i_m + u^i u_m/c^2$, the momentum balance

$$\left(\mu + p/c^2\right)\dot u^i + h^{im} p_{,m} = 0. \tag{21.85}$$

Equation (21.85) shows that the pressure too contributes to the inertia of the matter elements, the classical analogue of this equation being of course

$$\rho\frac{d\mathbf{v}}{dt} = -\mathbf{grad}\,p. \tag{21.86}$$

The equations of motion (21.84) and (21.85) must in each case be completed by the specification of an equation of state. One can regard as the simplest equation of state that of dust, $p = 0$. From this and the equations of motion follow

$$\dot u^m = Du^m/D\tau = 0 \quad \text{and} \quad (\mu u^n)_{;n} = 0; \tag{21.87}$$

that is, the stream-lines of the matter are geodesics, and the rest mass μ is conserved.

21.7 Other fundamental physical laws

Just as with the examples of particle mechanics, electrodynamics, thermodynamics and mechanics of continua which have been described in detail, so also one can carry over to Riemannian spaces other classical theories, for example, those of the Dirac equation, of the Weyl equation for the neutrino field, and for the Klein–Gordon equation. Although the foundations of these theories have been thoroughly worked out, convincing examples and applications within the theory of gravitation are still lacking, and we shall therefore not go into them further. For the Einstein gravitation theory only the following property of closed systems, that is, systems upon which act no forces whose origins lie outside the systems, will be important; namely, that their energy-momentum tensor T_{mn} is symmetric (expressing the law of conservation of angular momentum in Special Relativity) and its divergence vanishes (generalization of the law of conservation of energy-momentum in Special Relativity):

$$T_{mn} = T_{nm}, \qquad T^{mn}{}_{;n} = 0. \tag{21.88}$$

With the fundamental laws of quantum mechanics and quantum field theory things are rather different. Here indeed some work has been done, addressed to particular questions, but one cannot yet speak of a real synthesis between quantum theory and gravitation theory (see Chapter 38).

Exercises

21.1   In Minkowski space, the four-potentials are often gauged by $A^n{}_{;n} = 0$ (Lorentz gauge). Is that gauge also possible in a Riemannian space? And will this gauge decouple the wave equations?

21.2   For a perfect fluid in hydrostatic equilibrium, the metric does not depend on time in the fluid's rest system. Determine g_{tt} in terms of $p(\mu)$ and give the explicit expression for $p = \alpha\mu$. Is a surface $p = 0$ always possible in the latter case?

21.3   Show that for hydrostatic *and* thermal equilibrium (with $q_a = 0$) the temperature T satisfies $T\sqrt{-g_{tt}} = $ const.

Further reading for Chapter 21

Ehlers (1961, 1966, 1971), Neugebauer (1980), Synge (1960), Tolman (1934).

III. Foundations of Einstein's theory of gravitation

22
The fundamental equations of Einstein's theory of gravitation

22.1 The Einstein field equations

As we have already indicated more than once, the basic idea of Einstein's theory of gravitation consists of geometrizing the gravitational force, that is, mapping all properties of the gravitational force and its influence upon physical processes onto the properties of a Riemannian space. While up until the present we have concerned ourselves only with the mathematical structure of such a space and the influence of a given Riemannian space upon physical laws, we want now to turn to the essential physical question. Gravitational fields are produced by masses – so how are the properties of the Riemannian space calculated from the distribution of matter? Here, in the context of General Relativity, 'matter' means everything that can produce a gravitational field (i.e. that contributes to the energy-momentum tensor), for example, not only atomic nuclei and electrons, but also the electromagnetic field.

Of course one cannot derive logically the required new fundamental physical law from the laws already known; however, one can set up several very plausible requirements. We shall do this in the following and discover, surprisingly, that once one accepts the Riemannian space, the Einstein field equations follow almost directly.

The following requirements appear reasonable.

(a) The field equations should be tensor equations (independence of coordinate systems of the laws of nature).

(b) Like all other field equations of physics they should be partial differential equations of at most second order for the functions to be determined (the components of the metric tensor g_{mn}), which are linear in the highest derivatives.

(c) They should (in the appropriate limit) go over to the Poisson (potential) equation:

$$\Delta U = 4\pi f \mu \qquad (22.1)$$

of Newtonian gravitation theory (here U is the potential, f is the Newtonian gravitational constant, and μ is the mass density).

(d) Since the energy-momentum tensor T^{mn} is the special relativistic analogue of the mass density, it should be the cause (source) of the gravitational field.

(e) If the space if flat, T^{mn} should vanish.

We now want to see where these requirements lead us. Plainly we need a tensor (requirement (a)) that contains only derivatives of the metric up to second order (requirement (b)); as building blocks for this *Einstein tensor* G_{mn}, only the curvature tensor, the metric tensor and the ε-tensor are available, as we have already shown in Section 19.3. Requirement (d) means that the field equations have the structure

$$G_{mn} = \kappa T_{mn}, \qquad (22.2)$$

with a constant of nature κ which is still to be determined; this is consistent with the symmetry and vanishing divergence (21.88) of the energy-momentum tensor only if

$$G^{mn}{}_{;n} = 0 \quad \text{and} \quad G_{mn} = G_{nm}. \qquad (22.3)$$

There is now, as one can show, only one second rank tensor which is linear (requirement (b)) in the components of the curvature tensor and which satisfies (22.3); namely, $R_{mn} - \frac{1}{2}g_{mn}R$, which we have already met in (19.39) during the discussion of the Bianchi identities. Since the metric tensor itself also satisfies (22.3), G^{mn} has the form

$$G^{mn} = R^{mn} - \frac{1}{2}g^{mn}R + \Lambda g^{mn}. \qquad (22.4)$$

The natural constant Λ is the *cosmological constant*, introduced by Einstein (1917). If it does not vanish, a completely matter-free space ($T^{mn} = 0$) would always be curved, in contradiction to requirement (e), since because of (22.2) and (22.4) the Ricci tensor R^{mn} cannot vanish. This requirement (e) is, however, difficult to prove. It is only possible to distinguish the cases $\Lambda = 0$ and $\Lambda \neq 0$ by making observations and relating them to cosmological models. We shall assume that $\Lambda = 0$, but we shall also discuss for a series of examples the influence of the term Λg^{mn} by bringing it into the right-hand side of (22.2) and formally regarding it as part of the energy-momentum tensor.

The Einstein tensor thus has the form

$$G_{mn} = R_{mn} - \tfrac{1}{2}R g_{mn}. \tag{22.5}$$

For actual calculations, use of the explicit representation in terms of the components of the curvature tensor

$$
\begin{aligned}
G_1^1 &= -(R^{23}{}_{23} + R^{24}{}_{24} + R^{34}{}_{34}), \\
G_2^2 &= -(R^{13}{}_{13} + R^{14}{}_{14} + R^{34}{}_{34}), \\
G_2^1 &= -(R^{31}{}_{32} + R^{41}{}_{42}), \\
G_3^2 &= -(R^{12}{}_{13} + R^{42}{}_{43}), \quad \text{etc.,}
\end{aligned}
\tag{22.6}
$$

is often useful.

Our demands have led us in a rather unambiguous manner to the *Einstein field equations*

$$G_{mn} = R_{mn} - \tfrac{1}{2}R g_{mn} = \kappa T_{mn}. \tag{22.7}$$

Einstein (1915) himself derived them after about ten years of research. They were published nearly simultaneously by Hilbert (1915), who knew about Einstein's quest for the correct form of the equations and used the variational principle (see Section 22.4) to derive them. The natural law (22.7) shows how the space curvature (represented by the Ricci tensor R_{mn}) is related to the matter distribution (represented by the energy-momentum tensor T_{mn}).

The field equations (22.7) constitute a system of ten different equations to determine the ten metric functions g_{mn}. But even for fixed initial conditions this system has no unique solution; it must still always be possible to carry out arbitrary coordinate transformations. In fact precisely this under-determinacy in the system of field equations is guaranteed by the existence of the contracted Bianchi identities

$$G^{mn}{}_{;n} = \kappa T^{mn}{}_{;n} = 0. \tag{22.8}$$

They of course express the fact that the ten field equations (22.7) are not independent of each other.

The equations (22.8) permit a conclusion of great physical significance. Since the divergence of the Einstein tensor G^{mn} vanishes identically, the Einstein field equations are integrable and free of internal contradiction only if $T^{mn}{}_{;n} = 0$. The covariant derivative in this condition is, however, to be calculated with respect to the metric g_{mn}, which should be first determined from this very energy-momentum tensor! It is therefore in principle impossible first to specify the space-time distribution of the matter (the matter and its motion) and from this to calculate the space structure. Space structure (curvature) and motion of the matter in this space constitute a dynamical system whose elements are so closely coupled with one another that they can only be solved simultaneously. The space is not the stage for the physical event, but rather an aspect of the interaction and motion of the matter.

Sometimes one can assume to good approximation that the space structure is determined by a part of the energy-momentum tensor (for example, by the masses of the stars) and that the remainder (for example, the starlight) no longer alters the curvature. One then speaks of test fields. These are fields which do not cause gravitational fields, but are only influenced by the gravitational fields already existing and hence serve to demonstrate the properties of these fields; they do not appear on the right-hand side of the field equations (22.7).

22.2 The Newtonian limit

In every new physical theory the previous one is contained as a limiting case. This experience is confirmed also in the theory of gravitation. The purpose of this section is to bring out the connection between the Einstein equations (22.7) and the Newtonian theory of gravitation and thereby to clarify the physical meaning of the natural constant κ introduced in (22.7). First of all we must define what we mean by 'Newtonian limit'. In the Newtonian theory of gravitation the mass density μ is the only source of the field. In the applications in which its predictions have been verified, such as planetary motion, all velocities in the rest system of the centre of gravity of the field-producing masses, for example the Sun, are small compared with the velocity of light. Therefore the following characterization of the Newtonian limit is appropriate.

(a) There exists a coordinate system in which the energy density

$$T_{44} = \mu c^2 \tag{22.9}$$

is the effective source of the gravitational field and all other components of the energy-momentum tensor are ignorable.

(b) The fields vary only slowly; derivatives with respect to $x^4 = ct$, which of course contain the factor c^{-1}, are to be ignored.

(c) The metric deviates only slightly from that of a Minkowski space:

$$g_{mn} = \eta_{nm} + f_{nm}, \quad \eta_{mn} = \text{diag}\,(1,1,1,-1). \tag{22.10}$$

Terms which are quadratic in f_{mn} and its derivatives are ignored; the Einstein field equations are linearized (see Section 27.2).

We have now to incorporate these three ideas into the field equations (22.7). By contraction we have quite generally from (22.7) the relation

$$-R = \kappa T^a{}_a = \kappa T, \tag{22.11}$$

so that one can also write the Einstein equations in the form

$$R_{mn} = \kappa\big(T_{mn} - \tfrac{1}{2}g_{mn}T\big). \tag{22.12}$$

Of these ten equations only

$$R_{44} = \kappa\big(T_{44} - \tfrac{1}{2}\eta_{44}T\big) = \kappa\big(\mu c^2 - \tfrac{1}{2}\mu c^2\big) = \tfrac{1}{2}\kappa\mu c^2 \tag{22.13}$$

is of interest in the Newtonian approximation. In order to calculate R_{44} from the metric (22.10) we start from the defining equation (19.7) for the curvature tensor and ignore terms which are quadratic in the Christoffel symbols, that is, we use

$$R^a{}_{mbn} = \Gamma^a{}_{mn,b} - \Gamma^a{}_{mb,n} = \tfrac{1}{2}\eta^{as}(f_{sn,mb} + f_{mb,sn} - f_{mn,bs} - f_{bs,mn}). \tag{22.14}$$

Then we have

$$R_{44} = R^a{}_{4a4} = \tfrac{1}{2}\eta^{as}(f_{s4,a4} + f_{a4,s4} - f_{44,as} - f_{as,44}), \tag{22.15}$$

or, on ignoring all time derivatives,

$$R_{44} = -\tfrac{1}{2}\eta^{as}f_{44,as} = -\tfrac{1}{2}\eta^{a\sigma}f_{44,\alpha\sigma} = -\tfrac{1}{2}\Delta f_{44}, \tag{22.16}$$

and the field equation (22.13) simplifies to

$$\Delta f_{44} = -\kappa\mu c^2. \tag{22.17}$$

This equation has indeed the structure of a Poisson equation – but not every quantity which satisfies a Poisson equation necessarily coincides with the Newtonian gravitational potential! In order not to make a mistake in the physical interpretation of (22.17) we need one additional piece of information, which is furnished by the geodesic equation

$$\frac{d^2 x^n}{d\tau^2} = -\Gamma^n_{ab}\frac{dx^a}{d\tau}\frac{dx^b}{d\tau}. \tag{22.18}$$

For slowly moving particles (e.g. planets) proper time almost coincides with coordinate time $t = x^4/c$, and the four-velocity on the right-hand side of (22.18) can be replaced by $u^a = (0,0,0,c)$:

$$\frac{\mathrm{d}^2 x^\nu}{\mathrm{d}t^2} = -\Gamma^\nu_{44} c^2 = \tfrac{1}{2}\eta^{\nu\mu} g_{44,\mu} c^2 = \tfrac{1}{2}\eta^{\nu\mu} f_{44,\mu} c^2. \tag{22.19}$$

If we compare this equation of motion with that for a particle in the gravitational potential U, that is, with

$$\frac{\mathrm{d}^2 \mathbf{r}}{\mathrm{d}t^2} = -\mathbf{grad}\,U, \tag{22.20}$$

then we see that the Newtonian gravitational potential U is related to the metric by the relation

$$U = -c^2 f_{44}/2, \quad g_{44} = -(1 + 2U/c^2), \tag{22.21}$$

and that because of (22.1), (22.17) and (22.21) we have the relation

$$8\pi f/c^4 = \kappa = 2.07 \times 10^{-48}\,\mathrm{g}^{-1}\mathrm{cm}^{-1}\mathrm{s}^2 \tag{22.22}$$

between the Newtonian constant of gravitation f and the Einstein natural constant κ. This establishes the required connection between Newtonian and Einsteinian gravitational theories.

The relation (22.21) between g_{44} and the potential U is in agreement with equation (21.17), since for small spatial regions we certainly have $U(x^\nu) = U_{,\nu} x^\nu = a_\nu x^\nu$ ($a_\nu = +U_{,\nu}$, because we are dealing with components of the acceleration seen from a freely falling inertial system).

22.3 The equations of motion of test particles

Monopole particle It is one of its particular merits that, in the Einstein theory, the equations of motion are a consequence of the field equations. If we take, for example, the Maxwell theory, then charge conservation is of course a consequence of the field equations, but the motion of the sources and the distribution of the charges are arbitrarily specifiable. Also the field of two point charges at rest a finite distance apart is an exact solution of Maxwell's equations – although the charges exert forces upon one another and therefore would be immediately accelerated into motion.

Even after the Einstein field equations had been set up it was thought that one had to demand in addition that the geodesic equation be the equation of motion of a test particle; but eventually it was realized that this can be deduced from the relation

$$T^{mn}{}_{;n} = 0 \tag{22.23}$$

which is always valid in the Einstein theory, and is thus a consequence of the local energy-momentum conservation.

In order to show this we first of all need the energy-momentum tensor for a pointlike particle of constant rest mass m. We use the four-dimensional δ-function defined in Section 11.1, preferring here to write volume integrals as for example

$$\int F(x^i)\frac{\delta^4[x^n - a^n]}{\sqrt{-g}}\sqrt{-g}\,\mathrm{d}^4x = F(a^n), \tag{22.24}$$

and perform the transition from the tensor $T^{mn} = \mu u^m u^n$ of dust to that of a pointlike particle by

$$T^{ik}(y^n) = mc \int \frac{\delta^4[y^n - x^n(\tau)]}{\sqrt{-g(x^a)}} \frac{\mathrm{d}x^i}{\mathrm{d}\tau}\frac{\mathrm{d}x^k}{\mathrm{d}\tau}\,\mathrm{d}\tau. \tag{22.25}$$

In the local Minkowski rest-system ($\sqrt{-g} = 1$, $t = \tau$, $x^\nu = 0$) it corresponds precisely to the transition

$$\mu \to m\delta(x)\delta(y)\delta(z) \tag{22.26}$$

of a continuous distribution of matter to a point mass.

We now insert the energy-momentum tensor (22.25) into (22.23). Using (22.24) and (16.19) we can rewrite the partial derivatives as:

$$
\begin{aligned}
T^{ik}{}_{,k} = \frac{\partial T^{ik}}{\partial y^k} &= mc \int \frac{\partial\{\delta^4[y^n - x^n(\tau)]\}/\partial y^k}{\sqrt{-g(x^a)}} \frac{\mathrm{d}x^i}{\mathrm{d}\tau}\frac{\mathrm{d}x^k}{\mathrm{d}\tau}\,\mathrm{d}\tau \\
&= -mc \int \frac{\partial\{\delta^4[y^n - x^n(\tau)]\}/\partial x^k}{\sqrt{-g(x^a)}} \frac{\mathrm{d}x^k}{\mathrm{d}\tau}\frac{\mathrm{d}x^i}{\mathrm{d}\tau}\,\mathrm{d}\tau \\
&= mc \int \delta^4[y^n - x^n(\tau)]\frac{\mathrm{d}}{\mathrm{d}\tau}\left(\frac{\mathrm{d}x^i}{\mathrm{d}\tau}\frac{1}{\sqrt{-g}}\right)\mathrm{d}\tau \\
&= mc \int \frac{\delta^4[y^n - x^n(\tau)]}{\sqrt{-g}}\left(\frac{\mathrm{d}^2x^i}{\mathrm{d}\tau^2} - \Gamma^a_{ab}\frac{\mathrm{d}x^i}{\mathrm{d}\tau}\frac{\mathrm{d}x^b}{\mathrm{d}\tau}\right)\mathrm{d}\tau.
\end{aligned}
\tag{22.27}
$$

From (22.23) we thus obtain

$$
\begin{aligned}
0 = T^{ik}{}_{;k} = T^{ik}{}_{,k} + \Gamma^i_{nk}T^{in} \\
= mc \int \frac{\delta^4[y^n - x^n(\tau)]}{\sqrt{-g}}\left(\frac{\mathrm{d}^2x^i}{\mathrm{d}\tau^2} - \Gamma^a_{ak}\frac{\mathrm{d}x^i}{\mathrm{d}\tau}\frac{\mathrm{d}x^k}{\mathrm{d}\tau}\right. \\
\left. + \Gamma^i_{nk}\frac{\mathrm{d}x^n}{\mathrm{d}\tau}\frac{\mathrm{d}x^k}{\mathrm{d}\tau} + \Gamma^a_{ak}\frac{\mathrm{d}x^i}{\mathrm{d}\tau}\frac{\mathrm{d}x^k}{\mathrm{d}\tau}\right)\mathrm{d}\tau,
\end{aligned}
\tag{22.28}
$$

and hence

$$0 = mc \int \frac{\delta^4[y^n - x^n(\tau)]}{\sqrt{-g}}\left(\frac{\mathrm{D}^2x^i}{\mathrm{D}\tau^2}\right)\mathrm{d}\tau. \tag{22.29}$$

On the world line $y^n = x^n(\tau)$ of the particle this equation can only be satisfied if

$$\frac{\mathrm{D}^2 x^i}{\mathrm{D}\tau^2} = 0, \tag{22.30}$$

and so the particle must move on a geodesic.

At first sight it is perhaps not apparent where it has in fact been assumed in this rather formal derivation that we are dealing with a *test* particle. But the gravitational field produced by a pointlike particle will certainly not be regular at the position of the particle (the electrical field at the position of a point charge is also singular), so that the metric and Christoffel symbols do not exist there at all.

Spinning particle   An extended body, for example, a planet, will in general not move exactly along a geodesic. This is due not so much to the gravitational field caused by the body itself as to the action of 'tidal forces'. Because of the space curvature, the distance between neighbouring geodesics is not constant (see Section 1.4); that is, the gravitational forces (which try to move every point of the body along a geodesic) deform the body, change its state of rotation and thereby lead to a complicated orbit. We can take account of one part of this effect by starting off from the model of a pointlike body, but associating with it higher moments (dipole moment, spin) in addition to the mass. Mathematically we can do this by using for its description not just δ-functions, but also their derivatives.

As we shall be interested later on in the action of the gravitational field upon a top, the equations of motion of a spinning (monopole-dipole) particle will be described in brief; for details and proofs we refer to the extensive literature on this problem.

An extended body can be approximately described by its mass $m(\tau)$, the four-velocity $u^a(\tau)$ of a suitably chosen point, and the antisymmetric spin-tensor $S^{ab}(\tau)$. From the vanishing of the divergence of the energy-momentum tensor

$$T^{ik}(y^n) = \int \frac{c}{\sqrt{-g}} \Big[\delta^4[y^n - x^n(\tau)] \left(m u^i u^k + \tfrac{1}{2} u_n [\dot{S}^{ni} u^k + \dot{S}^{nk} u^i]/c^2 \right)$$

$$- \big\{ \delta^4[y^n - x^n(\tau)] \tfrac{1}{2}(S^{mi} u^k + S^{mk} u^i) \big\}_{;m} \Big] \,\mathrm{d}\tau \tag{22.31}$$

follow the equations of motion as

$$\mathrm{D}S^{ab}/\mathrm{D}\tau \equiv \dot{S}^{ab} = \left(u_n \dot{S}^{na} u^b - u_n \dot{S}^{nb} u^a \right)/c^2 \tag{22.32}$$

$$\mathrm{D}\big(m u^a + u_b \dot{S}^{ba}/c^2 \big)/\mathrm{D}\tau = \tfrac{1}{2} R^a{}_{bcd} S^{dc} u^b. \tag{22.33}$$

Of these equations only seven are independent, as contraction with u^a shows. Thus they do not suffice for the determination of the ten unknown functions (m, S^{ab}, and three components of u^a). The physical reason for this is that S^{ab}, like angular momentum or dipole moment in Newtonian mechanics, depends upon the reference point, and we have not yet fixed this point and its world line. We now define the reference world line $x^a(\tau)$ by the requirement that in the instantaneous rest frame of an observer moving on the world line, the dipole moment of the body is zero. Since the total mass is positive, such a line always exists. One possible version of this condition is, as one can show,

$$S^{ab}u_b = 0. \tag{22.34}$$

Because of this subsidiary condition the antisymmetric tensor S^{ab} has only three independent components, which can be mapped uniquely onto the *spin-vector (intrinsic angular-momentum vector)* S_a according to

$$S_a = \tfrac{1}{2}\varepsilon_{abmn}u^b S^{mn}/c, \qquad S^{mn} = \varepsilon_{aq}{}^{mn}S^a u^q/c, \tag{22.35}$$

with $S^n u_n = 0$.

Substitution of (22.34) and (22.35) into the equations of motion (22.32) to (22.33) yields

$$\frac{DS_a}{D\tau} = \frac{1}{c^2}u_a S_n \frac{Du^n}{D\tau}, \tag{22.36}$$

$$\frac{D}{D\tau}m = 0, \tag{22.37}$$

$$m\frac{D}{D\tau}u^a = -\frac{1}{c}\varepsilon^{arq}{}_p S_r u_q \frac{D^2 u^p}{D\tau^2} + \frac{1}{2c}R^a{}_{bcd}\varepsilon^{cdpq}S_p u_q u^b. \tag{22.38}$$

The first of these equations says that the spin-vector S_a is Fermi–Walker transported along the orbit (cp. (18.19)); its magnitude then remains constant:

$$S^a S_a = \text{const.} \tag{22.39}$$

From the third equation (22.38) we see that the point defined by (22.34) does not move on a geodesic; this effect will in general be ignorable.

The equation of motion for spin, (22.36), is also valid when additional non-gravitational forces act, provided only that these forces exert no couple on the body. An observer can therefore realize his Fermi–Walker transported triad in the directions of the axes of three tops which are suspended freely.

Further reading for Section 22.3

Taub (1965), Westpfahl (1967).

22.4 A variational principle for Einstein's theory

All known fundamental, physically significant equations of classical fields can be derived from a variational principle, including the Einstein field equations. What demands must one make regarding the Lagrangian density $\mathcal{L}$ in order that precisely the field equations (22.7) follow from

$$\delta \int \mathcal{L} \, \mathrm{d}^4 x = 0? \tag{22.40}$$

Of course the quantity in (22.40) must be an invariant; that is, $\mathcal{L}$ must be the product of a scalar L and $\sqrt{-g}$. But for the pure gravitational field there is only one unique scalar which is quadratic in first derivatives of the metric and linear in the second derivatives, namely, the scalar curvature R (there is no scalar which contains only first derivatives). Since the matter must also be represented in L, we couple it – as usual in field theory – by simply adding a part κL_{M} arising from the matter distribution (for example, from an electromagnetic field). Our variational principle reads, upon appropriate choice of numerical factors, thus,

$$\delta W = \delta \int \left(\tfrac{1}{2} R + \kappa L_{\mathrm{M}} \right) \sqrt{-g} \, \mathrm{d}^4 x = 0 \tag{22.41}$$

(Hilbert 1915). We shall now show that the Einstein field equations (22.7) really do follow from this ansatz.

As fundamental quantities of the gravitational field, which are to be varied independently of one another, we shall naturally take the components of the metric tensor g_{mn}. (If in (22.41) one varies the non-metrical field quantities contained in L_{M}, one obtains the corresponding field equations, for example, Maxwell's equations.) As usual with action integrals containing second derivatives, the variations δg_{mn} of the basic quantities and the variations of their first derivatives (combined into the variations $\delta \Gamma^n_{ab}$ of the Christoffel symbols) will be restricted so as to vanish on the bounding surfaces of the four-dimensional region of integration. Our first goal is to express the variations occurring in the equation

$$\delta W = \tfrac{1}{2} \int \left[R \, \delta \sqrt{-g} + \sqrt{-g} \, R_{mn} \, \delta g^{mn} + \sqrt{-g} \, g^{mn} \, \delta R_{mn} \right.$$
$$\left. + \delta \left(2 \kappa L_{\mathrm{M}} \sqrt{-g} \right) \right] \mathrm{d}^4 x = 0 \tag{22.42}$$

in terms of δg_{mn}.

From the properties of the metric tensor and its determinant described in Section 16.2 we obtain immediately

$$\delta \sqrt{-g} = \frac{\partial \sqrt{-g}}{\partial g_{mn}} \, \delta g_{mn} = \tfrac{1}{2} \sqrt{-g} \, g^{mn} \, \delta g_{mn} \tag{22.43}$$

and

$$\delta g^{mn} = -g^{ma} g^{nb} \delta g_{ab}. \tag{22.44}$$

The defining equations (19.7) and (19.27) for the curvature tensor and the Ricci tensor, respectively, lead to

$$\delta R_{mn} = -(\delta \Gamma^a_{ma})_{,n} + (\delta \Gamma^a_{mn})_{,a} - \delta(\Gamma^a_{rn}\Gamma^r_{ma} - \Gamma^a_{ra}\Gamma^r_{mn}). \tag{22.45}$$

The evaluation of the variational principle will be seen later to depend only upon the structure of the term containing δR_{mn}, which is not easily found by direct calculation from (22.45). We therefore give the result (and its structure)

$$g^{mn} \delta R_{mn} = \frac{1}{\sqrt{-g}} [\sqrt{-g} (g^{mn} \delta \Gamma^a_{mn} - g^{ma} \delta \Gamma^n_{mn})]_{,a} \equiv F^a_{;a} \tag{22.46}$$

without calculation and prove the correctness of this equation by showing that it is a tensor equation and that it is satisfied in a particular coordinate system.

The tensor property of equation (22.46) follows from the fact that not only the δR_{mn}, but also the difference $\delta \Gamma^a_{mn}$ of Christoffel symbols (the disturbing terms in (16.31) cancel when the difference is formed), are tensors. The equation (22.46) is clearly correct in locally geodesic coordinates; for $\sqrt{-g} = 1$, $g^{mn}{}_{,a} = 0$, $\Gamma^a_{bc} = 0$, (22.45) and (22.46) lead to the same equation, namely, to

$$g^{mn} \delta R_{mn} = (g^{mn} \delta \Gamma^a_{mn} - g^{ma} \delta \Gamma^n_{mn})_{,a} = g^{mn}(\delta \Gamma^a_{mn,a} - \delta \Gamma^a_{ma,n}). \tag{22.47}$$

We can only really work out the last term in (22.42) with exact knowledge of the Lagrangian L_M of the matter. In order to obtain L_M, we shall invoke the aid of the usual transcription formula. One starts from the corresponding special-relativistic Lagrangian, replaces the partial by covariant derivatives and now forms scalar products with g_{mn} instead of with η_{mn}. L_M may thus contain Christoffel symbols, but certainly no second derivatives of the metric, since it is in general constructed from the field quantities and their first derivatives only. We can therefore write generally

$$\delta (\sqrt{-g} L_M) = \frac{\delta (\sqrt{-g} L_M)}{\delta g_{mn}} \delta g_{mn} + \left(\sqrt{-g} \frac{\partial L_M}{\partial g_{mn,a}} \delta g_{mn}\right)_{,a}, \tag{22.48}$$

in which we have used the usual abbreviation

$$\frac{\delta (\sqrt{-g} L_M)}{\delta g_{mn}} = \frac{\partial (\sqrt{-g} L_M)}{\partial g_{mn}} - \left(\frac{\partial (\sqrt{-g} L_M)}{\partial g_{mn,a}}\right)_{,a} \tag{22.49}$$

for the so-called variational derivative.

If we now substitute (22.43), (22.44), (22.46) and (22.48) into the variational principle (22.42) then we obtain

$$\delta W = \tfrac{1}{2} \int \Big[\left(\tfrac{1}{2} g^{mn} R - R^{mn} + \frac{2\kappa}{\sqrt{-g}} \frac{\delta \left(\sqrt{-g}\, L_{\mathrm{M}} \right)}{\delta g_{mn}} \right) \delta g_{mn} + F^a{}_{;a}$$
$$+ \frac{2\kappa}{\sqrt{-g}} \left(\sqrt{-g}\, \frac{\partial L_{\mathrm{M}}}{\partial g_{mn,a}}\, \delta g_{mn} \right)_{,a} \Big] \sqrt{-g}\, \mathrm{d}^4 x = 0. \tag{22.50}$$

With the help of the Gauss law (20.34) we can reduce the last two terms of the sum to a surface integral; this surface integral vanishes, however, because we have demanded that $\delta g_{mn} = 0$ and $\delta \Gamma^a_{bc} = 0$ on the boundary. Hence (22.50) simplifies to

$$\delta W = \tfrac{1}{2} \int \left(\tfrac{1}{2} g^{mn} R - R^{mn} + \frac{2\kappa}{\sqrt{-g}} \frac{\delta \left(\sqrt{-g}\, L_{\mathrm{M}} \right)}{\delta g_{mn}} \right) \delta g_{mn} \sqrt{-g}\, \mathrm{d}^4 x = 0. \tag{22.51}$$

Because of the independence of the variations δg_{mn} the sum of the terms contained in the parentheses must vanish identically, so that from our variational principle we obtain precisely the Einstein equations

$$R^{mn} - \tfrac{1}{2} g^{mn} R = \kappa T^{mn}, \tag{22.52}$$

if we identify the energy-momentum tensor with the variational derivative according to

$$T^{mn} = \frac{2}{\sqrt{-g}} \frac{\delta \left(\sqrt{-g}\, L_{\mathrm{M}} \right)}{\delta g_{mn}}. \tag{22.53}$$

How can one justify this identification? Two standpoints are possible. On the one hand one can regard (22.53) as defining the energy-momentum tensor or, put more exactly, *the* energy-momentum tensor out of the many possible energy-momentum tensors of a classical field theory, which must stand on the right-hand side of the Einstein field equations. In this sense (22.53) is the construction principle for the symmetric energy-momentum tensor, which is remarkably complicated to find in many field theories. Although this procedure is quite natural in the Einstein theory, one can of course also corroborate it by comparison with the energy-momentum tensor T^{mn}, which is already known. We will do this for the example of the Maxwell field. Since its Lagrangian

$$L_{\mathrm{M}} = -\tfrac{1}{4} F_{ab} F^{ab} = -\tfrac{1}{4} (A_{b,a} - A_{a,b})(A_{s,r} - A_{r,s}) g^{ar} g^{bs} \tag{22.54}$$

does not depend at all upon derivatives of the metric, we have, because of (22.48) and (22.44),

$$\frac{\delta\left(\sqrt{-g}\,L_{\mathrm{M}}\right)}{\delta g_{mn}} = \frac{\partial\left(\sqrt{-g}\,L_{\mathrm{M}}\right)}{\partial g_{mn}} = \tfrac{1}{2}\sqrt{-g}\,(L_{\mathrm{M}}g^{mn} + F^{nb}F^{m}{}_{b}), \quad (22.55)$$

and from the definition (22.53)

$$T^{mn} = F^{nb}F^{m}{}_{b} - \tfrac{1}{4}g^{mn}F_{ab}F^{ab}, \qquad (22.56)$$

which is indeed the correct energy-momentum tensor of electrodynamics.

Finally, we want to draw attention to a peculiar property of the action function of the gravitational field. Although the Lagrangian contains derivatives of second order, it does not give rise to differential equations of fourth order for the field equations – as we would normally have expected. This has to do with the fact that the curvature scalar R contains derivatives of second order in precisely such a combination that a four-dimensional divergence can be formed from them,

$$R = \frac{1}{\sqrt{-g}}[\sqrt{-g}\,(g^{ma}\Gamma^{n}_{ma} - g^{mn}\Gamma^{a}_{ma})]_{,n} + F(g_{mn}, \Gamma^{a}_{mn}), \qquad (22.57)$$

which, by means of the Gauss law, can be turned into a surface integral and hence supplies no contribution to the variation. This is also the deeper reason why one part of the variation of the integrand appears as a divergence (see (22.46)). The non-covariant decomposition (22.57) of the Lagrangian of the gravitational field into a divergence and a remainder containing only first derivatives plays an important rôle in the attempt to define an energy-momentum tensor (energy-momentum complex) of the gravitational field in the context of the Lagrangian function.

23
The Schwarzschild solution

23.1 The field equations

The gravitational fields which are most important in our daily life, namely, that of the Earth and that of the Sun, are produced by slowly rotating, nearly spherical mass distributions; they are approximately spherically symmetric. Since, on the other hand, we may hope that spherically symmetric gravitational fields are especially simple, we dis-

cuss, as a first application of the Einstein field equations, the problem
of obtaining exact spherically symmetric solutions.

Line element We shall naturally try to introduce coordinates appropriate to the problem. Since a choice of coordinates always leads to requirements on the metric functions, we must proceed carefully in order not to
lose solutions by making the restrictions too strong. Spherical symmetry
evidently signifies that in three-dimensional space, $T = $ const., all radial
directions are equivalent and no perpendicular direction is singled out;
in spherical coordinates R, ϑ, φ we have

$$d^{(3)}s^2 = g_{11}(R, cT)\,dR^2 + f(R, cT)\big[d\vartheta^2 + \sin^2\vartheta\,d\varphi^2\big]. \tag{23.1}$$

The angular coordinates at different times can be so chosen that $g_{T\vartheta}$
and $g_{T\varphi}$ do not appear in the metric (they would single out tangential
directions). Our ansatz thus reads

$$\begin{aligned}
ds^2 &= g_{11}(R, cT)\,dR^2 + f(R, cT)\big[d\vartheta^2 + \sin^2\vartheta\,d\varphi^2\big] \\
&\quad + 2g_{14}(R, cT)\,dR\,dcT + g_{44}(R, cT)c^2\,dT^2.
\end{aligned} \tag{23.2}$$

For many calculations it is expedient to simplify ds^2 further. By the
coordinate transformation $r^2 = f(R, cT)$, where f is positive (for if not
ϑ, φ would be additional timelike coordinates), we bring the line element
into the form

$$\begin{aligned}
ds^2 &= h^2(r, T)\,dr^2 - 2a(r, T)b(r, T)c\,dT\,dr - b^2(r, T)c^2\,dT^2 \\
&\quad + r^2\big(d\vartheta^2 + \sin^2\vartheta\,d\varphi^2\big),
\end{aligned} \tag{23.3}$$

which already contains the usual two-dimensional spherical surface
element. Here we have assumed implicitly that r is a spacelike and
T a timelike coordinate. A further transformation

$$e^{\nu/2}\,d(ct) = b\,d(cT) + a\,dr \tag{23.4}$$

($e^{\nu/2}$ plays the rôle of an integrating factor) eliminates the undesired
non-orthogonal term. Thus we arrive at the Schwarzschild form,

$$ds^2 = e^{\lambda(r,t)}\,dr^2 + r^2\big(d\vartheta^2 + \sin^2\vartheta\,d\varphi^2\big) - e^{\nu(r,t)}\,d(ct)^2, \tag{23.5}$$

of the line element of a spherically symmetric metric.

Christoffel symbols The Christoffel symbols associated with a metric are
constructed most quickly by comparing the Euler–Lagrange equations

$$\frac{d}{d\tau}\frac{\partial L}{\partial\,(dx^i/d\tau)} - \frac{\partial L}{\partial x^i} = 0, \tag{23.6}$$

for the Lagrangian

$$L = \frac{1}{2}\left[e^\lambda\left(\frac{\mathrm{d}r}{\mathrm{d}\tau}\right)^2 + r^2\left(\frac{\mathrm{d}\vartheta}{\mathrm{d}\tau}\right)^2 + r^2\sin^2\vartheta\left(\frac{\mathrm{d}\varphi}{\mathrm{d}\tau}\right)^2 - e^\nu\left(\frac{\mathrm{d}x^4}{\mathrm{d}\tau}\right)^2\right] \quad (23.7)$$

with the geodesic equation

$$\frac{\mathrm{d}^2x^i}{\mathrm{d}\tau^2} + \Gamma^i_{mn}\frac{\mathrm{d}x^m}{\mathrm{d}\tau}\frac{\mathrm{d}x^n}{\mathrm{d}\tau} = 0. \quad (23.8)$$

The Christoffel symbols can then easily be read off. With the abbreviations $\dot{} \equiv \partial/\partial ct$ and $' \equiv \partial/\partial r$ equations (23.6) become

$$e^\lambda\left[\frac{\mathrm{d}^2r}{\mathrm{d}\tau^2} + \frac{1}{2}\left(\frac{\mathrm{d}r}{\mathrm{d}\tau}\right)^2\lambda' + \frac{\mathrm{d}r}{\mathrm{d}\tau}\frac{\mathrm{d}x^4}{\mathrm{d}\tau}\dot\lambda\right] - r\left(\frac{\mathrm{d}\vartheta}{\mathrm{d}\tau}\right)^2 - r\sin^2\vartheta\left(\frac{\mathrm{d}\varphi}{\mathrm{d}\tau}\right)^2$$
$$+ \frac{1}{2}\left(\frac{\mathrm{d}x^4}{\mathrm{d}\tau}\right)^2 e^\nu\nu' = 0,$$

$$r^2\frac{\mathrm{d}^2\vartheta}{\mathrm{d}\tau^2} + 2r\frac{\mathrm{d}r}{\mathrm{d}\tau}\frac{\mathrm{d}\vartheta}{\mathrm{d}\tau} - r^2\sin\vartheta\cos\vartheta\left(\frac{\mathrm{d}\varphi}{\mathrm{d}\tau}\right)^2 = 0, \quad (23.9)$$

$$r^2\sin^2\vartheta\frac{\mathrm{d}^2\varphi}{\mathrm{d}\tau^2} + 2r\sin^2\vartheta\frac{\mathrm{d}r}{\mathrm{d}\tau}\frac{\mathrm{d}\varphi}{\mathrm{d}\tau} + 2r^2\sin\vartheta\cos\vartheta\frac{\mathrm{d}\varphi}{\mathrm{d}\tau}\frac{\mathrm{d}\vartheta}{\mathrm{d}\tau} = 0,$$

$$e^\nu\left[\frac{\mathrm{d}^2x^4}{\mathrm{d}\tau^2} + \frac{\dot\nu}{2}\left(\frac{\mathrm{d}x^4}{\mathrm{d}\tau}\right)^2 + \nu'\frac{\mathrm{d}x^4}{\mathrm{d}\tau}\frac{\mathrm{d}r}{\mathrm{d}\tau}\right] + \frac{1}{2}e^\lambda\dot\lambda\left(\frac{\mathrm{d}r}{\mathrm{d}\tau}\right)^2 = 0.$$

Of the total of forty independent Christoffel symbols only the following twelve are non-zero (here $x^1 = r$, $x^2 = \vartheta$, $x^3 = \varphi$ and $x^4 = ct$):

$$\begin{array}{lll}
\Gamma^1_{11} = \tfrac{1}{2}\lambda', & \Gamma^1_{14} = \tfrac{1}{2}\dot\lambda, & \Gamma^1_{22} = -re^{-\lambda}, \\
\Gamma^1_{33} = -r\sin^2\vartheta\,e^{-\lambda}, & \Gamma^1_{44} = \tfrac{1}{2}e^{\nu-\lambda}\nu', & \Gamma^2_{12} = 1/r, \\
\Gamma^2_{33} = -\sin\vartheta\cos\vartheta, & \Gamma^3_{13} = 1/r, & \Gamma^3_{23} = \cot\vartheta, \\
\Gamma^4_{11} = \tfrac{1}{2}\dot\lambda e^{\lambda-\nu}, & \Gamma^4_{14} = \tfrac{1}{2}\nu', & \Gamma^4_{44} = \tfrac{1}{2}\dot\nu.
\end{array} \quad (23.10)$$

Ricci tensor From the general defining equation

$$R^a_{\ mbn} = \Gamma^a_{mn,b} - \Gamma^a_{mb,n} + \Gamma^a_{rb}\Gamma^r_{mn} - \Gamma^a_{rn}\Gamma^r_{mb} \quad (23.11)$$

we obtain, bearing in mind (23.10),

$$\begin{aligned}
R^1_{\ m1n} &= \Gamma^1_{mn,1} - \Gamma^1_{1m,n} + \Gamma^1_{11}\Gamma^1_{mn} + \Gamma^1_{14}\Gamma^4_{mn} - \Gamma^1_{rn}\Gamma^r_{m1}, \\
R^2_{\ m2n} &= \Gamma^2_{mn,2} - \Gamma^2_{2m,n} + \Gamma^2_{12}\Gamma^1_{mn} - \Gamma^2_{rn}\Gamma^r_{m2}, \\
R^3_{\ m3n} &= -\Gamma^3_{3m,n} + \Gamma^3_{13}\Gamma^1_{mn} + \Gamma^3_{23}\Gamma^2_{mn} - \Gamma^3_{rn}\Gamma^r_{m3}, \\
R^4_{\ m4n} &= \Gamma^4_{mn,4} - \Gamma^4_{4m,n} + \Gamma^4_{14}\Gamma^1_{mn} + \Gamma^4_{44}\Gamma^4_{mn} - \Gamma^4_{rn}\Gamma^r_{m4}.
\end{aligned} \quad (23.12)$$

Unless $m = n$ or $(m,n) = (1,4)$ these components vanish. Also $R_{1234} = 0$. Thus only the following components of the Ricci tensor differ from zero:

$$R_{11} = e^{\lambda-\nu}\left[\tfrac{1}{2}\ddot{\lambda} + \tfrac{1}{4}\dot{\lambda}^2 - \tfrac{1}{4}\dot{\lambda}\dot{\nu}\right] - \tfrac{1}{2}\nu'' - \tfrac{1}{4}\nu'^2 - \tfrac{1}{4}\nu'\lambda' + \lambda'/r,$$

$$R_{44} = e^{\nu-\lambda}\left[\tfrac{1}{2}\nu'' + \tfrac{1}{4}\nu'^2 - \tfrac{1}{4}\nu'\lambda' + \tfrac{1}{4}\nu'/r\right] - \tfrac{1}{2}\ddot{\lambda} - \tfrac{1}{4}\dot{\lambda}^2 + \tfrac{1}{4}\dot{\lambda}\dot{\nu},$$

$$R_{14} = \dot{\lambda}/r,$$

$$R_{22} = -e^{-\lambda}\left[1 + \tfrac{1}{2}r(\nu' - \lambda')\right] + 1 = R_{33}/\sin^2\vartheta. \qquad (23.13)$$

We have indicated here in detail how the necessary calculations could be done by hand. Of course, for really computing Christoffel symbols and Riemann tensor components one would use one of the many existing programs for algebraic computing, for example Maple or Mathematica.

Vacuum field equations Outside the field-producing masses the energy-momentum tensor vanishes, and, since it follows immediately by taking the trace of

$$R^{mn} - \tfrac{1}{2}g^{mn}R = 0 \qquad (23.14)$$

that $R = 0$, the field equations for the vacuum are simply

$$R_{mn} = 0. \qquad (23.15)$$

That is, all the components (23.13) of the Ricci tensor must vanish.

23.2 The solution of the vacuum field equations

Birkhoff theorem From $R_{14} = 0$ we have immediately that $\dot{\lambda} = 0$; thus λ and $\dot{\lambda}$ depend only upon the radial coordinate r. The equation $R_{22} = 0$ can then only be satisfied if ν' is also independent of time,

$$\nu = \nu(r) + f(t). \qquad (23.16)$$

Since ν occurs in the line element in the combination $e^{\nu(r)}e^{f(t)}\,\mathrm{d}(ct)^2$ one can always make the term $f(t)$ in (23.16) vanish by a coordinate transformation

$$\mathrm{d}t' = e^{f/2}\,\mathrm{d}t, \qquad (23.17)$$

so that in the new coordinates we have

$$\lambda = \lambda(r), \qquad \nu = \nu(r); \qquad (23.18)$$

that is, the metric no longer depends upon time. And thus we have proved the *Birkhoff theorem*: every spherically symmetric vacuum solution is independent of t. (The assumptions made in Section 23.1 may fail, so that t is not a timelike coordinate and r is not a spacelike coordinate, for example, in a black hole (see Section 35.3). However, the theorem still holds, although one would no longer describe the solution as static.)

If one considers the vacuum gravitational field produced by a spherically symmetric star, then this field remains static even if the material inside the star experiences a spherically symmetric radial displacement (explosion). Thus the Birkhoff theorem is the analogue of the statement in electrodynamics that a spherically symmetric distribution of charges and currents does not radiate, that there are no spherically symmetric electromagnetic waves.

Schwarzschild solution   For static vacuum fields the field equations (23.15) simplify to

$$\tfrac{1}{2}\nu'' + \tfrac{1}{4}\nu'^2 - \tfrac{1}{4}\nu'\lambda' - \lambda'/r = 0,$$
$$(\nu' + \lambda')/r = 0, \qquad (23.19)$$
$$e^{-\lambda}(1 - r\lambda') - 1 = 0.$$

The second of these equations is equivalent to

$$\lambda(r) = -\nu(r), \qquad (23.20)$$

since a possible additive constant in (23.20) is a special case of the $f(t)$ in (23.16) and can thus be made to vanish by a coordinate transformation.

Under the substitution $\alpha = e^{-\lambda}$, the third equation transforms into the differential equation

$$\alpha' + \alpha/r = 1/r, \qquad (23.21)$$

whose general solution is

$$\alpha = e^{-\lambda} = e^{\nu} = 1 - 2M/r, \qquad (23.22)$$

with $2M$ as a freely adjustable constant of integration. The spherically symmetric vacuum solution, found by Schwarzschild (1916) and Droste (1916), therefore has the line element

$$ds^2 = \frac{dr^2}{1 - 2M/r} + r^2(d\vartheta^2 + \sin^2\vartheta\, d\varphi^2) - \left(1 - \frac{2M}{r}\right)c^2\, dt^2. \quad (23.23)$$

One can verify by direct substitution that the first of the field equations (23.19) is also satisfied and furnishes no further conditions. We shall discuss the physical meaning of the constant of integration M in the following section.

23.3 General discussion of the Schwarzschild solution

In order to understand the physical properties of the Schwarzschild line element (23.23) we have first to clarify the physical significance of the integration parameter M. This is best done through a comparison with Newtonian theory. For large values of the coordinate r, (23.23) deviates only a little from the metric of a flat space, and, from the relation (22.21), which is valid in this limit and links the Newtonian gravitational potential U to the metric, we have

$$U = -c^2(1 + g_{44})/2 = -Mc^2/r. \qquad (23.24)$$

We have thus to interpret the Schwarzschild solution as the gravitational field present outside a spherically symmetric mass distribution whose (Newtonian) mass m is

$$m = \frac{Mc^2}{f} = \frac{8\pi M}{\kappa c^2} > 0. \qquad (23.25)$$

According to (23.25), the (positive) constant of integration $2M$ is a measure of the total mass; since it has the dimension of a length, one also calls $r_G = 2M$ the *Schwarzschild radius* or *gravitational radius* of the source. For normal stars or planets r_G is very small in relation to the geometrical radius. The Schwarzschild radius of the Sun, for example, has the value $r_G = 2.96$ km, that of the Earth $r_G = 8.8$ mm. Since the Schwarzschild metric describes only the gravitational field outside the matter distribution (we shall discuss the interior field in Chapter 26), whilst the Schwarzschild radius mostly lies far in the interior, we shall initially suppose that $r \gg 2M$ always. See, however, Chapter 35, where we shall investigate the Schwarzschild metric again, and in more detail.

In the discussion of physical properties of the Schwarzschild metric (23.23) one must always remember that r and t in particular are only coordinates and have no immediate physical significance. We therefore call t the *coordinate time*, to distinguish it, for example, from the proper time τ of an observer at rest in the gravitational field; in the Schwarzschild field these two quantities are related by

$$d\tau = \sqrt{1 - 2M/r}\, dt. \qquad (23.26)$$

The radial coordinate r is so defined that the surface area of a sphere $r = \text{const.}$, $t = \text{const.}$ has the value $4\pi r^2$. The infinitesimal displacement in the radial direction ($d\vartheta = d\varphi = dt = 0$) is given, however, by

$$ds = dR = \frac{dr}{\sqrt{1 - 2M/r}}, \qquad (23.27)$$

and is therefore always greater than the difference of the radial coordinates. One can illustrate the metrical relations in the surface $t = \text{const.}$,

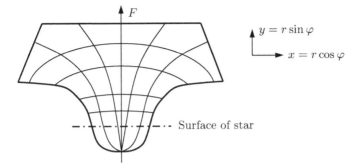

Fig. 23.1. Illustration of the section $t = \text{const.}$, $\vartheta = \pi/2$ of the Schwarzschild metric; $F = [8M(r - 2M)]^{1/2}$.

$\vartheta = \pi/2$ by means of a surface of revolution $F = F(r)$, which for $r \to \infty$ goes over to a plane and for small r has a bulge out of this plane (see Fig. 23.1). When discussing paths of motion (t variable) one must always remember that g_{44} is also dependent upon position.

23.4 The motion of the planets and perihelion precession

Figure 23.1 gives a qualitative idea of the planetary orbits if one imagines the planets as spheres which roll about on the surface under the influence of a downwardly directed gravitational field. According to the Newtonian gravitational theory the orbits of planets are ellipses (in the xy-plane of Fig. 23.1). Does the Einstein theory in any way change this well verified result?

To answer this question properly we should integrate the Lagrange equations of the second kind already set up in (23.9). But since one can always satisfy the initial conditions $\vartheta = \pi/2$ and $\mathrm{d}\vartheta/\mathrm{d}\tau = 0$ by a suitable rotation of the coordinate system, and (23.9) then implies that $\mathrm{d}^2\vartheta/\mathrm{d}\tau^2$ also vanishes, the orbit remains permanently in the plane $\vartheta = \pi/2$; as in Newtonian theory, the orbit of a planet runs in a 'plane' which passes through the middle of the Sun. We can therefore proceed from the simplified Lagrangian

$$L = \frac{1}{2}\left[\frac{1}{1 - 2M/r}\left(\frac{\mathrm{d}r}{\mathrm{d}\tau}\right)^2 + r^2\left(\frac{\mathrm{d}\varphi}{\mathrm{d}\tau}\right)^2 - \left(1 - \frac{2M}{r}\right)\left(\frac{\mathrm{d}x^4}{\mathrm{d}\tau}\right)^2\right] \quad (23.28)$$

which results from substitution of the Schwarzschild metric (23.23) and $\vartheta = \pi/2$ into (23.7).

Since φ and x^4 are cyclic coordinates, two conservation laws hold, namely, that of angular momentum,

$$r^2 d\varphi/d\tau = B, \qquad (23.29)$$

and that of energy,

$$(1 - 2M/r)\, dct/d\tau = A. \qquad (23.30)$$

In place of a third equation of motion, we use the defining equation

$$\frac{1}{1 - 2M/r}\left(\frac{dr}{d\tau}\right)^2 + r^2\left(\frac{d\varphi}{d\tau}\right)^2 - \left(1 - \frac{2M}{r}\right)\left(\frac{dct}{d\tau}\right)^2 = -c^2 \quad (23.31)$$

for the proper time τ, which like the energy law and the momentum law has the form of a first integral of the equations of motion.

From now on the procedure is analogous to that of Newtonian mechanics. In order to obtain the orbits $r = r(\varphi)$ we replace the variable τ by φ, with the aid of the angular-momentum law, and simplify the equation of motion by the substitution $u = r^{-1}$. Putting

$$r = \frac{1}{u}, \quad \frac{d\varphi}{d\tau} = Bu^2, \quad \frac{dct}{d\tau} = \frac{A}{1 - 2Mu}, \quad \frac{dr}{d\tau} = -B\frac{du}{d\varphi} \qquad (23.32)$$

into (23.31), we have

$$B^2 u'^2 + B^2 u^2(1 - 2Mu) - A^2 = -c^2(1 - 2Mu), \quad u' \equiv du/d\varphi. \quad (23.33)$$

This equation can in fact be integrated immediately, but it leads to elliptic integrals, which are awkward to handle. We therefore differentiate (23.33) and obtain the equation

$$u'' + u = Mc^2/B^2 + 3Mu^2, \qquad (23.34)$$

which is easier to discuss. The term $3Mu^2$ is absent in the Newtonian theory, where we have

$$u_0'' + u_0 = Mc^2/B^2. \qquad (23.35)$$

The solutions of this latter differential equation are, as is well known, the conics

$$u_0 = Mc^2(1 + \varepsilon \cos\varphi)/B^2. \qquad (23.36)$$

We can obtain an approximate solution u_1 to the exact orbit equation (23.34) valid for $M/r \ll 1$, if we substitute the Newtonian solution (23.36) into the term quadratic in u, that is, we solve

$$u_1'' + u_1 = \frac{Mc^2}{B^2} + 3Mu_0^2 = \frac{Mc^2}{B^2} + \frac{3M^3c^4}{B^4}\left(1 + 2\varepsilon \cos\varphi + \varepsilon^2 \cos^2\varphi\right). \qquad (23.37)$$

This differential equation is of the type due to a forced oscillation. As one can confirm by substitution, the first approximation sought for is

$$u_1 = u_0 + \frac{3M^3c^4}{B^4}\left[1 + \varepsilon\varphi\sin\varphi + \varepsilon^2\left(\tfrac{1}{2} - \tfrac{1}{6}\cos 2\varphi\right)\right]. \qquad (23.38)$$

The most important term on the right-hand side is the term linear in $\varepsilon\varphi$, because it is the only one which in the course of time (with many revolutions of the planet) becomes larger and larger. We therefore ignore the other corrections to u_0 and obtain (after substituting for u_0)

$$u_1 = \frac{Mc^2}{B^2}\left[1 + \varepsilon\cos\varphi + \varepsilon\frac{3M^2c^2}{B^2}\varphi\sin\varphi\right], \qquad (23.39)$$

or, since $r_0 = u_0^{-1}$ is large compared with M ($M^2c^2/B^2 \ll 1$),

$$u_1 = \frac{1}{r} = \frac{Mc^2}{B^2}\left[1 + \varepsilon\cos\left(1 - \frac{3M^2c^2}{B^2}\right)\varphi\right]. \qquad (23.40)$$

The orbit of the planet is thus only approximately an ellipse (see Fig. 23.2). The solution (23.40) is indeed still a periodic function, but no longer, however, with the period 2π. The point at which the orbit is closest to the Sun is reached again only after an additional rotation through the angle

$$\Delta\varphi_P = 6\pi M^2c^2/B^2. \qquad (23.41)$$

This effect is the famous *perihelion precession*. If, using the equation of the ellipse (23.36), we express the factor Mc^2/B^2 in (23.41) in terms of the semi-major axis a of the ellipse and of ε, so that

$$\Delta\varphi_P = 6\pi M/a\left(1 - \varepsilon^2\right), \qquad (23.42)$$

then we see that the precession of the perihelion is greatest for a large central mass M and an elongated ellipse ($\varepsilon \approx 1$) with a small (motion close to the centre). For circular orbits it disappears.

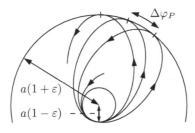

Fig. 23.2. Rosette motion of a planet due to perihelion precession ($\Delta\varphi_P$ exaggerated in magnitude).

23.5 The propagation of light in the Schwarzschild field

Light rays and deflection of light Light rays are null geodesics, that is, geodesics with $ds^2 = 0$. We can compute the corresponding geodesic equation as in the previous section and hence immediately take over a part of the results found there. However, we must use the affine parameter λ in place of the proper time τ and the relation

$$\frac{1}{1 - 2M/r}\left(\frac{dr}{d\lambda}\right)^2 + r^2\left(\frac{d\varphi}{d\lambda}\right)^2 - \left(1 - \frac{2M}{r}\right)\left(\frac{dct}{d\lambda}\right)^2 = 0 \qquad (23.43)$$

in place of (23.31), cp. (16.10). We then arrive at the statement that for suitable choice of coordinates light rays in the Schwarzschild metric travel in the surface $\vartheta = \text{const.} = \pi/2$ and satisfy the differential equation

$$\frac{d^2u}{d\varphi^2} + u = 3Mu^2, \quad u \equiv 1/r, \qquad (23.44)$$

which is analogous to (23.34).

In flat space ($M = 0$) the light rays are of course straight lines. With our choice of coordinates these straight lines are represented by

$$u_0 = \frac{1}{r} = \frac{1}{D}\sin(\varphi - \varphi_0). \qquad (23.45)$$

They run in the directions $\varphi = \varphi_0$ and $\varphi = \varphi_0 + \pi$ to infinity ($u = 0$) and have displacement D from the centre ($r = 0$).

To obtain an approximation solution u_1 to (23.44) we put the Newtonian value (23.45) into the term quadratic in u and solve

$$\frac{d^2u_1}{d\varphi^2} + u_1 = \frac{3M}{D^2}\sin^2(\varphi - \varphi_0). \qquad (23.46)$$

As one can verify by substitution, for suitable choice of φ_0,

$$u_1 = \frac{1}{r} = \pm\frac{\sin\varphi}{D} + \frac{M(1 + \cos\varphi)^2}{D^2} \qquad (23.47)$$

is a family of solutions. These curves come in parallelly from infinity (from the direction $\varphi = \pi$), see Fig. 23.3. The sign in (23.47) is always to be chosen so that $u_1 = 1/r$ is positive. Since a curve leaves the field in the direction in which u_1 again becomes zero (r infinite), its total deflection relative to a straight line is (ignoring terms quadratic in M)

$$\Delta\varphi = 4M/D. \qquad (23.48)$$

This effect is the familiar deflection of light in a gravitational field, one of the most important predictions of the Einstein theory. The deflection is inversely proportional to the (Newtonian) displacement

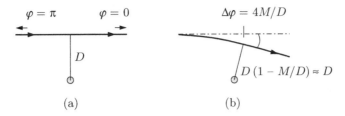

Fig. 23.3. Deflection of light: (a) flat space $u = u_0$, (b) Schwarzschild metric $u = u_1$.

from the centre; since $M/D \ll 1$ always holds in the Solar System (this was presupposed in the derivation) the effect is very small. In very strong gravitational fields (23.48) is no longer applicable (see Chapter 35).

Red shift When propagating in a gravitational field, light changes not only its direction, but also its frequency. Since the corresponding formulae can be derived for arbitrary static fields at no extra effort, we shall carry out this generalization and only substitute the special case of the Schwarzschild metric again in the final result. A more general relationship, valid for arbitrary (non-static) gravitational fields, will be described in Section 40.2.

In a static gravitational field, that is, in a metric g_{mn} which is independent of time and which satisfies the condition $g_{4\alpha} = 0$, it is possible by making the ansatz

$$A_\mu(x^n) = a_\mu(x^\alpha)e^{\mathrm{i}\omega t} \tag{23.49}$$

in the gauge

$$A_4 = 0, \qquad \left[\sqrt{-g}\, g^{44} A^\beta\right]_{,\beta} = 0, \tag{23.50}$$

to separate the Maxwell equations with respect to time and reduce them to the time-independent wave equation

$$\frac{1}{\sqrt{-g}}\left[\sqrt{-g}\, g^{\mu\alpha} g^{\nu\beta}(a_{\beta,\alpha} - a_{\alpha,\beta})\right]_{,\nu} + \frac{\omega^2}{c^2} g^{44} a^\mu = 0. \tag{23.51}$$

An observer who is at rest at the location P_1 of the transmitter will naturally use not the coordinate time t but the proper time τ_1 to measure the frequency ν_1 of the wave. Because of the general relation

$$\tau_P = \sqrt{-g_{44}(P)}\, t \tag{23.52}$$

between proper time and coordinate time, he will therefore associate the frequency

$$\nu_1 = \frac{\omega t}{2\pi \tau_1} = \frac{\omega}{2\pi \sqrt{-g_{44}(1)}} = \frac{\omega \sqrt{-g^{44}(1)}}{2\pi} \qquad (23.53)$$

with the monochromatic wave (23.49), in agreement with the interpretation of the factor $-g^{44}\omega^2/c^2$ in (23.51) as the square of a position-dependent wave number. Analogously, an observer at rest at the location P_2 of the receiver measures the frequency

$$\nu_2 = \frac{\omega}{2\pi \sqrt{-g_{44}(2)}} \qquad (23.54)$$

in his local Minkowski system. The frequencies ν_1 and ν_2 measured by observers at rest at the points P_1 and P_2 are thus related by

$$\frac{\nu_1}{\nu_2} = \sqrt{\frac{g_{44}(2)}{g_{44}(1)}} = 1 + z. \qquad (23.55)$$

Although ν_2 can just as well be larger as smaller than ν_1, in general one speaks of the effect of the redshift in the gravitational field and calls the quantity z defined by (23.55) the *redshift* at the position P_2.

For the Schwarzschild metric we have from (23.55) the relation

$$\frac{\nu_1}{\nu_2} = \sqrt{\frac{1 - 2M/r_2}{1 - 2M/r_1}} \approx 1 + M \left(\frac{1}{r_1} - \frac{1}{r_2} \right). \qquad (23.56)$$

Light reaching the Earth (P_2) from the Sun (P_1) ($r_1 < r_2$) is shifted to the red-wavelength region.

If we express g_{44} in terms of the Newtonian gravitational potential U according to (22.21) then we obtain to first approximation

$$\frac{\nu_1}{\nu_2} = 1 + \frac{U_2 - U_1}{c^2}. \qquad (23.57)$$

In the picture of light as particles (photons) the gravitational redshift corresponds to a change in the kinetic energy $E = h\nu$ by the gain or loss of potential energy $m\Delta U = (E/c^2)\Delta U$, in accordance with (23.57).

Light-travel times and the Fermat principle Here too we generalize the problem and permit arbitrary static gravitational fields ($g_{ab,4} = 0$, $g_{\alpha 4} = 0$), since in all these fields the equation

$$\frac{\mathrm{d}^2 x^a}{\mathrm{d}\lambda^2} + \Gamma^a_{mn} \frac{\mathrm{d}x^m}{\mathrm{d}\lambda} \frac{\mathrm{d}x^n}{\mathrm{d}\lambda} = 0 \qquad (23.58)$$

of a null geodesic can be brought into a form which allows a particularly simple physical interpretation. For this we introduce the coordinate time

t in place of the parameter λ and substitute for the Christoffel symbols the explicit expressions

$$\Gamma^{\alpha}_{\mu\nu} = \tfrac{1}{2}g^{\alpha\beta}(g_{\beta\mu,\nu} + g_{\beta\nu,\mu} - g_{\mu\nu,\beta}),$$

$$\Gamma^{4}_{4\nu} = 0, \quad \Gamma^{4}_{\mu\nu} = 0, \quad \Gamma^{4}_{\alpha 4} = \tfrac{1}{2}g^{44}g_{44,\alpha} = g_{44,\alpha}/2g_{44}, \qquad (23.59)$$

$$\Gamma^{\alpha}_{44} = -\tfrac{1}{2}g^{\alpha\beta}g_{44,\beta}, \quad \Gamma^{4}_{44} = 0.$$

The fourth of equations (23.58),

$$\frac{\mathrm{d}^2 t}{\mathrm{d}\lambda^2} + \frac{g_{44,\alpha}}{g_{44}}\frac{\mathrm{d}x^{\alpha}}{\mathrm{d}t}\left(\frac{\mathrm{d}t}{\mathrm{d}\lambda}\right)^2 = 0, \qquad (23.60)$$

enables us to eliminate λ from the three spatial equations

$$\frac{\mathrm{d}^2 x^{\alpha}}{\mathrm{d}t^2} + \Gamma^{\alpha}_{\mu\nu}\frac{\mathrm{d}x^{\mu}}{\mathrm{d}t}\frac{\mathrm{d}x^{\nu}}{\mathrm{d}t} + \Gamma^{\alpha}_{44}c^2 + \frac{\mathrm{d}^2 t}{\mathrm{d}\lambda^2}\left(\frac{\mathrm{d}\lambda}{\mathrm{d}t}\right)^2\frac{\mathrm{d}x^{\alpha}}{\mathrm{d}t} = 0, \qquad (23.61)$$

finding

$$\frac{\mathrm{d}^2 x^{\alpha}}{\mathrm{d}t^2} + g^{\alpha\beta}\tfrac{1}{2}(g_{\beta\mu,\nu} + g_{\beta\nu,\mu} - g_{\nu\mu,\beta})\frac{\mathrm{d}x^{\mu}}{\mathrm{d}t}\frac{\mathrm{d}x^{\nu}}{\mathrm{d}t}$$

$$- \frac{g_{44,\nu}}{g_{44}}\frac{\mathrm{d}x^{\nu}}{\mathrm{d}t}\frac{\mathrm{d}x^{\alpha}}{\mathrm{d}t} - \tfrac{1}{2}c^2 g^{\alpha\beta}g_{44,\beta} = 0. \qquad (23.62)$$

If we remember the property of null geodesics that $\mathrm{d}s = 0$, that is,

$$c^2\,\mathrm{d}t^2 = -g_{\alpha\beta}\,\mathrm{d}x^{\alpha}\,\mathrm{d}x^{\beta}/g_{44} \equiv \gamma_{\alpha\beta}\,\mathrm{d}x^{\alpha}\,\mathrm{d}x^{\beta} \equiv -\mathrm{d}l^2/g_{44}, \qquad (23.63)$$

then (23.62) can be reduced to

$$\frac{\mathrm{d}^2 x^{\alpha}}{\mathrm{d}t^2} + \hat{\Gamma}^{\alpha}_{\mu\nu}\frac{\mathrm{d}x^{\mu}}{\mathrm{d}t} = 0,$$

$$\hat{\Gamma}^{\alpha}_{\mu\nu} \equiv \tfrac{1}{2}\gamma^{\alpha\beta}(\gamma_{\beta\mu,\nu} + \gamma_{\beta\nu,\mu} - \gamma_{\mu\nu,\beta}), \quad \gamma^{\alpha\beta}\gamma_{\beta\nu} = \delta^{\alpha}_{\nu}. \qquad (23.64)$$

The curves described by this equation are, however, just the extremals which follow from the variational principle

$$\int \mathrm{d}t = \frac{1}{c}\int\sqrt{\gamma_{\alpha\beta}\frac{\mathrm{d}x^{\alpha}}{\mathrm{d}t}\frac{\mathrm{d}x^{\beta}}{\mathrm{d}t}}\,\mathrm{d}t = \frac{1}{c}\int\frac{\mathrm{d}l}{\sqrt{-g_{44}}} = \text{extremum}, \qquad (23.65)$$

see Section 14.3.

The variational principle (23.65) is the generalization of the Fermat principle, that light propagates in a three-dimensional space in such a way that the light-travel time t is an extremum.

The variational principle (23.65) can also be interpreted as saying that the three-dimensional space (metric $g_{\alpha\beta}$) has a refractive index $n = (-g_{44})^{-1/2}$, which is caused by the gravitational force (and which also contributes to the deflection of light), and that the velocity of light v in the gravitational field is decreased according to $c = nv$. But this latter interpretation is only to be used with the proviso that v is the velocity of light with respect to the coordinate time t and therefore has, like t itself, no immediate physical significance. Predictions about the numerical value of the velocity of light have little value in General Relativity; the only essential thing is that light propagates along null geodesics (and that in local inertial systems one can give the velocity of light the value c through choice of the unit of time).

23.6 Further aspects of the Schwarzschild solution

Isotropic coordinates   The Schwarzschild coordinates in which we have until now described the spherically symmetric gravitational field go over to spherical coordinates at a great distance from the centre (for $r \to \infty$). For many calculations or considerations it is more convenient to use coordinates which are related to Cartesian coordinates. We introduce them by the transformation

$$r = \bar{r}\left(1 + M/2\bar{r}\right)^2, \qquad \begin{aligned} \bar{x} &= \bar{r}\cos\varphi\sin\vartheta, \\ \bar{y} &= \bar{r}\sin\varphi\sin\vartheta, \end{aligned} \qquad \bar{z} = \bar{r}\cos\vartheta, \quad (23.66)$$

which turns (23.23) into

$$ds^2 = \left(1 + M/2\bar{r}\right)^4\left(d\bar{x}^2 + d\bar{y}^2 + d\bar{z}^2\right) - \left(\frac{1 - M/2\bar{r}}{1 + M/2\bar{r}}\right)^2 c^2\, dt^2. \quad (23.67)$$

Since in this form of the line element the three spatial directions enter on an equal footing, one speaks of isotropic coordinates.

Harmonic coordinates   Coordinates which are restricted by

$$\Box x^a \equiv \frac{1}{\sqrt{-g}}\left[\sqrt{-g}\, g^{nm} x^a{}_{,m}\right]_{,n} = \frac{1}{\sqrt{-g}}\left(\sqrt{-g}\, g^{an}\right)_{,n} = 0 \quad (23.68)$$

are called harmonic coordinates (of course (23.68) is not a covariant equation, rather is serves to pick out a coordinate system). Such coordinates are useful in approximation procedures for the solution of Einstein's equations (see Section 27.2). In such coordinates, the Schwarzschild metric has the form

$$ds^2 = \left[\left(1 + \frac{M}{\bar{r}}\right)^2\eta_{\alpha\beta} + \frac{\bar{r} + M}{\bar{r} - M}\frac{M^2}{\bar{r}^4}x_\alpha x_\beta\right]dx^\alpha\, dx^\beta - \frac{\bar{r} - M}{\bar{r} + M}c^2\, dt^2,$$
$$(23.69)$$

where

$$\bar{r} = r - M. \tag{23.70}$$

The interior field of a hollow (non-rotating) sphere The Schwarzschild line element (23.23) follows from the requirement of spherical symmetry alone, and it therefore holds also in the matter-free interior of a hollow, non-rotating sphere. But in this case the metric must be finite at $r = 0$, that is, M must vanish; the space inside a hollow sphere is field-free (flat) as in the Newtonian gravitation theory.

23.7 The Reissner–Nordström solution

The Reissner–Nordström solution (Reissner 1916, Nordström 1918) is the spherically symmetric, static, exterior field of a charged distribution of mass. We state without proof that the gravitational field is described by the metric

$$ds^2 = e^{\lambda(r)}dr^2 + r^2\left(d\vartheta^2 + \sin^2\vartheta\,d\varphi^2\right) - e^{-\lambda(r)}c^2\,dt^2,$$
$$e^{-\lambda(r)} = 1 - 2M/r + \kappa e^2/2r^2 \tag{23.71}$$

and the electromagnetic field by the four-potential

$$A_\alpha = 0, \quad U = -A_4 = e/r. \tag{23.72}$$

The potential (23.72) is a solution of the source-free Maxwell equations

$$\left[\sqrt{-g}\,g^{ma}g^{nb}(A_{b,a} - A_{a,b})\right]_{,n} = 0 \tag{23.73}$$

in the Riemannian space of the metric (23.71), and the metric (23.71) satisfies the Einstein field equations

$$R_{in} - \tfrac{1}{2}Rg_{in} = \kappa T_{in}, \tag{23.74}$$

with the energy-momentum tensor of the Maxwell field (23.72) on the right-hand side. The system (23.71) and (23.72) is thus an exact solution of the coupled Einstein–Maxwell equations.

Since for large r the term $\kappa e/2r^2$ in the metric can be ignored, an observer situated at great distance will interpret $m = 8\pi M/\kappa c^2$ as the total mass of the source (see (23.25)). From (23.72) one deduces that $Q = 4\pi e$ is the total charge of the source.

In practice celestial bodies are weakly charged or uncharged, and so the influence of the electromagnetic field on their metric can be ignored, the Reissner–Nordström solution being replaced by its special case, the Schwarzschild solution. Originally the hope was that in the Reissner–Nordström solution one had found a useful model of the electron. But even for the electron, the particle with the largest charge per unit mass, $\kappa e^2/M$ has the value of only 2.8×10^{-13} cm. The influence of the term $\kappa e^2/2r^2$ therefore only becomes important at such dimensions that the effects of Quantum Mechanics and Quantum Field Theory dominate, and the theories of General Relativity and Classical Electrodynamics are no longer adequate to describe the properties of matter.

The Reissner–Nordström solution thus has only slight physical significance. It deserves attention, however, as a simple example of an exact solution of the Einstein–Maxwell equations.

Exercises

23.1   A light ray from a distant star touches the Earth tangentially. By what angle will it be deflected?

23.2   Are there circular orbits $r = R = \text{const.} > 2M$, $\vartheta = \pi/2$ for any value of R?

24

Experiments to verify the Schwarzschild metric

24.1 Some general remarks

The gravitational fields of the Earth and the Sun constitute our natural environment and it is in these fields that the laws of gravity have been investigated and summed up by equations. Both fields are to good approximation spherically symmetric and, as a result, suitable objects to test the Einstein theory as represented in the Schwarzschild metric.

The Einstein theory contains the Newtonian theory of gravitation as a first approximation and in this sense is of course also confirmed by Kepler's laws. What chiefly interests us here, however, are the – mostly very small – corrections to the predictions of the Newtonian theory. In very exact experiments one must distinguish carefully between the following sources of deviation from the Newtonian spherically symmetric field:

(a) Relativistic corrections to the spherically symmetric field,

(b) Newtonian corrections, due to deviations from spherical symmetry (flattening of the Earth or Sun, taking into account the gravitational fields of other planets),

(c) Relativistic corrections due to deviations from spherical symmetry and staticity.

The Newtonian corrections (b) are often larger than the relativistic effects (a) which are of interest to us here, and can be separated from them only with difficulty. Except for the influence of the rotation of the Earth (Lense–Thirring effect, see Section 27.5), one can almost always ignore the relativistic corrections of category (c).

The discussion of measurements and experiments in spherically symmetric gravitational fields is often done by comparing the results for the Schwarzschild metric with those for a more general metric of the form (in isotropic coordinates, with higher order terms neglected)

$$
\begin{aligned}
\mathrm{d}s^2 = {} & \left(1 + \gamma\, 2M/\bar{r} + \cdots\right)\left(\mathrm{d}\bar{x}^2 + \mathrm{d}\bar{y}^2 + \mathrm{d}\bar{z}^2\right) \\
& - \left(1 - 2M/\bar{r} + \beta\, 2M^2/\bar{r}^2 + \cdots\right)c^2\, \mathrm{d}t^2.
\end{aligned}
\tag{24.1}
$$

The free parameters β and γ (two of the so-called PPN-parameters) are found as 'best-fit' parameters to the observational data and serve to measure the agreement between Einstein's theory and observation (they both have the value 1 for the Schwarzschild metric). Perihelion precession $\Delta\varphi_P$, light deflection $\Delta\varphi$ and light travel time Δt for the metric (24.1) differ from the Einstein values (shown with a suffix E) by

$$
\Delta\varphi_P = \tfrac{1}{3}(2 + 2\gamma - \beta)\Delta\varphi_{PE}, \quad \Delta\varphi = \tfrac{1}{2}(1+\gamma)\Delta\varphi_E, \quad \Delta t = \tfrac{1}{2}(1+\gamma)\Delta t_E.
\tag{24.2}
$$

24.2 Perihelion precession and planetary orbits

Einstein's theory predicts the following relativistic contribution to the perihelion precession per century:

$$
\begin{array}{lll}
\text{Mercury } 42.98'', & \text{Earth } 3.8'', & \\
\text{Venus } \quad 8.6'', & \text{Mars } 1.35'', & \text{Satellite} \le 1000'',
\end{array}
\tag{24.3}
$$

see (23.42). Because the deviation from spherical symmetry of the Earth's gravitational field is so large that one can determine the density distribution of the Earth from observed irregularities in satellite orbits, artificial satellites are far from ideal test objects.

In the first decade of relativity theory the most promising evidence came from data on Mercury's orbit. Astronomers before Einstein were already perturbed because although most of the observed perihelion precession, 5600″ per century, could be ascribed to the influence of other planets, there remained an unexplained 41″ now seen to be in good agreement with Einstein's theory.

The survey of the orbit of Mars on the Viking Mission (1976–82) and radar measurements of the distances to Venus and Mercury have furnished data of substantially increased accuracy. More measurements and a comprehensive computer analysis (for example the influence of the larger asteroids on Mars' orbit was taken into account) has produced the values:

$$\beta - 1 = 3\,(\pm 3.1) \times 10^{-3}, \qquad \gamma - 1 = 0.7\,(\pm 1.7) \times 10^{-3}. \qquad (24.4)$$

The unknown quadrupole moment of the Sun is the main source of uncertainty in the reduction of the data. The independently obtained data for Mars and Mercury suggest, however, that it can be neglected.

In 1974 a pulsar (PSR 1913+16) was discovered that forms a binary system with a smaller star whose nature (white dwarf, neutron star or black hole) is unknown. The elliptic orbit of the pulsar shows an unusually large rotation (periastron precession) of 4.22 (±0.04)° per year, that is, 271 times the total value for Mercury. It is highly probable that this is a purely relativistic effect.

24.3 Light deflection by the Sun

Maximum deflection occurs when the light ray grazes the surface of the Sun (see Fig. 24.1), giving

$$\Delta\varphi = 1.75''. \qquad (24.5)$$

To measure this effect for stars one compared the appearance of the night sky in some region with the appearance of the same region during a solar eclipse. (Without the eclipse the Sun's luminosity would swamp that of the stars.) The effect of the gravitational field of the Sun appears to move the closest stars away from it. Although experimental problems, caused, for example, by the distortion of the photographic emulsion during the developing process, have produced values between 1.43″ and

Fig. 24.1. How the Sun deflects light.

2.7″ for this effect, and Eddington's experimental verification in 1919 brought Einstein public recognition. The current deflection values give

$$\gamma - 1 = 0.1 \, (\pm 0.22). \tag{24.6}$$

Today's data for the reflection of electromagnetic waves by the Sun are much more accurate. Instead of stars from within our Galaxy, the sources are radio galaxies and quasars, and not their distances to the Sun, but the changes in their mutual distances due to the motion of the Earth are observed, using very long baseline interferometry. The current data confirm the Einsteinian value

$$\gamma - 1 = 0.000 \, (\pm 0.002). \tag{24.7}$$

24.4 Redshifts

The redshift produced by the Earth's gravitational field was measured first by Pound and Rebka (1960) using the Mössbauer effect. The ^{57}Fe source in the basement of a tower was set in motion so that the resultant Doppler shift corresponded exactly to the energy loss at the receiver, 22.5 m higher. The relation

$$\Delta\lambda/\lambda = gh \tag{24.8}$$

was confirmed with 1 per cent accuracy.

A somewhat more accurate confirmation, $7{\times}10^{-5}$, has been made in an experiment in which a hydrogen maser was taken, with the aid of a rocket, to a height of 10 000 km.

These results can also be considered as evidence for the assertion that atomic and molecular clocks measure proper rather than coordinate time.

24.5 Measurements of the travel time of radar signals (time delay)

The time taken by a radar signal that has been reflected by a planet (e.g. Mercury, Venus) or that has been emitted by a satellite can be compared with the relativistic formula

$$\Delta t = \frac{1}{c} \int \sqrt{-g_{\alpha\beta} \, \mathrm{d}x^\alpha \, \mathrm{d}x^\beta / g_{44}} \tag{24.9}$$

in order to verify Einstein's theory. This so-called fourth test of General Relativity was first proposed by Shapiro (1964). The main data come from the Viking mission to Mars (1977), giving

$$\gamma - 1 = 0 \ (\pm 0.002). \tag{24.10}$$

24.6 Geodesic precession of a top

The spin-vector S^a of a top which is transported along a geodesic (e.g. inside a satellite) satisfies, because of (22.34)–(22.39), the equations

$$\mathrm{D}S^a/\mathrm{D}\tau = 0, \qquad S^a S_a = \text{const.}, \qquad S^a u_a = 0. \tag{24.11}$$

Since the unit vectors $h^a{}_{(\nu)}$, which are used in the satellite and point towards the fixed stars, are not parallelly transported during the motion, the components $S_{(\nu)} = S_a h^a_{(\nu)}$ of the spin-vector in the rest system of the satellite change as

$$\frac{\mathrm{d}S_{(\nu)}}{\mathrm{d}\tau} = \frac{\mathrm{D}S_{(\nu)}}{\mathrm{D}\tau} = S_a \frac{\mathrm{D}h^a{}_{(\nu)}}{\mathrm{D}\tau}. \tag{24.12}$$

The components of the tetrad vectors in isotropic Schwarzschild coordinates are obtained from the three orthogonal unit vectors used in the rest system of the satellite by carrying out a Lorentz transformation with the speed $-v^\alpha = -u^\alpha \, \mathrm{d}\tau/\mathrm{d}t$ and a change of scale

$$\mathrm{d}x^{\alpha'} = \left(1 + M/2\bar{r}\right)^2 \mathrm{d}x^\alpha \approx \left(1 - U/c^2\right)\mathrm{d}x^\alpha,$$

$$\mathrm{d}t' = \left(\frac{1 - M/2\bar{r}}{1 + M/2\bar{r}}\right)\mathrm{d}t \approx \left(1 + U/c^2\right)\mathrm{d}t. \tag{24.13}$$

Including only terms of first order, we then have

$$h^\alpha{}_{(\nu)} = \left(1 + U/c^2\right)\delta^\alpha_\nu + v^\alpha v_\nu/2c^2, \quad h^4{}_{(\nu)} = v_\nu, \quad S_{(\alpha)} = S_\alpha. \tag{24.14}$$

Substituting this expression into (24.12), and, bearing in mind that

$$\mathrm{d}v_\alpha/\mathrm{d}\tau \approx -U_{,\alpha}, \tag{24.15}$$

we have finally

$$c^2 \, \mathrm{d}S_{(\nu)}/\mathrm{d}\tau \approx \tfrac{3}{2}S_{(\alpha)}, \ [U_{,(\alpha)}v_{(\nu)} - U_{,(\nu)}v_{(\alpha)}], \quad U_{,\nu} \approx U_{,(\nu)}, \quad v_\alpha = v_{(\alpha)}, \tag{24.16}$$

that is, the spin-vector rotates in the satellite's system with the angular velocity $\boldsymbol{\omega}$,

$$\mathrm{d}\mathbf{S}/\mathrm{d}t = \boldsymbol{\omega} \times \mathbf{S}, \quad \boldsymbol{\omega} = -\tfrac{3}{2}\mathbf{v} \times \mathbf{grad}\, U/c^2. \tag{24.17}$$

If the rotation of the Earth is taken into account, an additional contribution to $\boldsymbol{\omega}$ arises (Lense–Thirring effect, see Section 27.5).

For an Earth satellite ω amounts to about $8''$ per year; an experiment is in preparation. The Earth–Moon system can be considered as a gyroscope, with its axis perpendicular to the orbital plane. The theoretical geodesic precession of about $2''$ per century has been confirmed to about 2%. Similarly, the orbital spins of satellites undergo a precession, which has been verified with an accuracy of about 20%.

Further reading for Chapter 24

Will (1993), Ashby (1998), Schäfer (2000).

25
Gravitational lenses

25.1 The spherically symmetric gravitational lens

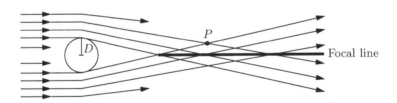

Fig. 25.1. The spherically symmetric gravitational field as a gravitational lens.

The family of curves (23.47) can obviously be interpreted as the family of light rays arriving from a very distant point source. If instead of just one ray (as in Fig. 23.3) the whole family is drawn (as in Fig. 25.1), one sees that the rays converge; a (spherically symmetric) gravitational field behaves like a lens. This gravitational lens is far from ideal, possessing two closely related peculiarities: it produces double images, and incoming parallel rays are focussed onto a focal *line* rather than a focal *point*. (In interpreting pictures Fig. 23.3 and 25.1 it should be noted that the r-φ-section of the Schwarzschild metric has been drawn as an r-φ-plane, which will be a good description only for large r.)

This picture may lead to the following predictions. An observer at P

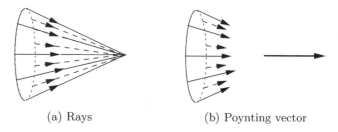

(a) Rays   (b) Poynting vector

Fig. 25.2. Einstein ring (a), and what really happens (b).

outside the focal line (see Fig. 25.1) will see two images, with different intensities, corresponding to the two rays arriving at P. At a point F on the focal line, all rays on the surface of a cone around the axis arrive. Consequently, an observer will have the impression that the light comes from a full circle: instead of a point, he will see a ring, the famous *Einstein ring*, see Fig. 25.2(a).

Unfortunately, these predictions are completely wrong, and *this* Einstein ring does not exist. The reason is that light is to be described by waves; as shown in Section 21.4, geometrical optics and rays are only an approximation, and it is exactly at regions where rays meet that this approximation breaks down. Here interference of the waves arriving along different rays takes place (note that light originating from a point-source, and travelling the same distance on the different paths, is always coherent).

Without any calculation, one can see what the result of that interference will be. At a given point, there is a unique Maxwell field, with a unique Poynting vector. Because of the rotational symmetry about the focal line, that Poynting vector must have the direction of that line; the source of the wave will be seen in the undisturbed direction, see Fig. 25.2(b).

The details of the interference pattern are rather complicated, see Herlt and Stephani (1976). Most impressive is the increase of intensity: for a wave of wavelength λ, the intensity on the focal line is increased by the (dimensionless) factor $4\pi M/\lambda$ (for a solar mass, and visible light, this factor is of the order 10^{10}). But also the apparent position of the double images differs from that found by geometrical optics.

Both approaches, wave optics and geometrical optics, predict that behind any pair of sources and lenses (stars, or galaxies), a focal line extends to infinity; the universe is full of such focal lines.

25.2 Galaxies as gravitational lenses

If we try to observe the image of a gravitational lens, the first idea may be to inspect the focal line of the Sun; but that focal line starts only at a distance of $d \approx 8 \cdot 10^{10}$ km from the Sun, practically outside the Solar System. So we have to look out for more distant lenses. Since the gravitational fields of stars are comparatively weak, galaxies are better candidates for lenses. Galaxies are not spherically symmetric, even less than stars; the model of a spherically gravitational lens does not apply. Focussing will take place, but the waves travelling along the different rays will in general no longer be coherent, their arrival times differing up to years. So for most applications geometrical optics suffices.

In a good approximation, the gravitational lens may be described by a transparent matter distribution $\mu(\mathbf{r})$ in a plane orthogonal to the line to the observer, and one can assume the deflection to take place only within that plane. Again in an approximation, any mass element $\mu(\mathbf{r}')$ will deflect the passing ray in the direction towards the mass as the Schwarzschild solution does, i.e. by $\mathrm{d}\varphi = 4M(\mathbf{r}' - \mathbf{r})/(\mathbf{r}' - \mathbf{r})^2$, with $M = f\mu(\mathbf{r}')/c^2$, see (23.48) and (23.25). Integrating over all masses gives the total deflection

$$\varphi(\mathbf{r}) = \frac{4f}{c^2} \int \frac{\mu(\mathbf{r}')(\mathbf{r}' - \mathbf{r})}{(\mathbf{r}' - \mathbf{r})^2} \mathrm{d}^2 r' \qquad (25.1)$$

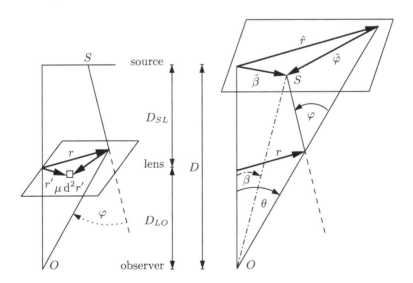

Fig. 25.3. Model of a gravitational lens.

(all angles are small, so that we can treat them as vectors in a plane tangent to the observer's unit sphere), see Fig. 25.3. Note that the distances D_{SL} and D_{LO} may need a careful definition if a cosmological background has to be taken into account.

If β is the direction to the unperturbed position of the source, and all angles are projected onto the source plane, one sees that

$$\widehat{\beta} - \widehat{\varphi} = \widehat{\mathbf{r}}, \quad \text{with} \ \widehat{\beta} = D\beta \ , \ \widehat{\varphi} = D_{SL}\varphi \ , \ \widehat{\mathbf{r}} = D\mathbf{r}/D_{LO}, \quad (25.2)$$

or

$$\widehat{\beta} = \mathbf{r}D/D_{LO} - D_{SL}\varphi(\mathbf{r}) \qquad (25.3)$$

holds.

Equation (25.3) (the 'lens equation') obviously describes a mapping of the lens plane ($\mathbf{r}$) onto the source plane ($\widehat{\beta}$). For a given position $\widehat{\beta}$ of the source there may be more than one solution $\mathbf{r}$; the source may be seen at several positions on the sky. The *magnification* caused by the deflection equals the ratio of the solid angles under which (a small part of) the source and its image, respectively, appear to the observer; it equals the Jacobian

$$J = \left| \det \frac{\partial \Theta}{\partial \beta} \right| = \left| \det \frac{\partial \beta}{\partial \Theta} \right|^{-1}, \quad \Theta = \mathbf{r}/D_{LO}, \qquad (25.4)$$

of the mapping (25.3). The zeros of J^{-1} correspond to caustics.

Although very few lens models have been treated rigorously, some general results can be obtained from the mathematical theory of mappings, including catastrophe theory, see Schneider *et al.* (1992) for further details.

Gravitational lensing, with both source and lens being galaxies, has been observed on many occasions. When the lens has a high symmetry, the images of the (extended) source may even be part of a ring, which is often called the Einstein ring.

Exercise

25.1   Use equations (25.1) and (25.3) to find the two directions in which the star is seen by the observer at P in the Schwarzschild field, cp. Fig. 25.1. What are the magnifications for the two images?

26

The interior Schwarzschild solution

26.1 The field equations

If we want to determine the gravitational field inside a celestial body then we need a model for this body, that is, we must say something about its energy-momentum tensor. Ignoring thermodynamic effects, such as heat conduction and viscosity, the ideal fluid medium (21.81)

$$T_{mn} = (\mu + p/c^2)u_m u_n + p g_{mn} \tag{26.1}$$

is a useful approximation.

We seek a spherically symmetric, static solution (ignoring radial matter currents in the stars), and thus require that the general line element (23.5) is independent of time, that is,

$$ds^2 = e^{\lambda(r)} dr^2 + r^2(d\vartheta^2 + \sin^2\vartheta \, d\varphi^2) - e^{\nu(r)}c^2 \, dt^2 \tag{26.2}$$

holds; the matter is at rest in this coordinate system,

$$u^m = (0, 0, 0, c \, e^{-\nu/2}), \tag{26.3}$$

and μ and p are functions purely of the radius r.

In setting up the field equations we can use the components of the Ricci tensor already calculated in (23.13), setting to zero all time derivatives. Because of

$$R = R_n^n = -2 e^{-\lambda}\left[\tfrac{1}{2}\nu'' + \tfrac{1}{4}\nu'^2 - \tfrac{1}{4}\nu'\lambda' - (\nu' - \lambda')/r + 1/r^2\right] + 2/r^2, \tag{26.4}$$

the Einstein equations

$$R_m^n - \tfrac{1}{2}R g_m^n = \kappa T_m^n, \tag{26.5}$$

hence assume the form

$$\kappa p = R_1^1 - \tfrac{1}{2}R = e^{-\lambda}[\nu'/r + 1/r^2] - 1/r^2, \tag{26.6}$$

$$\kappa p = R_2^2 - \tfrac{1}{2}R = e^{-\lambda}[\tfrac{1}{2}\nu'' + \tfrac{1}{4}\nu'^2 - \tfrac{1}{4}\nu'\lambda' + \tfrac{1}{2}(\nu' - \lambda')/r], \tag{26.7}$$

$$-\kappa\mu c^2 = R_4^4 - \tfrac{1}{2}R = -e^{-\lambda}[\lambda'/r - 1/r^2] - 1/r^2. \tag{26.8}$$

The four functions λ, ν, p and μ are to be determined from these three equations and an equation of state $f(\mu, p) = 0$ yet to be formulated.

26.2 The solution of the field equations

As we have discussed in detail in Section 22.1, the field equations are only integrable if the balance equations of energy and momentum $T^{mn}{}_{;n} = 0$ are satisfied. These conservation laws often give – analogously to the first integrals of Classical Mechanics – an important indication of how to solve the field equations. Since for static distributions of matter and pressure we have

$$\mu_{,n}\mu^n = 0, \quad p_{,n}u^n = 0, \quad u^n{}_{;n} = 0, \tag{26.9}$$

the equations

$$T^{mn}{}_{;n} = \left[pg^{mn} + \left(\mu - p/c^2\right)u^m u^n\right]_{;n} = 0 \tag{26.10}$$

simplify to the one equation

$$p' + \left(\mu + p/c^2\right)u_{1;4}u^4 = p' - \left(\mu + p/c^2\right)\Gamma_{14}^4 u_4 u^4 = 0, \tag{26.11}$$

which leads to

$$p' = -\tfrac{1}{2}\nu'\left(p + \mu c^2\right). \tag{26.12}$$

This equation is a consequence of the field equations (26.6)–(26.8) and can be used in place of one of these three equations.

The field equation (26.8) can be written in the form

$$\kappa\mu c^2 r^2 = -\left(e^{-\lambda}r\right)' + 1. \tag{26.13}$$

Assuming that $g^{11} = e^{-\lambda}$ is finite at $r = 0$, it is integrated by

$$2m(r) = -r\,e^{-\lambda} + r, \tag{26.14}$$

where $m(r)$ is defined by

$$m(r) = \tfrac{1}{2}\kappa c^2 \int_0^r \mu(x)x^2\,\mathrm{d}x. \tag{26.15}$$

The function $m(r)$ is called the *mass function*. Equation (26.15) can also be interpreted as showing that $m(r)$ is proportional to the total mass contained within a sphere of radius r, but then one must be careful to note that r is only the coordinate radius, the true radius $R(r)$ of the sphere being given by

$$R(r) = \int_0^r e^{\lambda(x)/2}\,\mathrm{d}x. \tag{26.16}$$

In (26.14) we have succeeded in specifying one of the metric functions,

$$e^{-\lambda(r)} = 1 - 2m(r)/r, \tag{26.17}$$

independently of a special equation of state. Taking now (26.12) and substituting ν' and e^λ by using (26.6) and (26.17), we get for p the differential equation

$$\frac{dp}{dr} = \frac{(\mu c^2 + p)[2m(r) + \kappa_0 pr^3]}{2r[2m(r) - r]} \tag{26.18}$$

(Tolman 1939, Oppenheimer and Volkoff 1939). To solve this equation we must fix the equation of state. The most simple possibility is to assume a constant rest-mass density,

$$\mu = \text{const.} \tag{26.19}$$

This equation of state certainly does not give a particularly good stellar model; a constant mass density is a first approximation only for small stars in which the pressure is not too large. The spherically symmetric, static solution with the special equation of state (26.19) is called the *interior Schwarzschild solution*.

For constant mass density (26.17) becomes

$$e^{-\lambda} = 1 - Ar^2, \quad A = \tfrac{1}{3}\kappa\mu c^2. \tag{26.20}$$

Instead of solving (26.18), we directly use (26.12) in the form

$$\left(p + \mu c^2\right)' = -\tfrac{1}{2}\nu'\left(p + \mu c^2\right), \tag{26.21}$$

and integrate it by

$$p + \mu c^2 = B\,e^{-\nu/2}. \tag{26.22}$$

As the third field equation to be solved, we choose the combination

$$\kappa(\mu c^2 + p) = e^{-\lambda}(\lambda' + \nu')/r = \kappa B\,e^{-\nu/2} \tag{26.23}$$

of (26.6) and (26.8). Upon substitution for $e^{-\lambda}$ it goes over to

$$e^{\nu/2}\left(2A - Ar\nu' + \nu'/r\right) = \kappa B, \tag{26.24}$$

and, through the intermediate step

$$2\left(1 - Ar^2\right)^{3/2}\left[e^{\nu/2}\left(1 - Ar^2\right)^{-1/2}\right]' = \kappa Br, \tag{26.25}$$

it can be easily solved to give

$$e^{\nu/2} = \kappa B/2A - D\sqrt{1 - Ar^2}. \tag{26.26}$$

The equations (26.20), (26.22) and (26.26) give the general solution for the case of constant mass density μ. They contain two constants of integration, B and D, which are to be determined by the matching conditions.

26.3 Matching conditions and connection to the exterior Schwarzschild solution

From the Maxwell theory one knows that matching conditions for certain field components must be satisfied at the interface between two media; these matching conditions follow from Maxwell's equations. In a completely analogous way, we must here ensure certain continuity properties of the metric at the surface of the star if we want to construct the complete gravitational field from the solution in the interior of the star (the interior Schwarzschild solution) and the solution in the exterior space (the Schwarzschild solution). As we shall show in Chapter 30, the appropriate matching conditions follow from the Einstein field equations. Since in our simple example physical plausibility considerations will answer the purpose, we shall limit ourselves here to some brief remarks on the matching conditions.

Continuity properties of the metric and its derivatives can obviously be destroyed by coordinate transformations and inappropriate choice of coordinates. We therefore formulate the matching conditions most simply in a special coordinate system in which the boundary is a coordinate surface $x^4 = $ const. (in our example: $r = r_0$) and Gaussian coordinates are employed in the neighbourhood of the boundary, so that

$$ds^2 = \varepsilon(dx^4)^2 + g_{\alpha\beta}\,dx^\alpha\,dx^\beta, \quad \varepsilon = \pm 1 \qquad (26.27)$$

($\varepsilon = +1$, if x^4 is a spacelike coordinate). Since second derivatives of the metric appear in the field equations, their existence must be ensured, that is, we demand that:

$$g_{\alpha\beta} \text{ and } g_{\alpha\beta,4} \text{ are continuous on } x^4 = \text{const.} \qquad (26.28)$$

By these requirements we have excluded layer structures in the surface (δ-function singularities in the energy-momentum tensor).

In order to connect interior and exterior Schwarzschild solutions to one another on the surface of the star $r = r_0$, we ought first of all to introduce (separate interior and exterior) Gaussian coordinates by

$$dx^4 = dr\,e^{\lambda(r)/2}, \qquad (26.29)$$

and try to satisfy the conditions (26.28) through choice of the still unspecified integration parameters B and D (the internal solution) and of M (the external solution). But here we want to deal with the problem in a more intuitive fashion; the reader may reflect on the equivalence of the two methods.

We require that the metric g_{mn} is continuous for $r = r_0$ and that the pressure p vanishes on the surface of the star. Because of (26.22) and (26.26), the pressure depends upon r according to

$$p = Be^{-\nu/2} - \mu c^2 = \frac{1}{\kappa} \frac{3AD\sqrt{1 - Ar^2} - \kappa B/2}{\kappa B/2A - D\sqrt{1 - Ar^2}}, \tag{26.30}$$

and hence these requirements correspond to the three equations

e^λ continous: $\qquad\qquad\qquad\qquad 1 - Ar_0^2 = 1 - 2M/r_0,$

e^ν continous: $\qquad \left(\kappa B/2A - D\sqrt{1 - Ar_0^2}\right)^2 = 1 - 2M/r_0, \tag{26.31}$

$p = 0: \qquad\qquad\qquad\qquad 3AD\sqrt{1 - Ar_0^2} = \kappa B/2.$

They have the solution

$$M = \tfrac{1}{2}Ar_0^3 = \tfrac{1}{6}\kappa\mu c^2 r_0^3 = \tfrac{1}{3}4\pi r_0^3 f\mu/c^2, \quad D = \tfrac{1}{2},$$
$$\kappa B = 3A\sqrt{1 - Ar_0^2} = \kappa\mu c^2\sqrt{1 - \tfrac{1}{3}\kappa\mu c^2 r_0^2}, \tag{26.32}$$

by means of which all the constants of integration are related to the mass density μ and the stellar radius r_0.

The spherically symmetric gravitational field of a star with mass density $\mu = \text{const.}$ and radius r_0 is thus described (Schwarzschild 1916) by the interior Schwarzschild solution

$$ds^2 = \frac{dr^2}{1 - Ar^2} + r^2\left(d\vartheta^2 + \sin^2\vartheta\, d\varphi^2\right)$$
$$- \left[\tfrac{3}{2}\sqrt{1 - Ar_0^2} - \tfrac{1}{2}\sqrt{1 - Ar^2}\right]^2 c^2\, dt, \tag{26.33}$$

$$\mu = \text{const.}, \quad A = \tfrac{1}{3}\kappa\mu c^2, \quad \kappa p = 2A\frac{\sqrt{1 - Ar^2} - \sqrt{1 - Ar_0^2}}{3\sqrt{1 - Ar_0^2} - \sqrt{1 - Ar^2}}$$

inside the star, and by the exterior (vacuum) Schwarzschild solution

$$ds^2 = \frac{dr^2}{1 - 2M/r} + r^2\left(d\vartheta^2 + \sin^2\vartheta\, d\varphi^2\right) - \left(1 - \frac{2M}{r}\right)c^2,$$
$$2M = Ar_0^3 \tag{26.34}$$

outside. We should point out that in the coordinates used here $\partial g_{rr}/\partial g_r$ is discontinuous on the boundary surface $r = r_0$, but that this discontinuity can be removed by making a coordinate transformation.

26.4 A discussion of the interior Schwarzschild solution

In interpreting the constant M, one must note that M is a measure of the total effective mass of the star as seen from outside (field-producing mass). As (26.32) shows, it is in fact proportional to the *coordinate volume* $4\pi r_0^3/3$, but not, however, to the true three-dimensional volume of the star.

While the mass density μ is constant, the pressure p increases inwards; the solution is non-singular as long as p is finite. At $r = 0$, where p takes its maximum value, this is only possible, because of (26.33), for

$$3\sqrt{1 - Ar_0^2} > 1; \tag{26.35}$$

that is, for

$$r_0 > \tfrac{9}{8}\, 2M. \tag{26.36}$$

This inequality is to be interpreted as saying that for given total mass M the interior solution is regular (exists) only if the stellar radius r_0 is large enough, and in any case larger than the Schwarzschild radius $2M$. For normal stars like our Sun this is always the case, but for stars with very dense matter (nuclear matter) it may not be possible to satisfy (26.36). There is then no interior Schwarzschild solution and, as we shall describe in detail later (Section 36.2), no stable interior solution at all.

The three-dimensional space

$$\mathrm{d}^{(3)}s^2 = \frac{\mathrm{d}r^2}{1 - Ar^2} + r^2\big(\mathrm{d}\vartheta^2 + \sin^2\vartheta\,\mathrm{d}\varphi^2\big) \tag{26.37}$$

of the interior Schwarzschild solution has an especially simple geometry. One sees this best by introducing a new coordinate χ via $r = A^{-1/2}\sin\chi$ and transforming the line element (26.37) into the form

$$\mathrm{d}^{(3)}s^2 = A^{-1}\big[\mathrm{d}\chi^2 + \sin^2\chi\big(\mathrm{d}\vartheta^2 + \sin^2\vartheta\,\mathrm{d}\varphi^2\big)\big]. \tag{26.38}$$

The metric (26.38) is that of a three-dimensional space of constant curvature, cp. Section 19.5 and equation (19.45). In this space all points are geometrically equivalent; of course the points in the star are physically 'distinguishable', because g_{44} (that is, in essence the pressure p) is position dependent.

The geometric optical behaviour of the interior Schwarzschild solution is exactly that of the Maxwell fish eye, truncated at some finite radius (Buchdahl 1983).

Exercises

26.1 Find the metric of the interior Schwarzschild solution in isotropic coordinates $ds^2 = e^{2F(\rho)}[d\rho^2 + \rho^2(d\vartheta^2 + \sin^2\vartheta\,d\varphi^2)] - e^{\nu[r(\rho)]}c^2\,dt^2$ by (i) performing a coordinate transformation with $r = 2a\rho/(a^2 + A\rho^2)$ and an appropriate a, or by (ii) setting up the field equations and solving them. Use the form (23.67) of the Schwarzschild metric to show that in these coordinates $g_{\rho\rho,\rho}$ is continous at the surface.

26.2 Show that the interior Schwarzschild solution is conformally flat.

26.3 Solve (26.18) for $\mu = $ const. and compare the result with (26.30).

26.4 How does the Newtonian limit of (26.18) read?

IV. Linearized theory of gravitation, far fields and gravitational waves

27

The linearized Einstein theory of gravity

27.1 Justification for a linearized theory and its realm of validity

One speaks of a linearized theory when the metric deviates only slightly from that of a flat space,

$$g_{mn} = \eta_{mn} + f_{mn}, \quad f_{mn} \ll 1, \tag{27.1}$$

and therefore all terms in the Einstein equations which are non-linear in f_{mn} or its derivatives can be discarded and the energy-momentum tensor T^{ik} can be replaced by its special-relativistic form.

This energy-momentum tensor then satisfies the special-relativistic equation

$$T^{ik}{}_{,k} = 0. \tag{27.2}$$

Since no covariant derivatives occur in (27.2), in the linearized theory the gravitational field has no influence upon the motion of the matter producing the field. One can specify the energy-momentum tensor arbitrarily, provided only that (27.2) is satisfied, and calculate the gravitational field associated with it. This apparently advantageous property of the linearized theory has, however, the consequence that the gravitational field corresponding to the exact solution can deviate considerably from that of the linearized theory if the sources of the field (under the influence of their own gravitational field) move in a manner rather dif-

ferent from that supposed. It is therefore quite possible that there is
no exact solution whose essential features agree with those of a particular solution of the linearized theory. Since, however, one would like to
use approximation procedures at precisely those places where the exact
solution is unknown, one must be careful with conclusions drawn from
the results of the linearized theory.

Statements made in the linearized theory will be reliable then if we
have a good knowledge of the motion of the sources and if these sources
are not too massive (if the field produces by them is weak). This is the
case, for example, in the planetary system. The linearized theory can
also be used to analyze the fields, due to sources regarded as known,
at great distance from these sources, or to describe the metric and the
gravitational field in the neighbourhood of a point at which we have
introduced a locally geodesic coordinate system. As (27.1) already
shows, the linearized theory is applicable only as long as one can introduce approximately Cartesian coordinates. From the standpoint of
the (curved) universe this means we shall always be dealing with local
applications.

27.2 The fundamental equations of the linearized theory

As we have already shown in equation (22.14), the curvature tensor
associated with the metric (27.1) has the form

$$R^a{}_{mbn} = \tfrac{1}{2}\eta^{as}(f_{sn,mb} + f_{mb,sn} - f_{mn,bs} - f_{bs,mn}), \qquad (27.3)$$

obtained by ignoring all non-linear terms in f_{mn}.

For the following calculations we use the convention that indices in the
f_{mn} and its derivatives are always moved up or down with the flat-space
metric η_{ab}, so that we have

$$f^{ab} = \eta^{am}\eta^{bn}f_{mn}, \quad f^a{}_a = \eta^{ab}f_{ab}, \dots. \qquad (27.4)$$

From (27.3) we thus obtain the linearized field equations

$$\begin{aligned}
R_{mn} - \tfrac{1}{2}R\eta_{mn} = -\tfrac{1}{2}\,[f_{mn}{}^{,a}{}_{,a} - \eta_{mn}f^i{}_i{}^{,a}{}_{,a} + \eta_{mn}f^{ab}{}_{,ab} \\
+ f^a{}_{a,mn} - f^a{}_{n,ma} - f^a{}_{m,na}] = \kappa T_{mn}.
\end{aligned} \qquad (27.5)$$

The following considerations lead, by means of suitable definitions and
subsidiary conditions (coordinate transformations), to a simpler and
mathematically clearer formulation of the linearized field equations.

We first of all introduce in the place of the quantities f_{mn} new field functions $\bar{f}_{mn}$, which occur in the expansion

$$\sqrt{-g}\, g^{mn} = \eta^{mn} - \bar{f}^{mn} \tag{27.6}$$

of the density of the metric tensor and which are linked to the old functions through the equations

$$\bar{f}_{mn} = f_{mn} - \tfrac{1}{2}\eta_{mn} f^a{}_a, \quad f_{mn} = \bar{f}_{mn} - \tfrac{1}{2}\eta_{mn} \bar{f}^a{}_a, \quad f^a{}_a = -\bar{f}^a{}_a. \tag{27.7}$$

The field equations (27.5) then read

$$\bar{f}_{mn}{}^{,a}{}_{,a} + \eta_{mn} \bar{f}^{ab}{}_{,ab} - \bar{f}^a{}_{n,ma} - \bar{f}^a{}_{m,na} = -2\kappa T_{mn}. \tag{27.8}$$

We shall now simplify them further by means of coordinate transformations of the form

$$\tilde{x}^n = x^n + b^n(x^i) \tag{27.9}$$

(these transformations are the analogue of the gauge transformations of electrodynamics). From (27.9) we obtain

$$\tilde{g}^{mn} = g^{as}(\delta_a^n + b^n{}_{,a})(\delta_s^m + b^m{}_{,s}), \quad \tilde{g} = |\tilde{g}^{mn}|^{-1} = g(1 + 2b^a{}_{,a})^{-1}, \tag{27.10}$$

and hence

$$\tilde{\bar{f}}^{mn} = \bar{f}^{mn} - b^{n,m} - b^{m,n} + \eta^{mn} b^a{}_{,a}. \tag{27.11}$$

The four functions $b^n(x^i)$ can be chosen arbitrarily; of course the transformation (27.9) must not take us outside the framework of the linearized theory, that is to say, $\bar{f}^{mn} \ll 1$ must hold. If we substitute (27.11) into the field equation (27.8), then we see, upon making the choice

$$\Box b^n = \eta^{rs} b^n{}_{,rs} = \bar{f}^{mn}{}_{,m}, \tag{27.12}$$

that the field equations become particularly simple. The field variables $\bar{f}_{mn}$ (we now drop the tilde) then satisfy the equation

$$\bar{f}^{mn}{}_{,n} = -(\sqrt{-g}\, g^{mn})_{,n} = 0 \tag{27.13}$$

(thus we use the harmonic coordinates defined in Section 23.6), and the Einstein field equations reduce to the inhomogeneous wave equation

$$\Box \bar{f}_{mn} = \bar{f}_{mn}{}^{,a}{}_{,a} = -2\kappa T_{mn}. \tag{27.14}$$

Of course, one must take only those solutions of the field equation (27.14) which satisfy the subsidiary conditions (27.13); the existence of such solutions is guaranteed by (27.2).

The linearized Einstein theory of gravity

27.3 A discussion of the fundamental equations and a comparison with special-relativistic electrodynamics

The fundamental equations (27.14) and (27.8) have just the usual form of the equations of a classical field theory in Minkowski space. They are linear and, after introduction of the subsidiary condition (27.13), they are even uncoupled. One can completely dispense with the idea of a Riemannian space and regard the $\bar{f}_{mn}$ as the components of a tensor field by means of which gravitation is described in a flat space. The action of this field upon a test particle is then given (according to the geodesic equation) by

$$\frac{\mathrm{d}^2 x^a}{\mathrm{d}\tau^2} = -\tfrac{1}{2}\eta^{ab}(f_{bm,n} + f_{bn,m} - f_{nm,b})\frac{\mathrm{d}x^n}{\mathrm{d}\tau}\frac{\mathrm{d}x^m}{\mathrm{d}\tau}. \tag{27.15}$$

Field equations and equations of motion are Lorentz invariant.

Although this kind of gravitational theory is very tempting (and has hence occasionally been interpreted as the correct theory of gravitation), it does however have a serious shortcoming; the gravitational force does not react back on the sources of the field. If one tries to correct this, one is led back to the Einstein theory.

The striking analogy between the linearized Einstein equations and electrodynamics is shown in Table 27.1.

Table 27.1. *Maxwell theory versus linearized Einstein theory*

	Maxwell theory	Linearized Einstein theory
Fundamental field variables	four-potential A_m	$\bar{f}_{mn}$
General field equations	$A_m{}^{,a}{}_{,a} - A^a{}_{,m,a}$ $= -j_m/c$	$\bar{f}_{mn}{}^{,a}{}_{,a} + \eta_{mn}\bar{f}^{ab}{}_{,ab}$ $+\bar{f}^a{}_{n,am} - \bar{f}^a{}_{m,na} = -2\kappa T_{mn}$
Field equations are invariant under	gauge transformations $\tilde{A}_m = A_m + b_{,m}$ (field tensor invariant)	coordinate transformations $\tilde{\bar{f}}_{mn} =$ $\bar{f}_{mn} - b_{n,m} - b_{m,n} + \eta_{mn}b^a{}_{,a}$ (Christoffel symbols changed, curvature tensor invariant)
Subsidiary conditions	$A^a{}_{,a} = 0$	$\bar{f}^{mn}{}_{,n} = 0$
Form of the field equations simplified by these conditions	$\Box A_m = -j_m/c$	$\Box \bar{f}_{mn} = -2\kappa T_{mn}$
Further possible gauge transformations are restricted by	$\Box b = 0$	$\Box b^n = 0$

In the linearized theory of gravitation, too, we can represent the solution to the field equations in terms of the sources, namely, in the form of a 'retarded potential'

$$\bar{f}_{mn}(\mathbf{r}, t) = \frac{\kappa}{2\pi} \int \frac{T_{mn}\left(\bar{\mathbf{r}}, t - |\mathbf{r} - \bar{\mathbf{r}}|/c\right)}{|\mathbf{r} - \bar{\mathbf{r}}|}\, \mathrm{d}^3 \bar{x}. \qquad (27.16)$$

To this particular solution one can still always add solutions of the homogeneous equations

$$\Box \bar{f}_{mn} = 0, \qquad \bar{f}^{mn}{}_{,n} = 0, \qquad (27.17)$$

and thus, for example, go over from the retarded to the advanced solutions.

Sometimes it is more convenient to simplify the metric by a coordinate transformation

$$\tilde{f}_{mn} = f_{mn} - b_{m,n} - b_{n,m} \qquad (27.18)$$

and *not* to use harmonic coordinates. Such a transformation can remove pure coordinate effects, that is, terms which give no contribution to the curvature tensor.

27.4 The far field due to a time-dependent source

In electrodynamics one learns that in general the following components dominate in the far field of an arbitrary distribution of charge and current (the characteristic r-dependence is in brackets): electrostatic monopole (r^{-1}), electrostatic and magnetostatic dipole (r^{-2}), electrostatic quadrupole (r^{-3}), oscillating electric and magnetic dipole and electric quadrupole (all r^{-1}). For charges which are not moving too quickly, the spacelike contribution to the four-current density is smaller than the timelike contribution by a factor of c, because $j^n = (\rho v, \rho c)$, and therefore the electromagnetic radiation emerging from a system is essentially that due to an oscillating electric dipole.

In a similar manner we now want to investigate and characterize the gravitational far field of a matter distribution. The calculations are simple, but somewhat tedious. To keep a clear view we divide them into three steps.

Step 1. Power series expansion of the integrand of (27.16) We assume that we are dealing with an isolated distribution of matter; that is, T_{mn} is non-zero only within a finite spatial region (Fig. 27.1). In the far field we can then replace $|\mathbf{r} - \bar{\mathbf{r}}|$ by the first terms of a power series expansion:

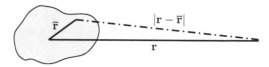

Fig. 27.1. Isolated matter distribution.

$$|\mathbf{r} - \bar{\mathbf{r}}| = \sqrt{\mathbf{r}^2 - 2\mathbf{r}\bar{\mathbf{r}} + \bar{\mathbf{r}}^2} = r - \frac{x^\alpha \bar{x}^\alpha}{r} - \frac{x^\alpha x^\beta}{2r^3}\left(\bar{x}^\alpha \bar{x}^\beta - \bar{r}^2 \delta^{\alpha\beta}\right) + \cdots,$$

$$\frac{1}{|\mathbf{r} - \bar{\mathbf{r}}|} = \frac{1}{r} + \frac{x^\alpha \bar{x}^\alpha}{r^3} + \frac{x^\alpha x^\beta}{2r^5}\left(3\bar{x}^\alpha \bar{x}^\beta - \bar{r}^2 \delta^{\alpha\beta}\right) + \cdots \tag{27.19}$$

Imagining the series (27.19) to have been substituted into the argument $t - |\mathbf{r} - \bar{\mathbf{r}}|/c$ of the energy-momentum tensor and an expansion of the components T_{mn} carried out, we then have

$$T_{mn}(\bar{\mathbf{r}}, t - |\mathbf{r} - \bar{\mathbf{r}}|/c) = T_{mn}(\bar{\mathbf{r}}, t - r/c) + \dot{T}_{mn}(\bar{\mathbf{r}}, t - r/c)\{r - |\mathbf{r} - \bar{\mathbf{r}}|\}/c$$

$$+ \tfrac{1}{2}\ddot{T}_{mn}(\bar{\mathbf{r}}, t - r/c)\{r - |\mathbf{r} - \bar{\mathbf{r}}|\}^2 c^{-2} + \cdots \tag{27.20}$$

For a motion of the matter periodic in time (frequency ω) it is justifiable to ignore higher time derivatives if the diameter of the matter distribution is small by comparison with c/ω, and thus small compared to the wavelength of the waves radiated out.

The integrand of (27.16) has, after substitutions of (27.19) and (27.20), the form

$$\frac{T_{mn}(\bar{\mathbf{r}}, t - |\mathbf{r} - \bar{\mathbf{r}}|/c)}{|\mathbf{r} - \bar{\mathbf{r}}|} = T_{mn}\left[\frac{1}{r} + \frac{x^\alpha \bar{x}^\alpha}{r^3} + \frac{x^\alpha x^\beta}{2r^5}\left(3\bar{x}^\alpha \bar{x}^\beta - \bar{r}^2 \delta^{\alpha\beta}\right)\right]$$

$$+ \frac{\dot{T}_{mn}}{c}\left[\frac{x^\alpha \bar{x}^\alpha}{r^2} + \frac{x^\alpha x^\beta}{2r^4}\left(3\bar{x}^\alpha \bar{x}^\beta - \bar{r}^2 \delta^{\alpha\beta}\right)\right]$$

$$+ \frac{\ddot{T}_{mn}}{c^2}\left[\frac{x^\alpha x^\beta}{2r^3}\bar{x}^\alpha \bar{x}^\beta\right], \tag{27.21}$$

where on the right-hand side the argument $t - r/c$ in T_{mn} and its derivatives have been suppressed.

Step 2. Definition of the moments of the energy-momentum tensor and simplification by Lorentz transformations and conservation laws For matter not moving too quickly, the component T_{44} dominates the energy-momentum tensor, and we have

$$|T_{44}| \gg |T_{4\alpha}| \gg |T_{\alpha\beta}|. \tag{27.22}$$

Accordingly, in substituting (27.21) into the integrand of formula (27.16) it is only necessary to evaluate the integrals

$$\int T_{44} \, \mathrm{d}^3\bar{x} \equiv m, \quad \int T_{44}\bar{x}^\alpha \, \mathrm{d}^3\bar{x} \equiv d^\alpha, \quad \int T_{44}\bar{x}^\alpha\bar{x}^\beta \, \mathrm{d}^3\bar{x} \equiv d^{\alpha\beta},$$
$$\int T_{4\nu} \, \mathrm{d}^3\bar{x} \equiv -p_\nu, \quad \int T_{4\nu}\bar{x}^\alpha \, \mathrm{d}^3\bar{x} \equiv b_\nu{}^\alpha, \quad \int T_{\alpha\beta} \, \mathrm{d}^3\bar{x} \equiv a_{\alpha\beta}. \tag{27.23}$$

All these quantities are in principle functions of the retarded time $t - r/c$. From the energy law

$$T^{n4}{}_{,4} = -T^{n\mu}{}_{,\mu} \tag{27.24}$$

and the law of angular momentum

$$(T^{4m}\bar{x}^s - T^{4s}\bar{x}^m)_{,4} = -(T^{\nu m}\bar{x}^s - T^{\nu s}\bar{x}^m)_{,\nu} \tag{27.25}$$

we obtain, upon integration over the matter distribution and application of the Gauss law (all operations are carried out in a flat space),

$$m = \mathrm{const} \, (>0), \qquad\qquad p^\nu = \mathrm{const.,}$$
$$b^{\nu\alpha} - b^{\alpha\nu} = B^{\nu\alpha} = \mathrm{const.,} \quad \dot{d}^\alpha/c = p^\alpha. \tag{27.26}$$

We can therefore transform the three-momentum p^α to zero by making a Lorentz transformation and then (because $m > 0$) transform away the matter dipole moment d^α by shifting the origin of the spatial coordinate system.

Further, from the two conservation laws (27.24) and (27.25), we obtain the equations

$$T^{4\alpha}\bar{x}^\beta + T^{4\beta}\bar{x}^\alpha = (T^{44}\bar{x}^\alpha\bar{x}^\beta)_{,4} + (T^{4\nu}\bar{x}^\alpha\bar{x}^\beta)_{,\nu} \tag{27.27}$$
$$(T^{44}\bar{x}^\alpha\bar{x}^\beta)_{,44} = (T^{\mu\nu}\bar{x}^\alpha\bar{x}^\beta)_{,\mu\nu} - 2(T^{\mu\alpha}\bar{x}^\beta + T^{\mu\beta}\bar{x}^\alpha)_{,\mu} + 2T^{\alpha\beta},$$

which upon integration lead to the relation

$$b^{\alpha\beta} + b^{\beta\alpha} = -\dot{d}^{\alpha\beta}/c, \qquad a_{\alpha\beta} = \ddot{d}_{\alpha\beta}/2c^2 \tag{27.28}$$

between the moments of the energy–momentum tensor.

Taken together, all the moments in which we are interested can thus be expressed in terms of the mass m, the angular momentum $B_{\nu\alpha}$ and the mass quadrupole moment $d_{\alpha\beta}$ according to

$$m = \mathrm{const.,} \qquad p^\nu = 0, \qquad d^\nu = 0, \qquad B_{\nu\alpha} = \mathrm{const.,}$$
$$b_{\nu\alpha} = (B_{\nu\alpha} - \dot{d}_{\nu\alpha}/c)/2, \qquad a_{\alpha\beta} = \ddot{d}_{\alpha\beta}/2c^2. \tag{27.29}$$

Step 3. Writing down the metric and simplifying it by coordinate trans-
formations Substituting the integrand (27.21) into the formula (27.16),

using the results (27.23), and remembering the relations (27.29), we have

$$\frac{2\pi}{\kappa}\bar{f}_{44} = \frac{m}{r} + \frac{x^\alpha x^\beta}{2r^5}(3d_{\alpha\beta} - \eta_{\alpha\beta}d_\sigma^\sigma)$$

$$+ \frac{x^\alpha x^\beta}{2r^4 c}(3\dot{d}_{\alpha\beta} - \eta_{\alpha\beta}\dot{d}_\sigma^\sigma) + \frac{x^\alpha x^\beta}{2c^2 r^3}\ddot{d}_{\alpha\beta}, \qquad (27.30)$$

$$\frac{2\pi}{\kappa}\bar{f}_{4\nu} = \frac{B_{\nu\alpha}x^\alpha}{2r^3} - \frac{\dot{d}_{\nu\alpha}x^\alpha}{2cr^3} - \frac{\ddot{d}_{\nu\alpha}x^\alpha}{2c^2 r^2}, \qquad \frac{2\pi}{\kappa}\bar{f}_{\nu\mu} = \frac{\ddot{d}_{\nu\mu}}{2c^2 r}.$$

The conversion to the f_{mn} yields relatively complicated expressions which we shall not give explicitly. The reader can verify by direct calculation that after a coordinate transformation (27.18) with the generating functions

$$\frac{2\pi}{\kappa}b_4 = \frac{x^\alpha x^\beta}{8r^4}(3d_{\alpha\beta} - \eta_{\alpha\beta}d_\sigma^\sigma) + \frac{x^\alpha x^\beta}{8cr^3}(\dot{d}_{\alpha\beta} + \eta_{\alpha\beta}\dot{d}_\sigma^\sigma), \qquad (27.31)$$

$$\frac{2\pi}{\kappa}b_\nu = -\frac{3d_{\nu\alpha}x^\alpha}{4r^3} - \frac{\dot{d}_{\nu\alpha}x^\alpha}{2cr^2} + \frac{3d_{\alpha\beta}x^\alpha x^\beta x^\nu}{4r^5} + \frac{(\dot{d}_{\alpha\beta} + \eta_{\alpha\beta}\dot{d}_\sigma^\sigma)x^\alpha x^\beta x^\nu}{8cr^4}$$

no time derivatives are contained in f_{44} and $f_{4\nu}$. The far field of an isolated matter distribution then has, in the linearized theory, the metric

$$g_{44} = -1 + f_{44} = -1 + \frac{2M}{r} + \frac{x^\alpha x^\beta}{r^5}(3D_{\alpha\beta} - \eta_{\alpha\beta}D_\sigma^\sigma) + O(r^{-4}),$$

$$g_{4\nu} = f_{4\nu} = \frac{2x^\alpha}{r^3}\varepsilon_{\mu\nu\alpha}P^\mu + O(r^{-3}),$$

$$g_{\mu\nu} = \eta_{\mu\nu}\left[1 + \frac{2M}{r} + \frac{x^\alpha x^\beta}{r^5}(3D_{\alpha\beta} - \eta_{\alpha\beta}D_\sigma^\sigma)\right] + \frac{2}{3c^2 r}(3\ddot{D}_{\nu\mu} - \eta_{\nu\mu}\ddot{D}_\lambda^\lambda)$$

$$+ (3\ddot{D}_{\alpha\beta} - \eta_{\alpha\beta}\ddot{D}_\lambda^\lambda)x^\alpha x^\beta\left[\frac{\eta_{\mu\nu}}{3c^2 r^3} + \frac{x^\nu x^\mu}{12c^2 r^5}\right] \qquad (27.32)$$

$$- \frac{(3\ddot{D}_{\nu\alpha} - \eta_{\nu\alpha}\ddot{D}_\lambda^\lambda)}{6c^2 r^3}x^\alpha x^\mu - \frac{(3\ddot{D}_{\mu\alpha} - \eta_{\mu\alpha}\ddot{D}_\lambda^\lambda)}{6c^2 r^3}x^\alpha x^\nu + O(r^{-2}),$$

with the abbreviations

$$mass:\quad M = \frac{\kappa}{8\pi}\int T_{44}(\bar{\mathbf{r}}, t - r/c)\,d^3\bar{x} = const.,$$

$$angular\ momentum:\quad P^\mu = \varepsilon^{\mu\nu}{}_\alpha\frac{\kappa}{8\pi}\int T_{4\nu}(\bar{\mathbf{r}}, t - r/c)\bar{x}^\alpha\,d^3\bar{x}$$

$$= const., \qquad (27.33)$$

$$quadrupole\ moment:\quad D^{\alpha\beta} = \frac{\kappa}{8\pi}\int T_{44}(\bar{\mathbf{r}}, t - r/c)\bar{x}^\alpha\bar{x}^\beta\,d^3\bar{x}$$

$$= D^{\alpha\beta}(t - r/c).$$

27.5 Discussion of the properties of the far field (linearized theory)

As we have already shown in Section 22.2, $f_{44} = g_{44} + 1$ is essentially the Newtonian potential of the matter distribution. Here it contains a mass term and a quadrupole term, but no dipole contribution (we have transformed this to zero by the choice of coordinate system). If we compare the linear approximation

$$g_{44} = -1 + 2M/r + x^\alpha x^\beta (3D_{\alpha\beta} - \eta_{\alpha\beta} D^\sigma_\sigma)/r^5 + O(r^{-4}) \qquad (27.34)$$

with the expansion

$$g_{44} = -1 + 2M/r - 2M^2/r^2 + O(r^{-3}) \qquad (27.35)$$

of the Schwarzschild metric (in isotropic coordinates) in powers of r^{-1}, see (23.67) and (24.1), then we see that retention of the quadrupole term in g_{44} is only justified in exceptional cases. The higher non-linear terms with M^2 will almost always dominate; (27.34) is a good approximation only up to terms in r^{-1}. The same restriction holds for the part of $g_{\mu\nu}$ proportional to $\eta_{\mu\nu}$.

The occurrence of the angular momentum in $g_{4\nu}$ is interesting. In the Newtonian theory there is no dependence of the gravitational field upon the rotation of a celestial body. To appreciate the physical meaning of this term in the metric, we recall that a metric of the form

$$ds^2 = \eta_{\alpha\beta}\, dx^\alpha\, dx^\beta - 2g_{4\beta}\, dx^4\, dx^\beta - \left(1 + 2a_\nu x^\nu/c^2 + \cdots\right)(dx^4)^2 \quad (27.36)$$

rotates with angular velocity

$$\omega^\alpha = -\tfrac{1}{2}c\varepsilon^{\alpha\beta\nu}g_{4\beta,\nu} \qquad (27.37)$$

with respect to a local inertial system, see Section 21.2. The coordinate system used here, upon which we have based the linearized gravitational theory, and which we have to identify with the system in which the fixed stars are at rest, is thus locally a rotating coordinate system, or, conversely, the local inertial systems rotate with angular velocity

$$\Omega^\sigma = \tfrac{1}{2}c\varepsilon^{\sigma\beta\nu}g_{4\beta,\nu} = -\left(\frac{cP^\sigma}{r^3} - \frac{3x^\sigma x_\nu P^\nu c}{r^5}\right) \qquad (27.38)$$

with respect to the fixed stars. This *Lense–Thirring* (1918) *effect* can be demonstrated by the precession of a top (which adds to that of the geodesic precession (24.17)). Experiments to check this are in preparation. By the way, the analogue in electrodynamics of the components $g_{4\nu}$ of the metric and their effect is the magnetic field, which is created by currents and exerts a couple upon dipoles (Ω^a has precisely the spatial structure of the force field of a dipole).

The most important terms for the far field (27.32) of a source are those strongest at infinity, namely, those proportional to r^{-1}, that is to say, the parts of the metric containing the mass M or the second derivative of the quadrupole moment $D_{\alpha\beta}(t - r/c)$. In electrodynamics the corresponding potentials

$$
\begin{aligned}
U &= \frac{Q}{4\pi r} + \frac{x^\alpha}{4\pi r^2}\left[\frac{\dot{p}_\alpha(t - r/c)}{c} + \frac{p_\alpha(t - r/c)}{r}\right] + O(r^{-2}), \\
A_\nu &= \frac{\dot{p}_\nu(t - r/c)}{4\pi r}
\end{aligned}
\tag{27.39}
$$

represent the far field of a charge Q and an oscillating electric dipole p_α, and the terms proportional to $\dot{p}_\alpha$ lead to the radiation of electromagnetic waves. We may therefore suppose that the occurrence in the metric of $\ddot{D}_{\alpha\beta}/r$ signifies that the system is emitting gravitational waves and that, in contrast to the possibility of dipole radiation from a charge distribution, the gravitational radiation is quadrupolar in character. Both suppositions can be to a certain extent verified. In order to be able to make more exact statements one must of course go beyond the linearized theory; we shall come back to this question in the next chapter. (In electrodynamics, too, the Poynting vector, which characterizes the radiation, is quadratic in the field strengths.)

27.6 Some remarks on approximation schemes

The linearized theory can be regarded as a first step in a systematic approximation procedure. Using harmonic coordinates one starts with

$$
\bar{f}^{mn} = \sqrt{-g}\, g^{mn} - \eta_{nm}, \quad \bar{f}^{mn}{}_{,n} = 0 \tag{27.40}
$$

and finds that the (exact) Einstein field equations can be written as

$$
\Box \bar{f}^{mn} = \tau^{mn}, \tag{27.41}
$$

where $\Box$ is the flat space wave operator, and τ^{mn} contains the energy-momentum tensor T^{mn} and all terms from G^{mn} which are non-linear in $\bar{f}^{mn}$. One then assumes a development of the sources and the field with respect to a parameter λ, which numbers the orders and may or may not have a physical meaning (such as the gravitational constant, or the velocity v of the sources divided by c):

$$
\bar{f}^{mn} = \sum \lambda^k \bar{f}^{mn}_{(k)}, \quad \tau^{mn} = \sum \lambda^k \tau^{mn}_{(k)}, \tag{27.42}
$$

so that

$$\Box \bar{f}^{mn}_{(k)} = \tau^{mn}_{(k)} \qquad (27.43)$$

holds. Since the $\tau^{mn}_{(k)}$ depend only on the $\bar{f}^{mn}_{(i)}$ of lower order, $i \leq k - 1$, the equations (27.43) can be solved successively using the retarded or advanced Green's functions of the wave operator. Since the field equations are integrable only if $\tau^{mn},_n = 0$ holds, these equations have to be satisfied in each step, thus giving the equations of motion for the sources of the field. Instead of using the wave operator $\Box$ one can also take the Poisson operator $\triangle$, shifting the time-derivatives with their c^{-2} to the right-hand side; this can be appropriate when considering corrections to Newtonian gravity. (Note the sometimes strange labelling: 'order 5/2' means for example that the terms contain $(c^{-2})^{5/2} = c^{-5}$).

The actual calculations are very long and cumbersome, but necessary to understand the motion of planets or binary pulsars including the back reaction from the outgoing quadrupole radiation.

Further reading for Section 27.6

Blanchet (2002).

Exercise

27.1   Insert (27.40) into (28.20) and show that the field equations really have the form (27.41).

28

Far fields due to arbitrary matter distributions and balance equations for momentum and angular momentum

28.1 What are far fields?

The linearized theory of gravitation is based on the presumption that over *whole* regions of space, at any rate in the vicinity of the sources of the field, the gravitational field is weak, and the metric deviates only slightly from that of a Minkowski space. In nature we often meet a situation in which a distribution of matter (a satellite near the Earth, the Earth, the planetary system, our Galaxy) is surrounded by vacuum, and the closest matter is so far away that the gravitational field is weak in an *intermediate region*. In the neighbourhood of the sources, however, the field can be strong.

If such an intermediate region exists, and far away sources are not present or their influence can be neglected, then we speak of the far field of the configuration in question (Fig. 28.1). Notice that here, by contrast, for example, to most problems in electrodynamics, we may not always assume an isolated matter distribution which is surrounded only by a vacuum. The assumption of a void (the 'infinite empty space') into which waves pass and disappear contradicts the basic conception of General Relativity; also the fact that we orient our local inertial system towards the fixed stars indicates that we must always in principle take into account the existence of the whole Universe whenever we examine the properties of a part of the Universe.

While in the linearized theory we investigated solutions to the *linearized* field equations, their dependence upon the structure of the sources and their behaviour at great distances from the sources, now we are interested in approximative and exact propertries of the solutions to the *exact* field equations in regions where the gravitational fields are weak. Our goal here is to obtain statements about the system from a knowledge of the far field.

The simplest examples are gravitational fields whose far fields are independent of time. We assume that to good approximation the metric can be written as

$$g_{mn} = \eta_{mn} + a_{mn}/r + b_{mn}/r^2 + O(r^{-3}). \tag{28.1}$$

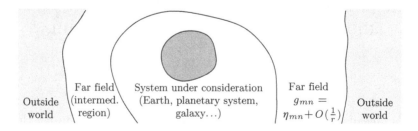

Fig. 28.1. How the far field is defined.

As the region of the far field is an annular or shell-like region and there-may be sources outside, we ought to add onto (28.1) a series with increasing powers of r. We drop these terms, however; this corresponds to the model of an isotropic external environment.

The functions a_{mn} and b_{mn}, which are independent of r and t, are to be determined by substitution of (28.1) into the vacuum field equations

$$R_{mn} = 0. \tag{28.2}$$

We can simplify these calculations by first obtaining solutions $\bar{f}_{mn}$ to the linearized field equations

$$\Delta \bar{f}_{mn} = 0, \qquad \bar{f}^{nm}{}_{,n} = 0, \tag{28.3}$$

and then calculating the non-linear corrections.

Every solution of the potential equation can be represented in the form of a multipole expansion. Thus, neglecting terms which are $O(r^{-3})$, we have

$$\bar{f}_{44} = A/r + A_\alpha x^\alpha/r^3, \qquad \bar{f}_{4\nu} = B_\nu/r + B_{\nu\alpha}x^\alpha/r^3,$$
$$\bar{f}_{\mu\nu} = \bar{f}_{\nu\mu} = C_{\nu\mu}/r + C_{\nu\mu\alpha}x^\alpha/r^3, \tag{28.4}$$

where, because of the subsidiary condition $\bar{f}^{mn}{}_{,n} = 0$, the constants are restricted by the algebraic conditions

$$B_\nu = 0, \qquad B_{\nu\alpha} = \eta_{\nu\alpha}B + \varepsilon_{\nu\alpha\beta}F^\beta,$$
$$C_{\nu\mu} = 0, \qquad C_{\nu\mu\alpha} = \delta_{\nu\mu}C_\alpha - \delta_{\mu\alpha}C_\nu - \delta_{\nu\alpha}C_\mu. \tag{28.5}$$

B and $C_{\mu\nu\alpha}$ can be eliminated by a coordinate transformation (27.9) with

$$b^4 = B/r, \qquad b^\alpha = -C^\alpha/r. \tag{28.6}$$

Experiment shows that in non-flat fields A is always non-zero (mass is always positive), so that by a shift of the origin of coordinates A_α can be transformed away as well. The linear approximation thus gives the metric

$$ds^2 = \left(\eta_{mn} + \bar{f}_{mn} - \tfrac{1}{2}\eta_{mn}\bar{f}^a_a\right) dx^n dx^m$$
$$= (1 + A/2r)\,\eta_{\alpha\beta}\,dx^\alpha\,dx^\beta + 2r^{-3}\varepsilon_{\nu\alpha\beta}x^\alpha F^\beta\,dx^\nu\,dx^4 \tag{28.7}$$
$$- (1 - A/2r)\,(dx^4)^2.$$

If we compare this expression with the metric (27.32), which we derived from the description of the fields in terms of the sources, then we see that the constants A and F^β can be identified with the mass M and the angular momentum P according to

$$A = 4M, \quad F^\beta = 2P^\beta. \tag{28.8}$$

This identification is not merely a repetition of the linearized theory. In the linearized theory mass and angular momentum were defined through the integrals (27.33) over the source distribution. Now, in the investigation of the far field of an (unknown) source, we take as definitions of the mass and the angular momentum of the source just those coefficients in the expansion of the far field which act upon a test body or a top in exactly the same way as the mass or angular momentum, respectively, of a weak source.

We have now to put into the metric (28.7) the corrections arising from the non-linearity of the Einstein equations; (28.7) is not of course a solution of the field equations (28.2), even if we ignore terms in r^{-3}. Since we are taking terms only up to r^{-2} and corrections due to the non-linearity are always weaker by at least one power of r than the original terms, we need to take into account terms quadratic in the mass parameter. However, we can obtain these by series expansion from the exact Schwarzschild solution (23.67) without performing additional calculations.

We thus obtain the result that the far field of an arbitrary, time-independent source has, in suitable coordinates, the form

$$ds^2 = \left(1 + 2M/r + 3M^2/2r^2\right)\eta_{\alpha\beta}\,dx^\alpha\,dx^\beta + 4r^{-3}\varepsilon_{\nu\alpha\beta}x^\alpha P^\beta\,dx^\nu\,dx^4$$
$$- \left(1 - 2M/r + 2M^2/r^2\right)(dx^4)^2 + O(r^{-3}), \tag{28.9}$$

in which M and P^α are regarded as the mass and the angular momentum because of the way they act upon test bodies and because of the analogy with the linearized theory.

28.2 The energy-momentum pseudotensor for the gravitational field

The problem Linearized gravitational theory and its strong analogy with electrodynamics leads one to suppose that time-varying gravitational systems emit gravitational waves. Is it possible by examining the far field to establish whether, and under what conditions, such waves exist?

In a special-relativistic field theory one would probably try to answer this question in the following fashion. The decisive factor in the existence of waves or radiation is not merely that the fields are time-dependent, but that energy, momentum and angular momentum are transported from one region of space to another. One therefore encloses the system under consideration in, for example, a spherical surface, writes down the balance equations for the above-mentioned quantities, and ascertains whether, for example, an energy current is flowing through the sphere. If this is the case, one can speak of radiation (provided that particles are not just flowing across the boundary surface and carrying with them energy, etc.). We therefore start out from the balance equations

$$T^{mn}{}_{,n} = 0, \qquad (T^{mn}x^a - T^{ma}x^n)_{,m} = 0 \qquad (28.10)$$

for momentum and angular momentum for the field under consideration.

If we want to perform a similar analysis in gravitation theory, we must construct an energy-momentum tensor of the gravitational field (not of the matter!) and derive balance equations from it. In General Relativity there is, however, just one energy-momentum tensor, namely, that due to the matter. Although its divergence vanishes,

$$T^{mn}{}_{;n} = 0, \qquad (28.11)$$

one cannot, however, deduce an integral conservation law from (28.11) in a Riemannian space, because integration is the reverse of partial, and not covariant, differentiation and one cannot apply the Gauss law to the divergence of a symmetric second-rank tensor (see Chapter 20).

Since we are supposing the existence of gravitational waves, this negative statement should not really surprise us. If energy can be transported in the form of gravitational waves, the energy of the sources alone cannot remain conserved. Rather, one would expect that in place of (28.11) there is a differential balance equation, formulated in terms of partial derivatives, of the structure

$$[(-g)(T^{mn} + t^{mn})]_{,n} = 0, \qquad (28.12)$$

which expresses the fact that a conservation law holds only for the sum of the matter (T^{mn}) and the gravitational field (t^{mn}).

The problem is, therefore, to construct a t^{mn} from the metric in such a way that (28.12) is satisfied alone as a consequence of the field equations

$$G_{nm} = R_{nm} - \tfrac{1}{2}Rg_{nm} = \kappa T_{nm}. \qquad (28.13)$$

Before turning to this problem, we want to formulate clearly the alternatives confronting us. Either we wish to deal only with tensors and allow only covariant statements, in which case we use (28.11) and can write down no balance equation for the energy transport by radiation. Or else we want such a balance equation (28.12), which can only be formulated in a non-covariant manner; as one can see from (28.12), t^{mn} is not a tensor, and we call it the *energy-momentum pseudotensor* of the gravitational field.

Since we pick out a Minkowski metric in the far field in a non-covariant fashion anyway, to begin with we shall accept the lack of covariance in (28.12), not going into its consequences until later. There are, however, good reasons for deciding to maintain covariance and to regard the question of energy transport by gravitational waves as inappropriate in the theory of gravitation, because the concept of energy has lost its meaning there (see Section 28.4).

Construction of the energy-momentum pseudotensor t^{mn} The program just described perhaps sounds plausible, but already in the initial equation (28.12) there is a flaw: t^{mn} is not uniquely determined. The addition of a term of the form $H^{ikl}{}_{,l}$ to $(-g)t^{ik}$ in no way affects the validity of (28.12), provided only that H^{ikl} is antisymmetric in k and l:

$$(-g)\tilde{t}^{ik} = (-g)t^{ik} + H^{i[kl]}{}_{,l}, \qquad \left[(-g)\tilde{t}^{ik}\right]_{,k} = \left[(-g)t^{ik}\right]_{,k}. \quad (28.14)$$

Thus, one finds in the extensive literature on this problem a whole series of different proposals, which finally, in the formulation of the conservation laws, give the same statements. We shall therefore not attempt to derive our preferred (Landau–Lifshitz) form of t^{mn}, but rather guess a trial substitution from seemingly plausible requirements and then verify its correctness.

In analogy to the properties of the energy-momentum tensors of all other fields, t^{mn} should be symmetric, it should be bilinear in the first derivatives of the metric, and it should contain no second derivatives. Furthermore, bearing in mind the field equations (28.13), equation (28.12) must be satisfied identically, that is,

$$[(-g)(G^{mn} + \kappa t^{mn})]_{,n} = 0 \tag{28.15}$$

must hold for every metric. Equation (28.15) can be satisfied most simply if we introduce a superpotential U^{mni} according to

$$U^{mni}{}_{,i} = (-g)(G^{mn} + \kappa t^{mn}), \quad U^{mni} = -U^{min}, \quad U^{mni}{}_{,i} = U^{nmi}{}_{,i}. \tag{28.16}$$

Since second derivatives of the metric occur in G^{mn}, U^{mni} should contain at most first derivatives. We can ensure this by writing U^{mni} as the divergence of a quantity U^{mnik}:

$$U^{mni} = U^{mnik}{}_{,k}, \tag{28.17}$$

which depends only upon the metric, not on its derivatives. From the symmetry requirement, the form of U^{mnik} is uniquely determined up to a factor; we make the choice

$$U^{mnik} = \tfrac{1}{2}(-g)(g^{mn}g^{ik} - g^{mi}g^{nk}).$$ (28.18)

While the validity of (28.15) is ensured because of our construction procedure, we must examine explicitly whether t^{mn} does contain no second derivatives of the metric. From (28.16), that is, from

$$\kappa(-g)t^{mn} = \tfrac{1}{2}[(-g)(g^{mn}g^{ik} - g^{mi}g^{nk})]_{ik} - (-g)G^{mn},$$ (28.19)

one obtains, after a rather long calculation,

$$
\begin{aligned}
2\kappa(-g)t^{mn} &= \tfrac{1}{8}(2g^{ml}g^{nk} - g^{mn}g^{lk})(2g_{ip}g_{qr} - g_{pq}g_{ir})\mathbf{g}^{ir}{}_{,l}\mathbf{g}^{pq}{}_{,k} \\
&\quad + \mathbf{g}^{mn}{}_{,k}\mathbf{g}^{ik}{}_{,i} - \mathbf{g}^{mi}{}_{,i}\mathbf{g}^{nk}{}_{,k} + \tfrac{1}{2}g^{mn}g_{li}\mathbf{g}^{lk}{}_{,p}\mathbf{g}^{pi}{}_{,k} \\
&\quad - g_{ik}\mathbf{g}^{ip}{}_{,l}(g^{mn}\mathbf{g}^{nk}{}_{,p} + g^{nl}\mathbf{g}^{mk}{}_{,p}) + g_{li}g^{kp}\mathbf{g}^{nl}{}_{,k}\mathbf{g}^{mi}{}_{,p}
\end{aligned}
$$ (28.20)

where the abbreviation

$$\mathbf{g}^{mn} \equiv \sqrt{-g}\, g^{mn}$$ (28.21)

has been used. The energy-momentum pseudotensor t^{mn} therefore really does have the desired properties. That we have succeeded so simply in expressing the second derivatives of the metric contained in G^{mn} by the derivatives of U^{mnik} is closely connected with the possibility mentioned in (22.48) of splitting up the Lagrangian of the gravitational field.

Properties of the energy-momentum pseudotensor The energy-momentum pseudotensor t^{mn} is not a tensor; one can see this property most clearly by noticing that at any point of space-time the energy-momentum pseudotensor can be made to vanish by the introduction of locally geodesic coordinates $g_{mn} = \eta_{mn}$, $g_{mn,a} = 0$. Therefore, if our idea of associating energy and momentum with the pure gravitational field is at all meaningful, then the gravitational energy is on no account to be thought of as localizable; it is at best a quantity which one can associate with a whole spatial region, its value at any one point being arbitrarily alterable through choice of the coordinate system.

On the other hand, the energy-momentum pseudotensor does transform like a tensor under coordinate transformations which have the formal structure of a Lorentz transformation:

$$x^{n'} = L^{n'}{}_n x^n, \quad t^{n'm'} = L^{n'}{}_n L^{m'}{}_m t^{nm}, \quad L^{n'}{}_n L_{n'}{}^a = \delta^a_n.$$ (28.22)

This property is important when the energy-momentum pseudotensor is used for the far field of an isolated matter distribution, where the space deviates only weakly from a Minkowski space and hence Lorentz transformations have a physical meaning.

28.3 The balance equations for momentum and angular momentum

We want now to use the energy-momentum pseudotensor and its super-potential to obtain global statements about the energy, momentum and angular momentum of the system under consideration, from the local balance equations for the four-momentum

$$[\kappa(-g)(T^{mn} + t^{mn})]_{,n} = U^{mni}{}_{,in} = U^{mnik}{}_{,kin} = 0, \tag{28.23}$$

and from the balance equation

$$\left\{[\kappa(-g)(T^{mn} + t^{mn})]x^a - [\kappa(-g)(T^{ma} + t^{ma})]x^n\right\}_{,m}$$
$$= (U^{mni}{}_{,i}x^a - U^{mai}{}_{,i}x^n)_{,m} = 0 \tag{28.24}$$

for angular momentum which follows from it.

To this end we integrate (28.23) and (28.24) over the region G_3 of the three-dimensional space $x^4 = \text{const.}$ indicated in Fig. 28.2, which contains the matter and which reaches into the far-field zone, and with the help of the Gauss law transform these integrals into surface-integrals over the surface Σ of G_3, giving

$$\frac{\mathrm{d}}{\mathrm{d}x^4}\left(\int_{G_3} U^{m4i}{}_{,i}\,\mathrm{d}^3x\right) = -\int_{G_3} U^{m\nu i}{}_{,i\nu}\,\mathrm{d}^3x = -\int_{\Sigma} U^{m\nu i}{}_{,i}\,\mathrm{d}f_\nu, \tag{28.25}$$

$$\frac{\mathrm{d}}{\mathrm{d}x^4}\left(\int_{G_3}(U^{4ni}{}_{,i}x^a - U^{4ai}{}_{,i}x^n)\,\mathrm{d}^3x\right) = -\int_{\Sigma}(U^{\nu ni}{}_{,i}x^a - U^{\nu ai}{}_{,i}x^n)\,\mathrm{d}f_\nu. \tag{28.26}$$

Because of the symmetry properties (28.16), $U^{m4i}{}_{,i}$ contains no time

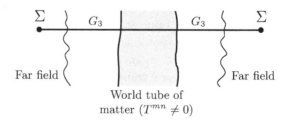

Fig. 28.2. Derivation of the balance equations.

derivative, so that the left-hand sides of these equations can also be transformed into surface integrals. We designate these integrals as

$$p^m \equiv \int_{G_3} [(-g)(T^{4m} + t^{4m})] \, \mathrm{d}^3 x = \kappa^{-1} \int_\Sigma U^{m4\nu} \, \mathrm{d}f_\nu, \quad (28.27)$$

$$
\begin{aligned}
I^{an} &\equiv \int_{G_3} [(-g)(T^{4n} x^a - T^{4a} x^n + t^{4n} x^a - t^{4a} x^n)] \, \mathrm{d}^3 x \\
&= \kappa^{-1} \int_\Sigma (U^{4n\nu} x^a - U^{4a\nu} x^n + U^{4na\nu} - U^{4an\nu}) \, \mathrm{d}f_\nu,
\end{aligned}
\quad (28.28)
$$

and notice that if T^{mn} vanishes on Σ then we obtain the relations

$$\mathrm{d}p^m/\mathrm{d}t = -c\kappa^{-1} \int_\Sigma U^{m\nu i}_{\ \ ,i} \, \mathrm{d}f_\nu = -c \int_\Sigma (-g) t^{m\nu} \, \mathrm{d}f_\nu, \quad (28.29)$$

$$
\begin{aligned}
\mathrm{d}I^{an}/\mathrm{d}t &= -c\kappa^{-1} \int_\Sigma (U^{\nu n i}_{\ \ ,i} x^a - U^{\nu a i}_{\ \ ,i} x^n) \, \mathrm{d}f_\nu \\
&= -c \int_\Sigma (-g)(t^{\nu n} x^a - t^{\nu a} x^n) \, \mathrm{d}f_\nu.
\end{aligned}
\quad (28.30)
$$

We can interpret these equations as balance equations for the momentum p^m and the angular momentum I^{an}. They state that the momentum and angular momentum of a source (of the region bounded by Σ) change when gravitational radiation is transported over the boundary surface Σ. Their particular advantage lies in the fact that all the quantities occurring need to be known only on Σ, that is, only in the far-field region.

In practice we shall identify the surface Σ with a 'sphere' $r = \mathrm{const.}$; since its surface element is given by

$$\mathrm{d}f^\nu = x^\nu r \sin \vartheta \, \mathrm{d}\vartheta \, \mathrm{d}\varphi + O(r), \quad (28.31)$$

in the far-field region (r very large) we need take into account only those contributions to the integrands which tend to zero no faster than as r^{-2}.

If we are to test the physical content of the balance equations in the example of the stationary metric (28.9), then, according to (28.27), for the calculation of the momentum we need retain only the terms in the metric proportional to r^{-1} (which give terms in r^{-2} in $U^{m4\nu}$); that is, we can use the relation

$$U^{mni} = \tfrac{1}{2} (\mathsf{g}^{mn} \mathsf{g}^{ik} - \mathsf{g}^{mi} \mathsf{g}^{nk})_{,k} = \tfrac{1}{2} (\eta^{nk} \bar{f}^{mi}_{\ \ ,k} - \eta^{ik} \bar{f}^{mn}_{\ \ ,k}), \quad (28.32)$$

valid in the linearized theory. The result of this simple calculation is

$$p^\nu = 0, \qquad p^4 = 8M\pi/\kappa = mc^2 = \mathrm{const.}; \quad (28.33)$$

that is, the spatial momentum is zero and the energy p^4 is connected with the mass m measured in the far field exactly as in the special-relativistic formula. On the other hand, only the term in $g_{4\nu}$ proportional to r^{-2} gives a contribution to the angular momentum I^{an}; one obtains

$$I^{4\nu} = 0, \qquad I^{\alpha\nu} = 8\pi\varepsilon^{\alpha\nu\beta}P_\beta/\kappa. \tag{28.34}$$

The results (28.33) and (28.34) thus confirm our ideas, in particular the interpretation of p^m as the momentum and I^{an} as the angular momentum of the system comprising matter plus gravitational field.

For time-dependent fields, momentum and angular momentum will not remain conserved. We examine this in the energy balance equation of the far field (27.32) of a time-dependent source. Since the energy-momentum pseudotensor is quadratic in the first derivatives of the metric (see (28.20)), we have to take into account in the energy law

$$\mathrm{d}p^4/\mathrm{d}t = -c\int(-g)t^4{}_\nu x^\nu r \sin\vartheta \,\mathrm{d}\vartheta \,\mathrm{d}\varphi \tag{28.35}$$

only the terms of the metric whose first derivatives go like r^{-1}, namely,

$$g^{44} \approx \eta^{44} - \bar{f}^{44} \approx -1 - \frac{\kappa}{4\pi} \frac{\dddot{D}_{\alpha\beta}x^\alpha x^\beta}{c^2 r^3},$$

$$g^{4\nu} \approx -\bar{f}^{4\nu} \approx \frac{\kappa}{4\pi} \frac{\dddot{D}_{\nu\alpha}x^\alpha}{c^2 r^3}, \qquad D_{\alpha\beta} = D_{\alpha\beta}\left(t - r/c\right), \tag{28.36}$$

$$g^{\mu\nu} \approx \eta^{\mu\nu} - \bar{f}^{\mu\nu} \approx \eta^{\mu\nu} - \frac{\kappa}{4\pi} \frac{\dddot{D}_{\nu\mu}}{c^2 r}.$$

After a simple, but rather lengthy, calculation one obtains

$$\frac{\mathrm{d}p^4}{\mathrm{d}t} = -\frac{1}{5c^2}\left(\dddot{D}_{\alpha\beta} - \tfrac{1}{3}\eta_{\alpha\beta}\dddot{D}^\sigma{}_\sigma\right)\left(\dddot{D}^{\alpha\beta} - \tfrac{1}{3}\eta^{\alpha\beta}\dddot{D}^\tau{}_\tau\right); \tag{28.37}$$

that is, the energy of the system always decreases. In the planetary system, this loss of energy through gravitational quadrupole radiation can certainly be ignored, since it is proportional to the sixth power of the frequency ω of the system. In the system of the binary pulsar 1913+16, however, this loss is significant.

The weakness of this application of the balance equations comes to light when one tries to calculate not the loss of energy but the total energy of the system emitting quadrupole waves: the corresponding integrals diverge for $r \to \infty$ if the system emits continuously (the whole space is filled with radiation). This diverging of the total energy is possible because in the linearized theory we have ignored the back reaction produced by the emission of radiation upon the motion of the sources, and consequently the system can give up energy continuously without exhausting the supply. Of course, one can put in this back reaction by hand, or better use an approximation scheme as sketched in Section 27.6, but it would be desirable to test the balance equations in the far field of an exact solution. Unfortunately, however, no exact solution is known which describes the emission of radiation by a physically reasonable system.

28.4 Is there an energy law for the gravitational field?

Because of the significance of the law of conservation of energy and (for systems which are not closed) the energy balance equation in many areas of physics, we shall examine their rôle in the theory of gravitation again, to some extent repeating the discussion of Section 28.3.

In Special Relativity, Electrodynamics, Thermodynamics, Quantum Mechanics and Quantum Field Theory it is always the case that a quantity 'energy' can be defined for a system which is constant if the system is isolated. If the system interacts with its surroundings, a balance equation can be written down so that the energy of the whole system (system plus surroundings) is again constant.

By analogy one would expect that, for example, electrical energy and energy of the gravitational field could transform into one another, their sum remaining constant (if there are no other types of interaction). The Einstein gravitational theory gives a completely different answer, however. In a *general* gravitational field there is indeed a conservation

$$T^{mn}{}_{,n} = 0 \tag{28.38}$$

for the energy and momentum of the field-producing matter law in the neighbourhood of a point, obtained upon introduction of the inertial coordinate system there (locally geodesic system). But it holds only so long as (in a region of space so small that) the curvature of the space, that is, the real gravitational effects, can be ignored. In this sense, and with this restriction, the theory of gravitation corroborates the conservation laws of special-relativistic physics.

Over larger spatial regions when the gravitational field is properly included there is no energy balance equation. It is incorrect to regard this as a violation of energy conservation; there exists in general no local covariant quantity 'energy' to which the property of conservation or non-conservation can be ascribed. None of the foundations of physics are thereby destroyed; energy is only a (very important) auxiliary quantity for describing interactions, but the interaction of all parts of the Universe is quite essential for the theory of gravitation.

The situation is rather more favourable if the gravitational field is not completely general, but possesses certain additional properties. Thus one can associate energy and momentum with a system that is separated from the rest of the Universe by a far-field zone, in the sense of Section 28.1, and for which the integrals (28.27) exist. Here these integrals assume an invariant significance through the use of Minkowski coordinates, which they do not have in a general system, in which, for example, the superpotential also exists. Then balance equations can be formulated, which for real systems, whose far fields of course do not reach to infinity, are only approximations. A localization of the energy in the interior of the system is in principle impossible.

Another important possibility for applying the concepts of energy, momentum and angular momentum occurs when the gravitational field possesses symmetries. While the local inertial system is invariant under the Lorentz group and possible translations (rotations) just correspond to the usual energy-momentum (angular-momentum) conservation law of physics, the whole space-time has symmetry properties only in exceptional cases. If, however, symmetries are present, they always correspond to conservation laws. We shall return to this problem in Chapter 33.

Further reading for Chapter 28

Misner *et al.* (1973), Landau and Lifschitz (1975).

29
Gravitational waves

29.1 Are there gravitational waves?

The existence of gravitational waves was disputed for a long time, but in recent years their existence has been generally accepted. As often in the history of a science, the cause of the variance of opinions is to be sought in a mixture of ignorance and inexact definitions. Probably in the theory of gravitation, too, the dispute will only be completely settled when a solution, for example, of the two-body problem, has been found, from which one can see in what sense such a double-star system in a Friedmann universe emits waves and in what sense it does not, and when the existence of such waves has been experimentally demonstrated.

Waves in the most general sense are time-dependent solutions of the Einstein equations; of course such solutions exist. But this definition of waves is, as we can see from experience with the Maxwell theory, rather too broad, for a field which changes only as a result of the relative motion of the source and the observer (motion past a static field) would not be called a wave. Most additional demands which a gravitational wave should satisfy lead, however, to the characterization 'radiation or transport of energy', and this is where the difficulties begin, as explained in the previous chapter, starting with the definition of energy.

In order to make the situation relatively simple, in spite of the non-linearity of the field equations, one can restrict attention to those solutions which possess a far-field zone in the sense of Section 28.1. Thus imagine the planetary system as seen from a great distance. Does this system emit gravitational waves as a consequence of the motion of the Sun and the planets? The linearized theory answers this question in the affirmative, but ignores the back reaction of the radiation upon the motion of the bodies. The general opinion of physicists is, however, that such a system tries to adjust its state (the Sun captures planets which have lost their kinetic energy by radiation) and thereby emits waves. There is little to be said against this supposition if one imagines the planetary system in an otherwise empty space. One may, however, regard the process also in the following way (see Fig. 29.1). From an initially non-spherically-symmetric field inside the far-field zone, and the external universe which (as a consequence) is also not spherically symmetric, there develops a Schwarzschild solution in the interior and a Friedmann universe in the exterior (see Chapter 41). Both parts of the universe strive to adjust their state, but whether, and in which direction, energy transport occurs through the far-field zone in unclear – neither of the two partners in the interaction is preferred in principle. It is therefore not at all certain whether a freely gravitating system (a system with exclusively gravitational interaction) emits gravitational waves.

The situation is clearer when the properties (the matter distribution) of a system are changed discontinuously by intervention from outside,

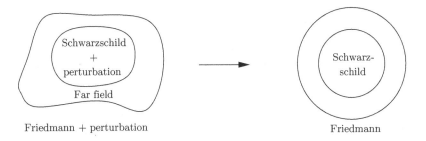

Fig. 29.1. Settling down of a perturbed gravitational system.

that is, by non-gravitational interaction, for example, by the explosion of a bomb or a supernova. The change thus produced in the gravitational field propagates out in the form of gravitational shock waves. We shall go into this again in the discussion of the initial value problem in Section 30.5.

This theoretical discussion of definitions will not interest the experimental physicist as much as the questions as to whether and how one can produce gravitational waves and demonstrate their existence. Because the gravitational constant is so small there seems no prospect at the present time of producing gravitational waves of measurable intensity by forced motion of masses. The question therefore reduces to whether stars, stellar systems or other objects in our neighbourhood are emitting gravitational waves and with what experimental arrangements one could detect these waves. We come back to this problem briefly in Section 29.4.

Exact solutions describing the interactions between the motion of the sources and the emission of radiation are not known. The considerations of the next few pages almost always deal with an analysis of the local properties of possible solutions. One introduces a local inertial system in the far field or in the neighbourhood of a point and considers (small) deviations from the Minkowski metric caused by the space curvature which have wavelike character. Even when, in Section 29.3, we discuss exact solutions, we are really dealing with an inadmissible idealization and generalization of local properties of the gravitational fields, just as for exactly plane electromagnetic waves in the Maxwell theory, which also of course can only be realized in an approximate fashion (locally).

29.2 Plane gravitational waves in the linearized theory

The waves and their degrees of freedom The simplest solutions of the linearized field equations in matter-free space,

$$\Box \bar{f}_{mn} = \eta^{ab}\bar{f}_{mn,ab} = 0, \quad \bar{f}^{mn}{}_{,n} = 0,$$
$$g_{mn} = \eta_{mn} + f_{mn}, \quad \bar{f}_{mn} = f_{mn} - \tfrac{1}{2}\eta_{mn}f^a{}_a \tag{29.1}$$

(see Section 27.2), are the plane, monochromatic gravitational waves

$$\bar{f}_{mn} = \mathrm{Re}[\hat{a}_{mn}\mathrm{e}^{\mathrm{i}k_r x^r}], \quad \hat{a}_{mn} = \mathrm{const.}, \quad k_r k^r = 0, \quad \hat{a}_{mn}k^m = 0, \tag{29.2}$$

from which (in the sense of a Fourier synthesis) all solutions of (29.1) can be obtained by superposition. (In this section indices are again shifted with the flat-space metric η_{mn}.)

The independent components $\hat{a}_{mn}$, ten in number because of the symmetry, are restricted by the four subsidiary conditions $\hat{a}_{mn}k^n = 0$. One might therefore suppose that a plane, monochromatic gravitational wave has six degrees of freedom (of polarization). But the waves (29.2) contain pure coordinate waves; these are waves whose curvature tensor vanishes identically, so that they can be eliminated by coordinate transformations. For many calculations it is convenient to get rid of these physically meaningless degrees of freedom. To this end we have at our disposal the coordinate transformations (27.9), whose generating functions,

$$b^n(x^m) = -i\hat{b}^n e^{ik_m x^m},$$ (29.3)

satisfy the wave equation

$$\Box b^n = 0,$$ (29.4)

and which effect a change of gauge

$$a_{nm} = \hat{a}_{nm} - \hat{b}_n k_m - \hat{b}_m k_n + \eta_{nm} \hat{b}_r k^r.$$ (29.5)

The four constants $\hat{b}_n$ can now be chosen so that, in addition to (29.2),

$$a_{4m} = 0 = a_{n4}$$ (29.6)

(because $a_{mn} k^n = 0$ these are three additional conditions) and

$$a_\mu{}^\mu = 0 = a_m{}^m$$ (29.7)

are satisfied. The remaining two independent components of a_{mn} cannot be transformed away, and therefore are of true physical significance.

The conditions (29.2), (29.6) and (29.7) have a simple visual interpretation. Let us choose the spatial coordinate system in such a manner that the wave propagates along the z direction, that is, k^r has only the components

$$k^r = (0, 0, \omega/c, \omega/c).$$ (29.8)

Then because of (29.2), (29.6) and (29.7) only the amplitudes a_{xx}, a_{xy} and a_{yy} of the matrix a_{mn} are non-zero, and in addition we have $a_{xx} = -a_{yy}$. The gravitational wave is therefore transverse and, corresponding to the two degrees of freedom of the wave, there are two linearly independent polarization states, which when (29.8) holds can be realized, for example, by the two choices ('linear polarization')

$$a_{xx} = -a_{yy}, \quad a_{nm} = 0 \quad \text{otherwise},$$ (29.9)

$$a_{xy} = a_{yx}, \quad a_{nm} = 0 \quad \text{otherwise}.$$ (29.10)

The result of this analysis is thus the following. Gravitational waves propagate with the speed of light (k^r is a null vector). They are transverse and possess two degrees of freedom of polarization. In the preferred coordinates (29.8) they have the metric

$$ds^2 = (1 + f_{xx})\,dx^2 + 2f_{xy}\,dx\,dy + (1 - f_{xx})\,dy^2 + dz^2 - c^2\,dt^2,$$
$$f_{xx} = a_{xx}\cos\left(wz/c - wt + \varphi\right), \quad f_{xy} = a_{xy}\cos\left(wz/c - wt + \psi\right). \tag{29.11}$$

The curvature tensor of plane gravitational waves Independent of any special gauge the curvature tensor

$$R_{ambn} = \tfrac{1}{2}(f_{an,mb} + f_{mb,an} - f_{mn,ba} - f_{ab,mn}) \tag{29.12}$$

of a plane gravitational wave always has the property

$$R_{ambn}k^n = 0, \tag{29.13}$$

as a result of the relations

$$f_{mn}k^n = \tfrac{1}{2}k_m f_a^a, \quad f_{mn,ab} = -k_a k_b f_{mn} \tag{29.14}$$

which follow from (29.1) and (29.2). The null vector k^n characterizing the wave is an eigenvector of the curvature tensor.

In the special gauge of the metric (29.11) all non-vanishing components of the curvature tensor can be expressed by

$$R_{\alpha4\beta4} = -\frac{1}{2}\frac{d^2 f_{\alpha\beta}}{c^2\,dt^2}. \tag{29.15}$$

The motion of test particles in a plane, monochromatic gravitational wave If one writes down the equation of motion

$$\frac{d^2 x^a}{d\tau^2} + \Gamma^a_{nm}\frac{dx^n}{d\tau}\frac{dx^m}{d\tau} = 0 \tag{29.16}$$

of a test particle in the coordinate system (29.11), then one finds that

$$x^\alpha = \text{const.}, \quad x^4 = c\tau \tag{29.17}$$

is a solution of the geodesic equation, because

$$\Gamma^a_{44} = \tfrac{1}{2}\eta^{ab}(2f_{b4,4} - f_{44,b}) = 0. \tag{29.18}$$

Particles initially at rest always remain at the same place; they appear to be completely uninfluenced by the gravitational wave. This initially surprising result becomes comprehensible when we remember that the curvature of space enters the relative acceleration of two test particles, and the action of the gravitational waves should therefore be detectable in this relative *acceleration* (and not in the relative *positions*).

Now which acceleration is measured by an observer at rest at the origin (O, ct) of the spatial coordinate system, who observes a particle which is at rest at the point (x^α, ct)? For the interpretation of his measurement

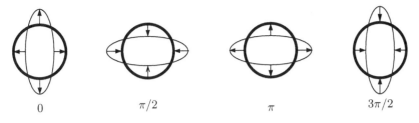

Fig. 29.2. Motion of a ring of test particles in a gravitational wave.

the observer will use not the coordinate system (29.11), but rather a local inertial system which he carries with him,

$$ds^2 = \eta_{\alpha\beta}\, d\hat{x}^\alpha\, d\hat{x}^\beta - c^2\, dt^2 + \text{higher terms}, \qquad (29.19)$$

which arises from (29.11) by the transformation

$$\hat{x}^\alpha = x^\alpha + \tfrac{1}{2} f^\alpha{}_\beta(O, ct)x^\beta. \qquad (29.20)$$

In this inertial system the test particle has the time-varying position $\bar{x}^\alpha$, and its acceleration is (see (29.15))

$$\frac{d^2 \bar{x}^\alpha}{dt^2} = \frac{1}{2}\frac{d^2 f^\alpha{}_\beta(O, ct)}{dt^2} x^\beta = -c^2 R^\alpha{}_{4\beta 4} x^\beta. \qquad (29.21)$$

Since $f_{\alpha\beta}$ has components only in the xy-plane, test particles also are only accelerated relative to one another in this plane, perpendicular to the direction of propagation of the wave. In this physical sense too, the gravitational wave is transverse. Figure 29.2 shows the periodic motion of a ring of test particles under the influence of the linearly polarized wave (29.9).

The energy-momentum pseudotensor of the plane wave In the case of the linearized plane wave the energy-momentum pseudotensor (28.20) has the simple form

$$t^{mn} = \tfrac{1}{4} a_{ir}\bar{a}^{ir} k^m k^n \sin^2 k_s x^s. \qquad (29.22)$$

Its proportionality to $k^m k^n$ is typical of a plane wave. It is found also for the electromagnetic wave (21.44) and expresses the fact that all the energy flows with the velocity of light, there is no static part.

29.3 Plane waves as exact solutions of Einstein's equations

Can one obtain exact solutions to Einstein's equations which have properties similar to those of the plane waves in the linearized theory? Before this question can be answered, it is necessary to characterize the required solutions in a covariant manner and thereby define in what sense one wants to make a generalization.

If one scrutinizes the results of the previous section, then one sees that only (29.13) is a covariant statement. One could try to start from there and characterize plane waves by

$$R_{mn} = 0, \qquad R_{ambn} k^n = 0, \qquad k_n k^n = 0. \qquad (29.23)$$

However, only the stronger conditions,

$$R_{mn} = 0, \qquad k_{a;n} = 0, \qquad k_n k^n = 0, \qquad (29.24)$$

actually give the restriction to plane waves. (One can convince oneself that (29.23) follows from (29.24)!) The requirement that $k_{a;n} = 0$ generalizes the property of plane waves in flat space of possessing parallel rays with which are associated a null vector k^n. These waves are therefore called *plane-fronted waves with parallel rays* (*pp*-waves).

We shall encounter the more general class of solutions (29.23) again in Chapter 32.

Choice of a suitable coordinate system Since $k_{[a;b]} = 0$ holds, the null vector k_a can be written as the gradient of a function u. If we identify u with the coordinate x^4, then we have

$$k_a = u_{,a}, \qquad u = x^4, \qquad k_a = (0,0,0,1). \qquad (29.25)$$

For a plane wave in flat space, which is propagating in the z-direction, u is proportional to $ct - z$. Since k_a is a null vector, g^{44} vanishes, and by coordinate transformations $x^{\alpha'} = x^{\alpha'}(x^a)$, $u' = u$ one can arrive at

$$g^{14} = g^{24} = g^{44} = 0, \qquad g^{34} = 1, \qquad (29.26)$$

and, because $g^{4a} g_{am} = \delta^4{}_m$,

$$g_{31} = g_{32} = g_{33} = 0, \qquad g_{34} = 1. \qquad (29.27)$$

The reader may verify for himself that this and the following transformations really do exist (existence theorems for partial differential equations) and do not destroy the form of the metric already obtained previously.

The null vector field k_a is covariantly constant, and from this and (29.23)–(29.27) it follows that

$$k_{a;n} = \Gamma^m_{an} k_m = -\tfrac{1}{2} g_{an,3} = 0. \qquad (29.28)$$

If we label the coordinates as $(x^n) = (x, y, v, u)$ and introduce conformally Euclidean coordinates into the two-dimensional xy-subspace (which is always possible on a surface) then we arrive at the metric

$$ds^2 = p^2(x, y, u)(dx^2 + dy^2)$$
$$+ 2m_1(x, y, u)\, dx\, du$$
$$+ 2m_2(x, y, u)\, dy\, du$$
$$+ 2m_4(x, y, u)\, du^2 + 2\, du\, dv,$$

$$g_{ab} = \begin{pmatrix} p^2 & 0 & 0 & m_1 \\ 0 & p^2 & 0 & m_2 \\ 0 & 0 & 0 & 1 \\ m_1 & m_2 & 1 & 2m_4 \end{pmatrix},$$

$$(29.29)$$

$$g^{ab} = \begin{pmatrix} p^{-2} & 0 & -m_1 p^{-2} & 0 \\ 0 & p^{-2} & -m_2 p^{-2} & 0 \\ -m_1 p^{-2} & -m_2 p^{-2} & -2m_4 + (m_1^2 + m_2^2)p^{-2} & 1 \\ 0 & 0 & 1 & 0 \end{pmatrix}.$$

Solution of the field equations Because we have

$$k_{m;b;n} - k_{m;n;b} = 0 = R^a{}_{mbn} k_a, \qquad (29.30)$$

the components $R^4{}_{mbn}$ and R_{3mbn} of the curvature tensor vanish identically, so that the field equations reduce to the five equations

$$R_{mn} = R^1{}_{m1n} + R^2{}_{m2n} = 0. \qquad (29.31)$$

Upon substitution of

$$\Gamma^1_{12} = \Gamma^2_{22} = -\Gamma^2_{11} = (\ln p)_{,2}, \quad \Gamma^2_{12} = \Gamma^1_{11} = -\Gamma^1_{22} = (\ln p)_{,1},$$
$$\Gamma^A_{3B} = 0, \quad \Gamma^4_{AB} = 0, \quad A, B = 1, 2, \qquad (29.32)$$

into the defining equation for the curvature tensor,

$$R^a{}_{mbn} = \Gamma^a_{mn,b} - \Gamma^a_{mb,n} - \Gamma^a_{rn}\Gamma^r_{mb} + \Gamma^a_{rb}\Gamma^r_{mn}, \qquad (29.33)$$

it follows that $R_{11} = 0$ and $R_{22} = 0$ are equivalent to

$$\Delta(\ln p) = \left(\frac{\partial^2}{\partial x^2} + \frac{\partial^2}{\partial y^2}\right)\ln p = 0. \qquad (29.34)$$

$\ln p$ is therefore the real part of an analytic function of $x + iy$, so that by a coordinate transformation in the xy-plane we can achieve

$$p = 1, \qquad \Gamma^A_{BC} = 0, \qquad A, B = 1, 2. \qquad (29.35)$$

If we now calculate the components $R^1{}_{412}$ and $R^2{}_{421}$ using (29.35) and

$$\Gamma^a_{3b} = 0, \qquad \Gamma^2_{14} = \tfrac{1}{2}(m_{2,1} - m_{1,2}) = -\Gamma^1_{24}, \qquad (29.36)$$

then we see that the relation

$$m_{1,2} - m_{2,1} = 2F'(u) \qquad (29.37)$$

follows from the field equations $R_{14} = R_{24} = 0$. With the aid of the coordinate transformations

$$v = \bar{v} - \int m_1 \, \mathrm{d}x + F'(u)xy,$$
$$\bar{x} = x \cos F(u) + y \sin F(u), \quad \bar{y} = -x \sin F(u) + y \cos F(u), \qquad (29.38)$$

this enables us to introduce the simplified form

$$\mathrm{d}s^2 = \mathrm{d}x^2 + \mathrm{d}y^2 + 2 \, \mathrm{d}u \, \mathrm{d}v + H(x, y, u) \, \mathrm{d}u^2 \qquad (29.39)$$

of the metric (the bar on the new coordinates has been dropped). The remaining field equation yet to be satisfied, $R_{44} = 0$, then reads

$$\Delta H = \left(\frac{\partial^2}{\partial x^2} + \frac{\partial^2}{\partial y^2} \right) H(x, y, u) = 0. \qquad (29.40)$$

Metrics (29.39) which satisfy this relation are the most general plane-fronted waves with parallel rays.

Properties of plane-fronted waves with parallel rays In the coordinates

$$u = (z - ct)/\sqrt{2}, \qquad v = (z + ct)/\sqrt{2} \qquad (29.41)$$

flat space has the line element

$$\mathrm{d}s^2 = \mathrm{d}x^2 + \mathrm{d}y^2 + 2 \, \mathrm{d}u \, \mathrm{d}v. \qquad (29.42)$$

Comparing this expression with the gravitational wave (29.39), one can see that the wave is plane also in the intuitive sense that the characteristic function H depends upon the time only in the combination $z - ct$.

The general manifold of solutions also contains special wave-packets

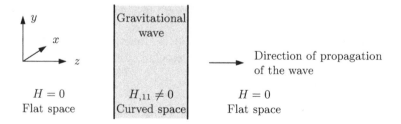

Fig. 29.3. A special wave-packet.

which are so constructed that the space before and after passage of the wave is flat, see Fig. 29.3.

Since there is no potential function which is regular over the whole xy-plane, H always possesses singularities (the only exception $H = H(u)$ leads to a flat four-dimensional space). To avoid such singularities it is in many cases convenient to use another coordinate system. For the simplest form of a wave

$$H = (x^2 - y^2)h(u),$$ (29.43)

for example, the coordinate transformation

$$x = \bar{x}a(u), \quad y = \bar{y}b(u), \quad v = \bar{v} - \tfrac{1}{2}a'a\bar{x}^2 - \tfrac{1}{2}bb'\bar{y}^2,$$
$$a''/a = h(u) = -b''/b,$$ (29.44)

leads to the line element (the dash on the coordinates has been suppressed after the transformation)

$$ds^2 = a^2(u)\,dx^2 + b^2(u)\,dy^2 + 2\,du\,dv, \quad a''b + ab'' = 0,$$ (29.45)

with a metric regular over the whole xy-plane.

To end this discussion we shall compare the exact solution with the plane waves of the linearized theory. If we once more go through the derivation of the metric (29.39) of the exact solution, or if we substitute this metric immediately into the field equations, then surprisingly we can establish that no quadratic expressions of any kind in H or its derivatives occur. The exact solution (29.39) is therefore also a solution of the linearized field equations, and it even satisfies the gauge conditions $(\sqrt{-g}\,g^{mn})_{,n} = 0$, but not always of course the requirement $|H| \ll 1$. If we want to compare exact solutions and approximate solutions in detail, then we must linearize the exact solution; in the case of the solution (29.45) this can be done by carrying out the substitution

$$a = 1 + \alpha/2, \quad b = 1 - \beta/2, \quad \alpha \ll 1, \ \beta \ll 1,$$ (29.46)

and ignoring higher terms in α and β. Because of (29.45) we have $\alpha'' = \beta''$, and hence $\alpha = \beta + c_1 u + c_2$, but then c_1 must be zero (the coordinate u can become arbitrarily large!) and c_2 can be eliminated by a coordinate transformation. Therefore the linearized form of (29.45) is

$$ds^2 = [1 + \alpha(u)]dx^2 + [1 - \alpha(u)]dy^2 + 2\,du\,dv.$$ (29.47)

As a comparison with (29.9)–(29.11) shows, we are dealing with a linearly polarized packet of plane waves of differing frequencies.

29.4 The experimental evidence for gravitational waves

Gravitational waves change the curvature of space-time. They can in principle be detected by the change in the trajectories of particles (mirrors, satellites, planets, ...) or the oscillations they produce in mechanical or electromagnetic systems. Only when large masses are rapidly accelerated does one expect that the resulting gravitational waves are of detectable strength. Such processes could occur, for example, in gravitational collapse (see Chapter 36), in a supernova explosion of a star, in a rapidly moving binary stellar systems or in processes near to black holes.

The first experimental search for gravitational waves was initiated by J. Weber in 1961. His 'aerial' consisted of an aluminium cylinder 1.53 m long and with radius 0.33 m; waves arriving at the cylinder transversally would cause length oscillations. The initially observed 'events' could not be reproduced, in spite of a greatly improved experimental procedure and a cryogenic environment. Current technology of this type can measure relative displacements $h \approx 10^{-18}$ (for millisecond pulses), corresponding to length perturbations of 10^{-16}cm, that is, a thousandth of the radius of the nucleus of an atom.

Most gravitational detectors now built or under consideration use laser interferometry: they measure the displacement of freely suspended mirrors in a Michelson-type interferometer. To achieve the planned sensitivity of $h \approx 10^{-21}$, the arm-lengths of the interferometers have to be large: 600 m in the recently built GEO 600 detector near to Hannover, 4 km in LIGO (Laser Interferometer Gravitational-wave Observatory) in the USA, 3 km in the French–Italian VIRGO project, and $5 \cdot 10^6$ km in the space-borne LISA (Laser Interferometer Space Antenna) project, see Rowan and Hough (2000) for details.

Although in the last 40 years the sensitivity of the receivers has been improved by an order of 10^6, in the same time the theoretical predictions about the wave magnitude to be expected have been revised and say that – as in the 1960s – the receivers are just one order of magnitude less sensitive than they ought to be. Gravitational waves have not yet been detected.

An indirect proof of the existence of gravitational waves arises from the very precise data from observations of the pulsar PSR 1913+16. This rapidly rotating binary system should emit appreciable amounts of gravitational quadrupole radiation, thereby lose energy and hence rotate faster. The observed relative change in period of $-2.422\,(\pm0.006)\cdot10^{-12}$ is in remarkable agreement with the theoretical value.

Exercises

29.1   Show that (29.24) implies (29.23).

29.2   Mr. X claims to have found a particularly interesting but simple solution of Einstein's equations, namely the wave $ds^2 = dy^2 + [1 - \sin(z - ct)]dz^2 + 2\sin(z - ct)dz\,dct - [1 + \sin(z - ct)]c^2\,dt^2$. Is he right?

30
The Cauchy problem for the Einstein field equations

30.1 The problem

The basic physical laws mostly have a structure such that from a knowledge of the present state of a system its future evolution can be determined. In mechanics, for example, the trajectory of a point mass is fixed uniquely by specifying its initial position and initial velocity; in quantum mechanics, the Schrödinger equation determines the future state uniquely from the present value of the ψ function.

As we shall see in the following sections, the equations of the gravitational field also have such a causal structure. In order to appreciate this we must first clarify what we mean by 'present' and 'present state'. As a preliminary to this we examine the properties of a three-dimensional surface in a four-dimensional space. In the later sections we shall concern ourselves with the initial value problem mainly in order to gain a better understanding of the structure of the field equations.

30.2 Three-dimensional hypersurfaces and reduction formulae for the curvature tensor

Metric and projection tensor   Suppose we are given a three-dimensional hypersurface in a four-dimensional Riemannian space which can be imagined as an element of a family of surfaces; the normal vectors n^a to this family of surfaces must not be null:

$$n_a n^a = \varepsilon = \pm 1. \tag{30.1}$$

Let us take these surfaces as the coordinate surfaces $x^4 = $ const. of a coordinate system that is not necessarily orthogonal and denote the components of the normal vectors by

$$n_a = (0, 0, 0, \varepsilon N), \quad n^a = (-N^\alpha/N, 1/N),$$
$$a, b, \ldots = 1, \ldots, 4, \quad \alpha, \beta, \ldots = 1, \ldots, 3. \tag{30.2}$$

Then the metric tensor $g_{\alpha\beta}$ of the hypersurface,

$$\overset{(3)}{\mathrm{d}s}{}^2 = g_{\alpha\beta}\,\mathrm{d}x^\alpha\,\mathrm{d}x^\beta, \tag{30.3}$$

and the metric tensor g_{ab} of the four-dimensional space are related by

$$\overset{(4)}{\mathrm{d}s}{}^2 = g_{ab}\,\mathrm{d}x^a\,\mathrm{d}x^b = g_{\alpha\beta}(\mathrm{d}x^\alpha + N^\alpha\,\mathrm{d}x^4)(\mathrm{d}x^\beta + N^\beta\,\mathrm{d}x^4) + \varepsilon(N\,\mathrm{d}x^4)^2, \tag{30.4}$$

from which we obtain for the inverse tensors

$$g^{ab} = \begin{pmatrix} g^{\alpha\beta} + \varepsilon N^\alpha N^\beta/N^2 & -\varepsilon N\alpha/N^2 \\ -\varepsilon N^\beta/N^2 & \varepsilon/N^2 \end{pmatrix}, \quad \begin{matrix} g_{\alpha\beta}g^{\beta\nu} = \delta_\alpha^\nu, \\ N_\alpha = g_{\alpha\beta}N^\beta. \end{matrix} \tag{30.5}$$

With the help of the projection tensor $h_{ab} = g_{ab} - \varepsilon n_a n_b$, which has the properties

$$h_{ab}h^b{}_c = h_{ac}, \quad h_{ab}n^a = 0, \quad h_{\alpha\beta} = g_{\alpha\beta}, \quad h^{\alpha\beta} = g^{\alpha\beta}, \quad h^4_b = 0, \tag{30.6}$$

we can decompose every tensor into its components parallel or perpendicular to the vector normal to the hypersurface.

The extrinsic curvature tensor K_{ab} In making the splitting

$$n_{a;b} = n_{a;i}(\varepsilon n^i n_b + h^i_b) \tag{30.7}$$

of the covariant derivative of the normal vector we encounter the tensor K_{ab} defined by

$$K_{ab} = -n_{a;i}h^i{}_b = -n_{a;b} + \varepsilon \dot{n}_a n_b. \tag{30.8}$$

Since n_a is a unit vector and is proportional to the gradient of a family of surfaces, K_{ab} is symmetric; it has of course no components in the direction of the normal to the surface:

$$K_{ab} = K_{ba}, \quad K_{ab}n^a = 0. \tag{30.9}$$

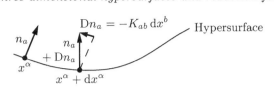

Fig. 30.1. Extrinsic curvature of a hypersurface.

Its components are linear combinations of the Christoffel symbols $\overset{4}{\Gamma}{}^4_{\alpha\beta}$ of the four-dimensional space

$$
K_{ab} = -h^i_a h^k_b n_{i;k} = \varepsilon N h^\alpha_a h^\beta_b \overset{4}{\Gamma}{}^4_{\alpha\beta},
$$

$$
K_{\alpha\beta} = \varepsilon N \overset{4}{\Gamma}{}^4_{\alpha\beta} = (N_{\alpha,\beta} + N_{\beta,\alpha} - 2N_\nu \overset{3}{\Gamma}{}^\nu_{\alpha\beta} - g_{\alpha\beta,4})/2N.
$$

(30.10)

Figure 30.1 shows that the tensor K_{ab} has a simple geometrical meaning; under a shift of the normal vector along the hypersurface we have

$$
\mathrm{D}n_a = n_{a;\beta}\,\mathrm{d}x^\beta = n_{a;i}h^i_\beta\,\mathrm{d}x^\beta = -K_{ab}\,\mathrm{d}x^b.
$$

(30.11)

K_{ab} is therefore a measure of the *extrinsic* curvature of the surface, that is, of the curvature in relation to the surrounding space (in contrast to the *intrinsic* curvature, which is characterized by the three-dimensional curvature tensor $\overset{3}{R}{}_{\alpha\beta\gamma\delta}$ of the surface alone.) In the theory of surfaces the tensor K_{ab} is associated with the *second fundamental form*.

Decomposition of the derivative of a vector perpendicular to n_a For the covariant derivative of an arbitrary vector T_a orthogonal to the normal vector n_a which obeys

$$
T_a n^a = 0, \qquad T^a = (T^\alpha, 0), \qquad T_a = (T_\alpha, T_\beta N^\beta),
$$

(30.12)

one obtains, after a short calculation using (30.9) and $n_{a;b}T^a = -n^a T_{a;b}$, the decomposition

$$
T_{a;b} = h^i_a h^k_b T_{i;k} + \varepsilon n_b \dot{T}_i h^i_a + \varepsilon n_a T^i K_{ib} - n_a n_b T^i \dot{n}_i.
$$

(30.13)

Because $h^4_a = 0$ and

$$
h^i_a h^k_b T_{i;k} = h^\alpha_a h^\beta_b (T_{\alpha,\beta} - T_r \overset{4}{\Gamma}{}^r_{\alpha\beta})
$$
$$
= h^\alpha_a h^\beta_b \left[T_{\alpha,\beta} - \tfrac{1}{2}T^\rho(g_{\rho\beta,\alpha} + g_{\rho\alpha,\beta} - g_{\alpha\beta,\rho})\right],
$$

(30.14)

the first term of this decomposition, which is wholly orthogonal to the normal vector, can be expressed in terms of the covariant derivative of the three-vector T_α with respect to the three-dimensional metric $g_{\alpha\beta}$:

$$h_a^i h_b^k T_{i;k} = h_a^\alpha h_b^\beta T_{\alpha||\beta}, \quad T_{\alpha||\beta} \equiv T_{\alpha,\beta} - \overset{3}{\Gamma}{}^\rho_{\alpha\beta} T_\rho. \tag{30.15}$$

In the derivation of this relation only the definition of the covariant derivative and the orthogonality of T_i to n_i have been used, and therefore analogous equations hold for the projections of the derivative of arbitrary tensors of higher rank perpendicular to the normal vector.

Reduction formulae for the curvature tensor The aim of the following calculations is to set up relations between the curvature tensor of the four-dimensional space and the properties of the hypersurface, that is, between $\overset{4}{R}{}_{abmn}$ on the one hand, and $\overset{3}{R}{}_{\alpha\beta\mu\nu}$ and the quantities n_a and K_{ab} on the other.

Because of (30.15), (30.13) and the equation

$$h_{ab;i} = -(\dot{n}_a n_b + n_a \dot{n}_b)n_i + \varepsilon(K_{ai} n_b + K_{bi} n_a), \tag{30.16}$$

which follows from (30.8), we have

$$\begin{aligned}
(T_{\beta||\mu||\nu} &- T_{\beta||\nu||\mu})h_b^\beta h_m^\mu h_p^\nu \\
&= (T_{r;s} h_i^r h_k^s)_{;q} h_b^i (h_m^k h_p^q - h_p^k h_m^q) \\
&= (T_{r;s;q} - T_{r;q;s})h_b^r h_m^s h_p^q + T_{r;s}(h_i^r h_k^s)_{;q} h_b^i (h_m^k h_p^q - h_p^k h_m^q)
\end{aligned} \tag{30.17}$$

or

$$\overset{3}{R}{}_{\alpha\beta\mu\nu} T^\alpha h_b^\beta h_m^\mu h_p^\nu = \overset{4}{R}{}_{\alpha r s q} T^\alpha h_b^r h_m^s h_p^q + \varepsilon(K_{pb}K_{m\alpha} - K_{bm}K_{p\alpha})T^\alpha. \tag{30.18}$$

Since this equation holds for every vector T^α, the relation

$$\overset{4}{R}{}_{\alpha\beta\mu\nu} = \overset{3}{R}{}_{\alpha\beta\mu\nu} + \varepsilon(K_{\beta\mu}K_{\alpha\nu} - K_{\beta\nu}K_{\mu\alpha}) \tag{30.19}$$

holds between the curvature tensors (remember that $h_\sigma^\beta = \delta_\sigma^\beta$, $h_a^4 = 0$). In the theory of surfaces one refers to the analogous relation between the intrinsic and extrinsic curvatures of a surface as *Gauss's equation*.

We obtain expressions for the remaining components of the four-dimensional curvature tensor by making similar transformation of the second derivatives of the normal vector. From (30.8), (30.15) and (30.16) we have first of all

$$\begin{aligned}
(n_{q;r;s} - n_{q;s;r})h_b^q h_m^r h_p^s &= \left[(n_{q;i} h_r^i)_{;s} - (n_{q;i} h_s^i)_{;r}\right] h_b^q h_m^r h_p^s \\
&= (K_{qs;r} - K_{qr;s})h_b^q h_m^r h_p^s \\
&= (K_{\beta\nu||\mu} - K_{\beta\mu||\nu})h_b^\beta h_m^\mu h_p^\nu,
\end{aligned} \tag{30.20}$$

and from this follows the *Codazzi equation*:

$$\overset{4}{R}{}^a{}_{\beta\mu\nu}n_a = K_{\beta\nu||\mu} - K_{\beta\mu||\nu}. \tag{30.21}$$

Analogously, from

$$(n_{q;r;s} - n_{q;s;r})h_b^q n^r h_p^s$$
$$= \left[(\varepsilon\dot{n}_q n_r - K_{qr})_{;s} - (\varepsilon\dot{n}_q n_s - K_{qs})_{;r}\right]n^r h_b^q h_p^s \tag{30.22}$$
$$= \dot{n}_{q;s}h_b^q h_p^s + K_{qr}n^r{}_{;s}h_b^q h_p^s - \varepsilon\dot{n}_b\dot{n}_p + K_{bp;r}n^r - K_{qs}(h_b^q h_p^s)_{;r}n^r,$$

we obtain finally

$$\overset{4}{R}{}^a{}_{\beta m\nu}n_a n^m = \dot{n}_{(\beta;\nu)} + K_{\beta\mu}K^\mu{}_\nu - \varepsilon\dot{n}_\beta\dot{n}_\nu + \mathcal{L}_\mathbf{n}K_{\beta\nu}. \tag{30.23}$$

The reduction formulae (30.19), (30.21) and (30.23) are frequently used for expressing the curvature tensor of a metric

$$ds^2 = g_{\alpha\beta}\,dx^\alpha\,dx^\beta + \varepsilon N^2 (dx^4)^2, \quad \varepsilon = \pm 1, \tag{30.24}$$

in terms of the three-dimensional subspace (the metric $g_{\alpha\beta}$) and the function N. In this special case $(N^\alpha = 0)$ the equations simplify to

$$\begin{aligned}
K_{\alpha\beta} &= -g_{\alpha\beta,4}/2N, \\
\overset{4}{R}{}_{\alpha\beta\mu\nu} &= \overset{3}{R}{}_{\alpha\beta\mu\nu} + \varepsilon(K_{\beta\mu}K_{\alpha\nu} - K_{\beta\nu}K_{\alpha\mu}), \\
\overset{4}{R}{}^4{}_{\beta\mu\nu} &= \varepsilon(K_{\beta\nu||\mu} - K_{\beta\mu||\nu})/N, \\
\overset{4}{R}{}^4{}_{\beta4\nu} &= \varepsilon K_{\beta\nu,4}/N - N_{,\beta||\nu}/N + \varepsilon K_{\alpha\beta}K^\alpha{}_\nu.
\end{aligned} \tag{30.25}$$

30.3 The Cauchy problem for the vacuum field equations

We are now in a position to be able to answer the following question. Given a spacelike surface, that is, a surface with a timelike normal vector n_a: which initial values of a metric *can* one specify on this surface and which *must* one prescribe in order to be able to calculate the subsequent evolution of the system with the aid of the vacuum field equations?

It is clear *ab initio* that, independent of the choice of the initial values, the *metric* of the space-time cannot be determined uniquely; we can carry out arbitrary coordinate transformations on the initial surface as in the whole four-dimensional space. Only certain characteristic geometrical properties will be specifiable which then evolve with time in a way which can be determined. For example, one can show that the quantities N^α of the metric (30.4) must be given not just on the initial surface $x^4 = \text{const.}$, but in the whole space, in order to fix the metric uniquely. To simplify the calculations, in the following we shall start from $N^\alpha = 0$; that is, we shall restrict ourselves to the time-orthogonal coordinates (30.24) with $\varepsilon = -1$. The essential results of the analysis of the initial value problem are unaffected by this specialization.

Vacuum field equations Let us write down the Einstein field equations $R_{bm} = 0$ for the metric (30.24), that is, for

$$\mathrm{d}s^2 = g_{\alpha\beta}\,\mathrm{d}x^\alpha\,\mathrm{d}x^\beta - N^2\big(\mathrm{d}x^4\big)^2,\tag{30.26}$$

using the reduction formulae (30.25) and

$$\overset{4}{R}_{bm} = (h^{ra} - n^r n^a)\,\overset{4}{R}_{abrm} = \overset{4}{R}{}^4{}_{b4m} + g^{a\nu}\,\overset{4}{R}_{ab\nu m}.\tag{30.27}$$

Then, after making a useful rearrangement, we obtain

$$\begin{aligned}
\overset{4}{R}{}^4_4 - g_{\beta\mu}\,\overset{4}{R}{}^{\beta\mu} &= -\overset{3}{R} - K^\beta_\beta K^\mu_\mu + K_{\beta\mu}K^{\beta\mu} = 0,\\[4pt]
N\overset{4}{R}{}^4_\mu &= K^\beta{}_{\mu||\beta} - K^\beta{}_{\beta||\mu} = 0,
\end{aligned}\tag{30.28}$$

and

$$\overset{4}{R}_{\beta\mu} = \overset{3}{R}_{\beta\mu} - 2K_{\alpha\mu}K^\alpha{}_\mu + K^\alpha_\alpha K_{\beta\mu} - K_{\beta\mu,4}/N - N_{,\beta||\mu}/N = 0.\tag{30.29}$$

Initial values and dynamical structure of the field equations The Einstein field equations are second-order differential equations; accordingly one would expect to be able to specify the metric and its first derivatives with respect to time (x^4) on an initial surface $x^4 = 0$ and hence calculate the subsequent evolution of the metric with time. As examination of the field equations (30.28) and (30.29) shows, this surmise must be made precise in the following way.

(1) In order to be able to calculate the highest time derivatives occurring in equation (30.29), namely, $K_{\beta\mu,4}$, one must know the metric $(g_{\alpha\beta}, N)$ and its first time derivatives $(K_{\alpha\beta})$; that is, one must specify these quantities on the hypersurface $x^4 = 0$.

(2) The field equations (30.28) contain only spatial derivatives of $g_{\alpha\beta}$ and $K_{\alpha\beta}$, and consequently the initial values $g_{\alpha\beta}(x^\nu, 0)$ and $K_{\alpha\beta}(x^\nu, 0)$ cannot be freely chosen. The equations (30.28) thus play the rôle of subsidiary conditions ('*constraints*'), limiting the degrees of freedom contained in the initial value data, namely, the intrinsic and the extrinsic curvature of the three-dimensional space.

(3) It is not possible to determine the time derivative of N with the aid of the field equations from the initial values $g_{\alpha\beta}(x^\nu, 0)$, $K_{\alpha\beta}(x^\nu, 0)$

and $N(x^\nu, 0)$. Rather, the function $N = \sqrt{-g_{44}}$ must be specified for all times (had we used the N_α of the general form (30.4) of the metric, we would have found the same for them). Since one can always achieve $N = 1$ by coordinate transformations (introduction of Gaussian coordinates), this special rôle played by N becomes understandable: N does not correspond to a true dynamical degree of freedom.

(4) If one has specified N and the initial values of $g_{\alpha\beta}$ and $K_{\alpha\beta}$, bearing in mind the four subsidiary conditions (30.28), then from the six field equations (30.29) one can calculate the subsequent time evolution of the metric. The equations (30.29) are therefore also called the true dynamical field equations.

The Bianchi identities,

$$\overset{4}{R}{}^a{}_{b;a} = \left[(h^a_i - n^a n_i) \overset{4}{R}{}^i{}_b \right]_{;a} = 0, \tag{30.30}$$

ensure, because of the equation

$$-(n_i \overset{4}{R}{}^i{}_b)_{;a} n^a = (N \overset{4}{R}{}^4{}_b)_{;4}/N = N K^\nu_\nu \overset{4}{R}{}^4{}_b - (\overset{4}{R}{}^\alpha{}_b)_{;\alpha} \tag{30.31}$$

that follows from them, that the subsidiary conditions (30.28) are satisfied not just for $x^4 = 0$ but for all times. That is to say, if the dynamical equations $\overset{4}{R}{}_{\beta\mu} = 0$ are satisfied for all times and if $R^4{}_b = 0$ for $x^4 = 0$, then because of (30.31) the time derivative of $R^4{}_b$ also vanishes (and with it all higher time derivatives), and so $R^4{}_b = 0$ holds always.

The splitting of the field equations into subsidiary conditions and dynamical equations, and the questions of which variables of the gravitational field are independent of one another, play an important part in all attempts at quantizing the gravitational field.

30.4 The characteristic initial value problem

From the initial values of the metric and its first derivatives in the direction normal to the surface we could in principle calculate the metric in the whole space-time, because the vacuum field equations gave us the *second* derivatives in the direction normal to the surface as functions of the initial value data. In this context it was of only secondary importance that the surface normal was timelike ($\varepsilon = -1$).

The situation is completely different, however, if the initial surface $u = x^4 = \text{const.}$ is a null surface, that is, a surface whose normal $k_a = u_{,a}$ is a null vector. Because

$$k_a = u_{,a} = (0, 0, 0, 1), \quad k_a k^a = g^{ab} k_a k_b = 0 \tag{30.32}$$

we have $g^{44} = 0$ and furthermore, by the coordinate transformation $x^{\nu'} = x^{\nu'}(x^{\alpha}, u)$, $u' = u$, we can also achieve $g^{41} = 0 = g^{42}$. In these preferred coordinates we therefore have (remembering that $k^a = (0, 0, k^3, 0)$ and $g_{ab}g^{bi} = \delta^i_a$)

$$g^{44} = g^{41} = g^{42} = 0, \quad g_{31} = g_{32} = g_{33} = 0, \quad g_{34}g^{34} = 1. \qquad (30.33)$$

Since second derivatives enter the curvature tensor only in the combination

$$\overset{4}{R}_{mabn} = \tfrac{1}{2}(g_{ab,mn} + g_{mn,ab} - g_{an,mb} - g_{mb,an}) + \cdots, \qquad (30.34)$$

second derivatives with respect to $u = x^4$ occur only in the field equation

$$\overset{4}{R}_{44} = -\tfrac{1}{2}g^{\mu\nu}g_{\mu\nu,44} + \cdots = 0, \qquad (30.35)$$

while the remaining nine field equations,

$$\overset{4}{R}_{4\alpha} = 0, \quad \overset{4}{R}_{\alpha\beta} = 0, \qquad (30.36)$$

contain at most first derivatives with respect to x^4.

Although through the choice (30.33) of coordinates we have more or less eliminated the unphysical degrees of freedom tied up with possible coordinate transformations, the field equations are in no way sufficient to calculate all second derivatives of the metric from the metric and its first derivatives. The *characteristic* initial value problem, that is, the initial value problem for a null hypersurface, differs fundamentally from the usual Cauchy problem, that is, from the initial value problem for a spacelike surface.

We shall not go into details here, but instead just clarify the physical reasons for this difference by reference to the example of plane waves

$$ds^2 = dx^2 + dy^2 + 2\,du\,dv + H\,du^2,$$

$$\left(\frac{\partial^2}{\partial x^2} + \frac{\partial^2}{\partial y^2}\right)H = 0, \quad H_{,n}k^n = \frac{\partial H}{\partial v} = 0 \qquad (30.37)$$

discussed in Section 29.3. If for $u = 0$ we were to specify the initial values of $H = g_{44}$ as a function $H(x, y, v, 0)$ which is initially arbitrary, then we could not determine the subsequent behaviour of the function H from these values; nor would the additional specification of derivatives with respect to u change anything. The field equations (30.37) only give conditions for the initial values, the dependence of the metric upon u remaining undetermined. Physically this indeterminacy is connected with the possible occurrence of gravitational shock waves, and thus of waves whose amplitude H is zero outside a finite region of u. For an observer in an inertial system a null surface $u = z - ct = 0$ is of course a two-dimensional surface which moves at the speed of light. A wave-front of a gravitational shock wave which is parallel to this surface will not be noticed on the surface (there is no point of intersection; see Fig. 30.2). An observer who knows only the metric on the surface $u = 0$ cannot predict the arrival of the shock wave. A spacelike surface, on the other hand, would intersect the shock wave somewhere; that is to say, from the initial data on such a surface the subsequent course of the wave can be determined (if one knows H on the surface $t = 0$ for *all* values of z, then H is known as a function of $u = z - ct$).

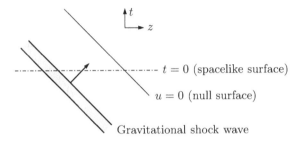

Fig. 30.2. A gravitational shock wave and the characteristic and usual initial value problems.

30.5 Matching conditions at the boundary surface of two metrics

In solving field equations one is often faced in practice with the problem of joining together two metrics obtained in different regions of space-time; for example, of joining a solution of the field equations $R_{ab} - Rg_{ab}/2 = \kappa T_{ab}$, valid inside a star, with that of the vacuum equations $R_{ab} = 0$, appropriate to the region outside.

Clearly it is not necessary for all components of the energy-momentum tensor to be continuous on the boundary surface. But what continuity properties must the energy-momentum tensor and the metric and its derivatives have in order that one can meaningfully speak of a solution to Einstein's equations?

We shall now deal with this problem under two restrictions: the boundary surface should not be a null surface (where even in the vacuum, pure discontinuities of the metric, that is, gravitational shock waves, can occur), and the energy-momentum tensor may indeed be discontinuous but should contain no δ-function singularities (a surface layer structure should not occur). Further, we want to simplify the formulae by using a coordinate system of the form

$$\mathrm{d}s^2 = g_{\alpha\beta}\,\mathrm{d}x^\alpha\,\mathrm{d}x^\beta + \varepsilon N^2 (\mathrm{d}x^4)^2 \tag{30.38}$$

on both sides of the boundary surface $x^4 = \mathrm{const}$.

We can obtain a qualitative statement about the results to be expected by the following consideration. If certain components of the energy-momentum tensor are discontinuous, then, because of the field equations, the components of the curvature tensor are at worst discontinuous. But if the second derivatives of the metric are at most discontinuous, then the metric and its first derivatives must be continuous.

When making this statement quantitative, one must note that by a clumsy choice of coordinates artificial discontinuities can be produced in the metric. The boundary surface between the two spatial regions I and II should of course be a reasonable surface, that is, whether it be approached from I or II it must always show the same metrical properties. To avoid unnecessary singularities we shall introduce the same coordinate system on both sides of the boundary surface, that is, on this surface we demand that

$$[g_{\alpha\beta}] \equiv \underset{I}{g_{\alpha\beta}} - \underset{II}{g_{\alpha\beta}} = 0. \tag{30.39}$$

Clearly all derivatives $g_{\alpha\beta,\nu\mu...}$ of this metric should also be continuous in the surface, particularly the curvature tensor $\overset{3}{R}{}_{\alpha\beta\mu\nu}$. We can make the function $N = \sqrt{\varepsilon g_{44}}$ continuous as well by suitable coordinate transformations, or even transform it to unity; here, however, we shall allow discontinuities, but no singularities.

Further statements about the continuity behaviour of the metric can be obtained from the field equations. As the reduction formulae (30.25) show, second derivatives of the metric in the direction of the surface normal are contained only in the components $\overset{4}{R}{}^4{}_{\beta4\nu}$ of the curvature tensor; they consequently enter the spatial part of the field equations in the combination

$$G^\alpha_\beta = \varepsilon(K^\alpha_\beta - \delta^\alpha_\beta K^\nu_\nu)_{,4}/N + \hat{G}^\alpha_\beta(K_{\mu\nu}, g_{\mu\nu}, g_{\mu\nu,\lambda}, N, N_{,\lambda}, \ldots) = \kappa T^\alpha_\beta. \tag{30.40}$$

Since in (30.40) neither T^α_β nor $\hat{G}^\alpha_\beta$ will be singular on the boundary surface, $K^\alpha_\beta - \delta^\alpha_\beta K^\nu_\nu$, and hence $K_{\alpha\beta}$ itself, must be continuous:

$$[K_{\alpha\beta}] = \underset{I}{K_{\alpha\beta}} - \underset{II}{K_{\alpha\beta}} = 0. \tag{30.41}$$

While (30.39) ensures the equality of the intrinsic curvature on both sides of the boundary surface, (30.41) demands equality of the extrinsic curvature too.

When the two matching conditions (30.39) and (30.41) are satisfied, then, because of (30.25), $\overset{4}{R}{}_{\alpha\beta\mu\nu}$ and $N\overset{4}{R}{}^4{}_{\beta\mu\nu}$ are also continuous; because of the field equations we must then also have

$$[T^4_4] = 0, \qquad [NT^4_\alpha] = 0. \tag{30.42}$$

To summarize: if on the boundary surface $x^4 = $ const. of two metrics of the form (30.38) the energy-momentum tensor is non-singular, then the metric $g_{\alpha\beta}$ and the extrinsic curvature $K_{\alpha\beta} = -g_{\alpha\beta,4}/2N$ of the surface, as well as the components $T^4_4 = \varepsilon n_a n^b T^a_b$ and $NT^4_\alpha = \varepsilon n_a T^a_\alpha$ of the energy-momentum tensor, must all be continuous there. While a possible discontinuity of $N = \sqrt{\varepsilon g_{44}}$ can be eliminated by a coordinate transformation, $T_{\alpha\beta}$ can be completely discontinuous; although we must of course have

$$[G^\alpha_\beta] = \kappa[T^\alpha_\beta]. \tag{30.43}$$

When in Section 26.3 we joined together the interior and exterior Schwarzschild solutions, we satisfied (30.39) by requiring continuity of the metric and (30.42) by the condition $p = 0$; the matching conditions (30.41) are then automatically satisfied. N also turns out to be continuous in this case, whilst $N_{,4}$ (note that $x^4 = r$!) is discontinuous.

V. Invariant characterization of exact solutions

Suppose that a solution of the Einstein equations is offered with the request to test it and establish whether it is already known, what physical situation it describes, what symmetries are present, and so on. Because of the freedom in the choice of coordinate system, such questions cannot usually be answered by merely looking at the solution. Thus one only establishes with certainty that

$$ds^2 = dx^2 - x \sin y \, dx \, dy + x^2(\tfrac{5}{4} + \cos y) dy^2$$
$$+ x^2(\tfrac{5}{4} + \cos y - \tfrac{1}{4}\sin^2 y)\sin^2 y \, dt^2 - dz^2 \tag{31.1}$$

describes flat Minkowski space (in inappropriate coordinates) by determining the curvature tensor. There exists, however, a series of methods for characterizing solutions invariantly (independently of the choice of coordinates), by means of which it has been possible to provide insight into the structure of solutions and hence often find ways of obtaining new solutions.

These methods, the most important of which we shall discuss in the following chapters, are at first sight of a purely mathematical nature. But, as often in theoretical physics, understanding of the mathematical structure simultaneously makes possible a deeper insight into the physical properties.

31
Preferred vector fields and their properties

31.1 Special simple vector fields

With many problems and solutions in General Relativity, preferred vector fields occur. Their origin may be of a more physical nature (velocity field of a matter distribution, light rays) or of a more mathematical nature (eigenvectors of the Weyl tensor, Killing vectors). One can use a knowledge of the properties of such vector fields for the purpose of classifying solutions or to simplify calculations by the introduction of coordinate systems which are adapted to the preferred vector field. We shall now discuss some special vector fields and coordinates appropriate to them.

Congruences of world lines The vector fields investigated in the following should have the property that at every point precisely one vector is defined. A family of world lines (congruence of world lines) is equivalent to such a vector field $a^n(x^i)$, its tangent vectors having the direction of a^n and covering the region of space under consideration smoothly and completely. This association is not unique, since not only a^n but also λa^n points in the direction of the tangent.

One obtains an especially simple form of the vectors $a^n(x^i)$ by taking these world lines as coordinate lines (for example, $x^\alpha = $ const., x^4 variable); the vector field then has the normal form

$$a^n(x^i) = \big(0, 0, 0, a^4(x^i)\big). \tag{31.2}$$

By means of a coordinate transformation $x^{4\prime} = x^{4\prime}(x^i)$ one can set $a^4 = 1$. If a^n is the four-velocity of the matter, then in (31.2) we are dealing with *comoving coordinates*.

Hypersurface-orthogonal fields A vector field is called hypersurface-orthogonal (or rotation-free) if it is possible to construct a family of surfaces $f(x^i) = $ const. across the congruence of world lines in such a way that the world lines, and with them the vectors of the field, are perpendicular to the surfaces (Fig. 31.1).The vector field a^n must therefore point in the direction of the gradient to the family of surfaces,

$$f_{,n} = \lambda a_n, \tag{31.3}$$

and hence must also satisfy the equations

$$a_{n;m} - a_{m;n} = (\lambda_{,n} a_m - \lambda_{,m} a_n)/\lambda, \tag{31.4}$$

which give

$$\omega^i \equiv \tfrac{1}{2}\varepsilon^{imnr} a_{[m;n]} a_r = 0. \tag{31.5}$$

A vector field a^n can be hypersurface-orthogonal only if its rotation ω^i as defined in (31.5) vanishes. One can show that this condition is also sufficient: a vector field is hypersurface-orthogonal if (31.5) holds.

While the contravariant components a^n of a vector can always be

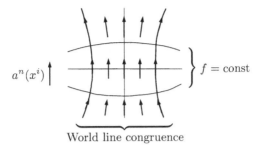

World line congruence

Fig. 31.1. Hypersurface-orthogonal vector field.

transformed to the normal form (31.2), a corresponding transformation of the covariant components to the form

$$a_n(x^i) = (0, 0, 0, a_4(x^i)) \qquad (31.6)$$

is only possible in a region of space if the vector field is hypersurface orthogonal. One can see this immediately from the fact that (31.6) is equivalent to $a_n = a_4 x^4_{,n}$. If a^n is not a null vector, one can take the surfaces $f = $ const. as coordinate surfaces, and simultaneously with (31.2) and (31.6) one can bring the metric to the form

$$ds^2 = g_{\alpha\beta}\, dx^\alpha\, dx^\beta + g_{44}(dx^4)^2. \qquad (31.7)$$

We shall examine the case of a null vector in Section 31.3.

Geodesic vector fields A vector field a^n is called geodesic when the world lines $x^i(s)$ of the associated congruence satisfy the geodesic equation:

$$t_{i;n}t^n = 0, \qquad t^i \equiv dx^i/ds. \qquad (31.8)$$

Since $t^i = \lambda(x^m)a^i$ should certainly hold, this implies

$$a_{[m}a_{i];n}a^n = 0. \qquad (31.9)$$

This condition is also sufficient; that is to say, if it is satisfied, then one can always determine a function λ which, when multiplied by a^i, gives a t^i which satisfies (31.8).

If the vector field is hypersurface-orthogonal *and* geodesic, then because of (31.2) and (31.6) we have in the metric (31.7)

$$a_{a;n}a^n = 0 = \Gamma^4_{\alpha 4} = \tfrac{1}{2}g_{44,\alpha}g^{44}, \qquad \alpha = 1, 2, 3; \qquad (31.10)$$

that is, g_{44} depends only upon x^4 and can be brought to the value ± 1

by a coordinate transformation $x^{4\prime} = x^{4\prime}(x^4)$:

$$ds^2 = g_{\alpha\beta}(x^i)\,dx^\alpha\,dx^\beta \pm (dx^4)^2. \tag{31.11}$$

Killing vector fields Killing vector fields $a_i(x^n)$ satisfy the condition

$$a_{i;n} + a_{n;i} = 0. \tag{31.12}$$

Because of their great importance we shall discuss them in more detail in Chapter 33.

Covariantly constant vector fields A vector field is covariantly constant if its covariant derivative vanishes:

$$a_{i;n} = 0. \tag{31.13}$$

From the definition (19.9) of the curvature tensor we have immediately

$$a^k R_{kinm} = 0. \tag{31.14}$$

The curvature tensor and with it the metric are restricted if such a vector field exists.

If a^i is not a null vector, then in the metric (31.11) we have

$$a_{\alpha;\beta} = 0 = \Gamma^4_{\alpha\beta}; \tag{31.15}$$

that is, $g_{\alpha\beta}$ is independent of x^4. Because of (30.10), the tensor $K_{\alpha\beta}$ of the extrinsic curvature of the surface $x^4 = \text{const.}$ vanishes, and the reduction formulae (30.19) and equation (31.14) lead to

$$\overset{4}{R}_{4\beta\nu\mu} = 0, \qquad \overset{3}{R}_{\alpha\beta\nu\mu} = \overset{4}{R}_{\alpha\beta\nu\mu}. \tag{31.16}$$

For vacuum solutions of the Einstein field equations we have accordingly

$$\overset{4}{R}_{\alpha\beta} = \overset{3}{R}_{\alpha\beta} = 0, \tag{31.17}$$

and, since the curvature tensor of the three-dimensional subspace can be constructed from its Ricci tensor alone, according to (19.32), then the curvature tensor of the four-dimensional space completely vanishes.

We thus have shown that, if a vacuum solution of the Einstein field equations possesses a covariantly constant vector field, then either space-time is flat or else we are dealing with a null vector field.

The vacuum solutions with a covariantly constant null vector field are just the plane gravitational waves investigated in Section 29.3.

31.2 Timelike vector fields

The invariant decomposition of $u_{m;i}$ and its physical interpretation One of the most important examples of a timelike vector field is the velocity field $u^i(x^n)$ of a matter distribution; for example, that of the matter inside a star or that of the stars or galaxies (imagined distributed continuously) in the universe. The properties of this velocity field are best recognized by examining the covariant derivative $u_{i;n}$. The idea consists essentially in decomposing that portion of the covariant derivative which is perpendicular to the four-velocity u^i, namely, the quantity $u_{i;n} + u_{i;m}u^m u_n/c^2$ (notice that the relation $u^i u_{i;n} = 0$ follows from $u^i u_i = -c^2$), into its antisymmetric part, its symmetric trace-free part, and the trace itself:

$$
\begin{aligned}
u_{i;n} &= -\dot{u}_i u_n/c^2 + \omega_{in} + \sigma_{in} + \Theta h_{in}/3, \\
\dot{u}_i &= u_{i;n}u^n = \mathrm{D}u_i/\mathrm{D}\tau, & \dot{u}_i u^i &= 0, \\
\omega_{in} &= u_{[i;n]} + \dot{u}_{[i}u_{n]}/c^2, & \omega_{in}u^n &= 0, \\
\sigma_{in} &= u_{(i;n)} + \dot{u}_{(i}u_{n)}/c^2 - \Theta h_{in}/3, & \sigma_{in}u^n &= 0, \\
\Theta &= u^i{}_{;i}, \\
h_{in} &= g_{in} + u_i u_n/c^2, & h_{in}u^n &= 0.
\end{aligned}
\tag{31.18}
$$

Since this splitting is covariant, the individual components characterize the flow field invariantly; they have the names:

$$
\begin{aligned}
\dot{u}_i &: \text{acceleration}, & \omega_{in} &: \text{rotation (twist)}, \\
\sigma_{in} &: \text{shear}, & \Theta &: \text{expansion}.
\end{aligned}
\tag{31.19}
$$

We shall now clarify the physical meaning of these quantities, and thereby also justify the names (31.19). The congruence of world lines

$$
x^a = x^a(y^\alpha, \tau),
\tag{31.20}
$$

which is associated with the velocity field

$$
u^a(x^i) = \partial x^a/\partial \tau,
\tag{31.21}
$$

obviously has the physical significance of being a family of streamlines (see Fig. 31.2). Along the world line of every particle (every volume element) the y^α are constant and τ varies; y^α labels the different world lines. Keeping the parameter τ fixed one passes from the world line (y^α) to the neighbouring world line $(y^\alpha + \delta y^\alpha)$ on advancing by

$$
\delta x^a = \frac{\partial x^a}{\partial y^\alpha}\delta y^\alpha.
\tag{31.22}
$$

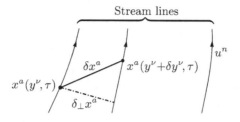

Fig. 31.2. Stream kinematics.

Since we have

$$\frac{\mathrm{D}}{\mathrm{D}\tau}\delta x^a = \frac{\mathrm{d}}{\mathrm{d}\tau}\delta x^a + \Gamma^a_{bc}\frac{\mathrm{d}x^b}{\mathrm{d}\tau}\delta x^c = \frac{\partial^2 x^a}{\partial\tau\,\partial y^\alpha}\delta y^\alpha + \Gamma^a_{bc}u^b\delta x^c$$

$$= \frac{\partial u^a}{\partial y^\alpha}\delta y^\alpha + \Gamma^a_{bc}u^b\delta x^c = \frac{\partial u^a}{\partial x^n}\delta x^n + \Gamma^a_{bc}u^b\delta x^c, \tag{31.23}$$

this difference vector changes with advance along the world line by

$$(\delta x^a)^{\cdot} = u^a{}_{;n}\delta x^n. \tag{31.24}$$

An observer comoving with the flow, however, will define as displacement to the neighbouring fluid elements not δx^a, but rather the projection of this quantity into his three-dimensional space, that is,

$$\delta_\perp x^a = (g^a_b + u^a u_b/c^2)\delta x^b = h^a_b\delta x^b. \tag{31.25}$$

Since this observer will use as his 'natural' comoving local coordinate system one whose axes are Fermi–Walker transported (cp. Sections 18.4 and 21.2), he will define as the velocity of the neighbouring matter elements the Fermi derivative of $\delta_\perp x^a$. Using (31.24) and (31.25), and remembering that $(\delta_\perp x^a)u_a = 0$, we obtain for this velocity

$$\mathrm{D}(\delta_\perp x^a)/\mathrm{D}\tau - c^{-2}(\delta_\perp x^n)(u^a\dot{u}_n - \dot{u}^a u_n) = (\delta_\perp x^n)^{\cdot}h^a_n, \tag{31.26}$$

and from this, with (31.18), finally

$$(\delta_\perp x^n)^{\cdot}h^a_n = (u^a{}_{;n} + \dot{u}^a u_n/c^2)(\delta_\perp x^n) = (\omega^a{}_n + \sigma^a{}_n + \Theta h^a_n/3)(\delta_\perp x^n). \tag{31.27}$$

Equation (31.27) gives the connection between the velocity $(\delta_\perp x^n)^{\cdot}h^a_n$ of the neighbouring particle to the observer (velocity relative to the observer) and the (infinitesimal) position vector $\delta_\perp x^n$ pointing from the observer to the particle. From it we can deduce the following.

(a) The expansion Θ leads to a radially directed velocity field whose magnitude is independent of direction; a volume element is thereby magnified ($\Theta > 0$) or diminished ($\Theta < 0$) in size with its form preserved.

(b) Since the antisymmetric tensor ω_{mi} of the rotation can be mapped into the vorticity vector ω^a according to (31.5) through

$$\omega^a = \tfrac{1}{2}\varepsilon^{abmi} u_b \omega_{mi}, \qquad \omega_{mi} = \varepsilon_{miab}\omega^a u^b, \tag{31.28}$$

the velocity field described by it has the form

$$(\delta_\perp x^n)^{\cdot}\, h_n^a = \varepsilon^a{}_{nmi}\omega^m u^i \delta_\perp x^n. \tag{31.29}$$

The velocity is perpendicular to the position vector $\delta_\perp x^n$ and to the vorticity vector ω^m, and thus we are dealing with a rotation about the axis defined by ω^m.

(c) The symmetric tensor σ_{an} of the shear leads to a direction-dependent velocity field which produces an ellipsoid out of a sphere of particles. Since the trace $\sigma^n{}_n$ vanishes, this ellipsoid has the same volume as the original sphere, and thus we here have a change in shape at constant volume.

Special cases and statements about possible coordinate systems When performing calculations one often uses the comoving coordinate system

$$u^i = (0,0,0,u^4). \tag{31.30}$$

If the rotation ω_{mi} vanishes, then the flow given by u^i is hypersurface-orthogonal and the metric can be brought to the form

$$ds^2 = g_{\alpha\beta}(x^i)\, dx^\alpha\, dx^\beta - u_4^2\, dt^2, \qquad u_i = (0,0,0,u_4). \tag{31.31}$$

If one writes down the covariant derivative $u_{a;b}$ explicitly in this metric and compares the result with (31.18), then one can show that:

(a) for $\omega_{mi} = 0$ and $\sigma_{mi} = 0$ the metric $g_{\alpha\beta}$ of the three-space contains the time only in a factor common to all elements:

$$g_{\alpha\beta}(x^i) = V^2(x^\nu, t)\bar{g}_{\alpha\beta}(x^\mu); \tag{31.32}$$

(b) for $\omega_{mi} = 0$ and $\Theta = 0$ the determinant of the three-metric $g_{\alpha\beta}$ does not depend upon the time;

(c) for $\omega_{mi} = 0$ and $\dot{u}_i = 0$ one can transform u_4 to c.

If the expansion and the shear vanish ($\Theta = 0$ and $\sigma_{mi} = 0$), but not the rotation ($\omega_{mi} \neq 0$), then for the comoving observer the distances to neighbouring matter elements do not change, and we have a *rigid rotation.* In the comoving coordinate system (31.30) one can see this from the fact that, because

$$u^4 h_{ab,4} = \mathcal{L}_{\mathbf{u}}(g_{ab} + u_a u_b/c^2) = u_{a;b} + u_{b;a} + (\dot{u}_a u_b + u_a \dot{u}_b)/c^2 = 0,$$

$$(31.33)$$

the purely spatial metric h_{ab} does not change with time.

31.3 Null vector fields

Null vector fields $k^n(x^i)$ can be characterized in a similar fashion to timelike vector fields by the components of their covariant derivative $k_{i;n}$. In this case some peculiarities arise from the fact that because $k_n k^n = 0$ one cannot simply decompose a vector, for example, into its components parallel and perpendicular to k^n; if we put $a_n = \lambda k_n + \hat{a}_n$, then λ and $\hat{a}_n$ cannot be uniquely determined. It is therefore preferable to use projections onto a two-dimensional subspace associated with the vector k^n (which is spanned by the vectors m_a and $\overline{m}_a$).

Geodesic null congruences and decomposition of $k_{a;b}$ Light rays and, as we shall see later, also the null vector fields induced by gravitational fields, lead to *geodesic* null vector fields. In the following we shall consider therefore only such fields.

We can describe a family of null geodesics by

$$x^a = x^a(y^\nu, v).$$

$$(31.34)$$

Here y distinguishes the different geodesics and v is an affine parameter along a fixed geodesic, that is, a parameter under the use of which the tangent vector

$$k^a = \partial x^a / \partial v, \quad k^a k_a = 0,$$

$$(31.35)$$

satisfies the equation

$$\dot{k}_a \equiv k_{a;b} k^b = 0.$$

$$(31.36)$$

The affine parameter v is not determined uniquely by this requirement; a linear transformation

$$v' = A^{-1}(y^\alpha)v + D(y^\alpha)$$

$$(31.37)$$

is still possible along every geodesic, corresponding to a transformation

$$k'^a = Ak^a.$$

$$(31.38)$$

Using the null tetrad of Section 9.1, we now decompose the covariant derivative $k_{i;n}$ of a geodesic null vector field with the help of the projection tensor

$$p_{ab} = m_a \overline{m}_b + \overline{m}_a m_b = g_{ab} + l_a k_b + l_b k_a \qquad (31.39)$$

into the components in the plane spanned (locally) by m_a and $\overline{m}_a$ and the perpendicular components. Taking into account (31.35) and (31.36), we obtain

$$k_{i;n} = A_{in} + a_i k_n + k_i b_n \qquad (31.40)$$

with

$$A_{in} = k_{a;b} p_i^a p_n^b = 2\mathrm{Re}\left[(\theta + i\omega)\overline{m}_i m_n - \sigma \overline{m}_i \overline{m}_n\right],$$

$$a_i k^i = 0, \quad b_i k^i = 0, \quad \theta, \omega \text{ real}, \sigma \text{ complex}. \qquad (31.41)$$

In spite of the non-uniqueness of the vectors m_a and $\overline{m}_a$ for fixed k^a (see (9.9)), the invariants

$$\omega = \sqrt{\tfrac{1}{2} A_{[nr]} A^{nr}} = \sqrt{\tfrac{1}{2} k_{[n;r]} k^{n;r}}, \quad \theta = \tfrac{1}{2} A^n{}_n = \tfrac{1}{2} k^i{}_{;i},$$

$$|\sigma| = \sqrt{\tfrac{1}{2} \left[A_{(nm)} A^{nm} - \tfrac{1}{2}(A^n_n)^2\right]} = \sqrt{\tfrac{1}{2}\left[k_{(n;r)} k^{n;r} - \tfrac{1}{2}(k^i{}_{;i})^2\right]} \qquad (31.42)$$

formed from the antisymmetric part, the symmetric trace-free part, and the trace of A_{in}, respectively, characterize the vector field in a unique fashion, since they can be expressed solely in terms of $k_{n;i}$. For a fixed congruence of world lines, under a gauge transformation of the associated null vector field according to (9.9), the invariants (31.42) will contain the factor A too.

The physical interpretation of the decomposition of $k_{i;n}$ – the optical scalars θ, ω and σ We fix attention on one element of the family of null geodesics (31.34), which we shall now call light rays, and consider the connecting vector

$$\delta x^a = \frac{\partial x^a(y^\nu, v)}{\partial y^\alpha} \delta y^\alpha \qquad (31.43)$$

to neighbouring light rays. The neighbouring light rays clearly form a three-parameter family. From this family we single out a two-parameter family by the condition

$$k_a \delta x^a = 0. \qquad (31.44)$$

Equation (31.44) expresses the fact that δx^a is a spacelike vector. In the rest system of this vector we have, because $\delta x^a = (\delta\mathbf{r}, 0)$ and $k^a = (\mathbf{k}, k^4)$,

$$\mathbf{k}\delta\mathbf{r} = 0; \qquad (31.45)$$

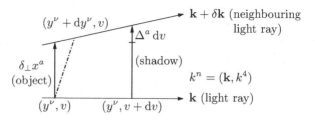

Fig. 31.3. Three-dimensional sketch for interpreting the optical scalars.

that is, the three-dimensional light rays of this family are perpendicular to the connecting vector $\delta\mathbf{r}$. In an arbitrary coordinate system we can define the component of the connecting vector restricted by (31.44) which is perpendicular to the light rays by

$$\delta_\perp x^a = p^a_b \delta x^b. \tag{31.46}$$

We shall now calculate how $\delta_\perp x^a$ changes along the light rays, which we can visualize as how the shadow which the light rays throw onto a screen at right-angles to them differs from the 'object' $\delta_\perp x^a$ (see Fig. 31.3).

The required quantity

$$\Delta^a = p^a_b (p^b_i \delta x^i)_{;n} k^n \tag{31.47}$$

can be easily calculated by using the equation

$$(\delta x^i)_{;n} k^n = k^i{}_{;n} \delta x^n, \tag{31.48}$$

which follows from

$$(\delta x^i)_{,n} k^n = \frac{\partial \delta x^i}{\partial v} = \frac{\partial^2 x^i}{\partial y^\alpha \, \partial v} \delta y^\alpha = \frac{\partial k^i}{\partial y^\alpha} \delta y^\alpha = \frac{\partial k^i}{\partial x^n} \delta x^n, \tag{31.49}$$

remembering the relations (17.44), (31.36), (31.40), (31.41) and (31.44). We obtain

$$\Delta^a = A^a{}_i \delta_\perp x^i = 2\mathrm{Re}\big[(\theta + \mathrm{i}\omega)\overline{m}_i m^a - \sigma \overline{m}_i \overline{m}^a\big]\delta_\perp x^i$$
$$= \theta \delta_\perp x^a + \mathrm{i}\omega\big(\overline{m}_i m^a - m_i \overline{m}^a\big)\delta_\perp x^i - 2\mathrm{Re}\big[\sigma \overline{m}_i \overline{m}^a\big]\delta_\perp x^i. \tag{31.50}$$

The three optical scalars θ, ω and σ can thus be visualized in the following way

(a) The antisymmetric part of A_{ai}, associated with ω, produces a difference vector Δ^a which is perpendicular to $\delta_\perp x^a$. Since the shadow is then rotated with respect to the object, ω is called the *torsion* or the

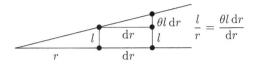

Fig. 31.4. How distance is defined with the aid of θ.

rotation of the light rays. Within the realm of validity of geometrical optics k_a is always hypersurface-orthogonal, so that ω vanishes; systems with $\omega \neq 0$ can therefore be realized in a simple manner only by a twisted bundle of light rays.

(b) The symmetric trace-free part of A_{ai}, associated with σ, produces as shadow of a circle an ellipse of equal area. The shrinking (stretching) of the axes is determined by $|\sigma|$, whilst the direction of the axes of the ellipse follows from the phase of σ. σ is called the *shear* of the null congruence.

(c) The trace of A_{ai}, which is associated with θ, produces a shadow which is diminished or magnified in size with respect to the object independently of direction. θ is therefore called the *expansion* of the light rays. The light rays emitted from a pointlike source of light constitute the standard example of a family with $\theta \neq 0$. Since in flat space we have $\theta = 1/r$ for these rays (see Fig. 31.4), one uses θ in curved space to define a parallax r_{P} distance by

$$\theta = 1/r_{\mathrm{P}}. \tag{31.51}$$

Special cases and appropriate coordinate systems For making calculations with null vector fields one often uses coordinate systems in which

$$k^i = (0, 0, k^3, 0). \tag{31.52}$$

If k_i is hypersurface-orthogonal (ω vanishes), then as well as (31.52) one can set

$$
\begin{aligned}
k_i &= \lambda u_{,i} = (0, 0, 0, k_4), \\
\mathrm{d}s^2 &= g_{AB}\,\mathrm{d}x^A\,\mathrm{d}x^B + 2m_i\,\mathrm{d}x^i\,\mathrm{d}u, \quad A, B = 1, 2.
\end{aligned}
\tag{31.53}
$$

A comparison of (31.53) with the form $k_i = (0, 0, \lambda, -\lambda)$ of a null vector in Minkowski space and Minkowski coordinates shows that $-u$ signifies a retarded time, for example, $u = z - ct$.

For plane waves with $k_{a;b} = 0$ all three optical scalars vanish in agreement with the intuitive interpretation of these quantities.

Exercises

31.1   Show that in a spherically symmetric metric (23.5) any radially
        directed vector field is hypersurface-orthogonal.

31.2   Show that in a Robinson–Trautman metric (34.4) the null vector
        $k_n = u_{,n}$ is hypersurface-orthogonal and shearfree, but has a
        non-zero divergence.

32

The Petrov classification

32.1 What is the Petrov classification?

The Petrov classification is the classification of Riemannian spaces
according to the algebraic properties of the Weyl tensor (conformal curvature tensor) defined by

$$C^{ai}{}_{sq} = R^{ai}{}_{sq} - \tfrac{1}{2}(g^a_s R^i_q + g^i_q R^a_s - g^i_s R^a_q - g^a_q R^i_s) + \tfrac{1}{6}(g^a_s g^i_q - g^i_s g^a_q). \quad (32.1)$$

From other areas of physics, one knows that algebraic properties of
tensors are linked with important physical properties. Thus, for example, in crystal optics the classification of media according to the number
of distinct eigenvalues of the ε-tensor leads to the division into optically
biaxial, uniaxial or isotropic crystals. Therefore we may also hope to find
physically interesting relations by investigating the algebraic structure
of the curvature tensor.

The examination of the conformal tensor does not suffice of course if
one wants to determine all algebraic properties of the curvature tensor.
The information lacking is hidden in the Ricci tensor or (because of the
field equations) in the energy-momentum tensor. Here, however, we shall
restrict our discussion to the Weyl tensor, which anyway coincides with
the curvature tensor for vacuum fields; the Petrov classification is the
classification of vacuum gravitational fields according to the algebraic
properties of the curvature tensor.

The Petrov classification of gravitational fields is the analogue of the
algebraic classification of electromagnetic fields performed in Sections 9.2

and 9.3, both formally and as regards physical content. The reader is advised to consult these sections, since because of those analogies we shall describe the Petrov classification relatively briefly.

32.2 The algebraic classification of gravitational fields

The expansion of the Weyl tensor in terms of self-dual bivectors With the Weyl tensor C_{arsq} of a gravitational field can be associated the tensor

$$\overset{*}{C}_{arsq} = C_{arsq} + \mathrm{i}\tilde{C}_{arsq} = C_{arsq} + \tfrac{1}{2}\mathrm{i}\varepsilon_{sqmn}C_{ar}{}^{mn} \qquad (32.2)$$

which is analogous to the complex field-strength tensor Φ_{ab}. This tensor is clearly self-dual with respect to the last two indices:

$$\overset{*}{C}\overset{\sim}{}_{arsq} = \tfrac{1}{2}\varepsilon_{sqmn}\overset{*}{C}_{ar}{}^{mn} = -\mathrm{i}\,\overset{*}{C}_{arsq} \,. \qquad (32.3)$$

Because of the definition (32.1) of the Weyl tensor its contraction vanishes:

$$C^{ar}{}_{aq} = 0. \qquad (32.4)$$

Hence the Weyl tensor (in contrast to the curvature tensor of an arbitrary space) has the property that the dual tensors formed with respect to the first and to the last pairs of indices coincide:

$$\tilde{C}\overset{*}{}_{arsq} = \tfrac{1}{4}\varepsilon_{arik}\varepsilon_{sqmn}C^{ikmn} = -C_{arsq} \qquad (32.5)$$

(use equations (6.17) and (17.30)!). The tensor $\overset{*}{C}_{arsq}$ is therefore automatically also self-dual with respect to the front index pair and can be expanded entirely in terms of the self-dual bivectors (9.15):

$$\overset{*}{C}_{arsq} = \Psi'_0 U_{ar}U_{sq} + \Psi'_4 V_{ar}V_{sq} + \Psi'_1(U_{ar}W_{sq} + U_{sq}W_{ar}) \qquad (32.6)$$
$$+ \Psi'_2(U_{ar}V_{sq} + U_{sq}V_{ar} + W_{ar}W_{sq}) + \Psi'_3(V_{ar}W_{sq} + V_{sq}W_{ar}).$$

In this expansion account has already been taken of the symmetry properties of the Weyl tensor; the ten algebraically independent components are described by the five complex coefficients Ψ'_A.

The original Petrov (1969) classification consists essentially of classifying the types of the self-dual tensor $\overset{*}{C}_{arsq}$ according to the number of eigenbivectors defined by

$$\tfrac{1}{2}\overset{*}{C}_{arsq}Q^{sq} = \lambda Q_{ar}, \quad \tilde{Q}_{ar} = -\mathrm{i}Q_{ar}. \qquad (32.7)$$

274 *The Petrov classification*

Here we shall take a rather different approach and therefore state the Petrov result without proof. Type I (special cases D, O) occurs when there are three eigenbivectors, type II (special case N) possesses two eigenbivectors and type III only one eigenbivector.

The classification of gravitational fields – first formulation By adapting the null tetrad, and along with it also the self-dual bivectors to the Weyl tensor under investigation, one can simplify the expansion (32.6) and set $\Psi_0 = 0$. To this end, because of (9.11) and (9.23), one has to line up the direction k^a (determine E) so that

$$\Psi_0 = \Psi_0' - 4\Psi_1'E + 6\Psi_2'E^2 - 4\Psi_3'E^3 + \Psi_4'E^4 = 0. \qquad (32.8)$$

Equation (32.8) has, as an equation of fourth degree, precisely four roots E (this is true with corresponding interpretation also in the special cases: for $\Psi_4' = 0$ and $\Psi_0' \neq 0$ one obtains by the exchange of labels $l_a \leftrightarrow -k_a$ an equation of fourth degree, for $\Psi_4' = \Psi_0' = 0$ then $E = 0$ is a double root, and so on). To these four roots correspond four directions k^a (eigenvectors k^a) with $\Psi_0 = 0$. According to the multiplicity of these roots one can divide Riemannian spaces into the following types:

Non-degenerate: Type I :   four distinct roots,
Degenerate: Type II :   one double root and two simple roots,
Type D :   two double roots,
Type III :   one triple root and one simple root,
Type N :   one four-fold root,
Type O :   the Weyl tensor and all Ψ_A are zero.

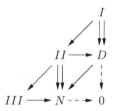

Fig. 32.1. The Penrose diagram.

The Penrose diagram (Fig. 32.1) provides a summary of the successive growth in degeneracy; every arrow signifies *one* additional degeneracy.

The classification of gravitational fields – second formulation It is possible to avoid the detour through the tensor $\overset{*}{C}_{arsq}$ and pick out the types of gravitational fields directly from the Weyl tensor and its null eigenvectors k^a.

First of all one expresses the coefficients Ψ_A directly in terms of products of the Weyl tensor with the null-tetrad vectors, using (32.2) and (9.16):

$$\Psi_0 = \tfrac{1}{8}\overset{*}{C}_{arsq} V^{ar} V^{sq} = \tfrac{1}{4}C_{arsq}V^{ar}V^{sq} = C_{arsq}k^a m^r k^s m^q,$$

$$\Psi_1 = -\tfrac{1}{16}\overset{*}{C}_{arsq} V^{ar} W^{sq} = C_{arsq}k^a m^r k^s l^q,$$

$$\Psi_2 = \tfrac{1}{8}\overset{*}{C}_{arsq} U^{ar} V^{sq} = -C_{arsq}k^a m^r l^s \bar{m}^q, \qquad (32.9)$$

$$\Psi_3 = -\tfrac{1}{16}\overset{*}{C}_{arsq} U^{ar} W^{sq} = C_{arsq}l^a \bar{m}^r l^s k^q,$$

$$\Psi_4 = \tfrac{1}{8}\overset{*}{C}_{arsq} U^{ar} U^{sq} = C_{arsq}l^a \bar{m}^r l^s \bar{m}^q.$$

Ψ_0 vanishes for null eigenvectors k^a; the real symmetric tensor

$$S_{aq} = C_{arsq}k^r k^s \qquad (32.10)$$

therefore contains no terms proportional to $m^r m^q$ and $\bar{m}^r \bar{m}^q$. Further, as a consequence of the symmetry properties of the Weyl tensor, the relations

$$S_{aq}k^q = 0, \qquad S^a_a = C^a{}_{rsa}k^r k^s = 0 \qquad (32.11)$$

hold, and therefore it has the structure

$$S_{aq} = \alpha k_a k_q + \text{Re}\big[\beta(k_a \bar{m}_q + k_q \bar{m}_a)\big]. \qquad (32.12)$$

Eigenvectors of the Weyl tensor therefore have the property

$$k_{[b}C_{a]rs[q}k_{n]}k^r k^s = 0 \quad \leftrightarrow \quad \Psi_0 = 0. \qquad (32.13)$$

If two eigenvectors coincide ($E = 0$ is a double root of (32.8)), then Ψ_1 must also vanish, besides Ψ_0. Because of (32.9), it then follows that $\beta = 0$, and hence that

$$k_{[b}C_{a]rsq}k^r k^s = 0 \leftrightarrow \Psi_0 = \Psi_1 = 0. \qquad (32.14)$$

By pursuing these considerations further one finally arrives at the results presented in Table 32.1 (in each case the last two columns hold for the null eigenvectors of highest degeneracy).

By rotations of the null tetrad (for fixed k^a) one can in addition set $\Psi_2 = 0$ for type I, $\Psi_3 = 0$ for type II, $\Psi_3 = \Psi_4 = 0$ for type D (for which $3\Psi_2\Psi_4 = 2\Psi_3{}^2$ always holds), and $\Psi_4 = 0$ for type III.

Table 32.1. *The Petrov types and their properties*

Type	Multiplicity of the null eigenvectors	Vanishing coefficients	Criterion satisfied by C_{abcd}
I	$(1,1,1,1)$	Ψ_0	$k_{[b}C_{a]rs[q}k_{n]}k^r k^s = 0$
II	$(2,1,1)$	$\Psi_0,\ \Psi_1$	$k_{[b}C_{a]rsq}k^r k^s = 0$
D	$(2,2)$		
III	$(3,1)$	$\Psi_0,\ \Psi_1,\ \Psi_2$	$k_{[b}C_{a]rsq}k^r = 0$
N	(4)	$\Psi_0,\ \Psi_1,\ \Psi_2,\ \Psi_3$	$C_{arsq}k^a = 0$
O	there are none	all Ψ_A	$C_{arsq} = 0$

In order to determine the Petrov type of a given space one must first compute the Weyl tensor and hence, using (32.9) with an arbitrary null tetrad, the Ψ'_As. Equation (32.8) then gives the possible transformations E and thus the multiplicity of the null eigenvectors.

32.3 The physical interpretation of degenerate vacuum gravitational fields

For vacuum solutions of Einstein's field equations the Weyl tensor and the curvature tensor are identical, so that all statements of the previous section also hold for the curvature tensor.

The simplest example of a degenerate vacuum solution is provided by the plane gravitational waves dealt with in Section 29.3. Because

$$R_{abmn}k^n = 0, \quad k^n k_n = 0, \tag{32.15}$$

they are of type N. One might therefore suppose that degenerate vacuum solutions are connected with gravitational radiation and gravitational waves. In fact one can show that the curvature tensor of an isolated matter distribution allows, under certain assumptions about the sources, at large distance an expansion

$$R_{abmn} = \frac{N_{abmn}}{r} + \frac{III_{abmn}}{r^2} + \frac{D_{abmn}}{r^3} + \cdots, \tag{32.16}$$

where the symbols N, III and D refer to tensors of the respective algebraic types. The far field of every source of gravitational radiation (if such a field exists) is therefore a plane wave (type N) locally; if one approaches closer to the source, then the four initially coincident directions of the null eigenvectors separate (*peeling theorem*). Unfortunately this result is not as fruitful as the corresponding one in electrodynamics, because in general the relation between N_{abmn}, III_{abmn} and D_{abmn} and the properties of the sources of the field are not known.

Two simple properties of the null eigenvector fields of degenerate vacuum fields can be deduced from the Bianchi identities, which it is best to use here in the form

$$\tilde{R}_{ar}{}^{sq}{}_{;q} = 0 \tag{32.17}$$

or, since $R_{ab} = 0$,

$$\overset{*}{C}{}^{arsq}{}_{;q} = 0. \tag{32.18}$$

We shall do this explicitly for type II and type D ($\Psi_0 = \Psi_1 = 0$, $\Psi_2 \neq 0$); the calculations for the other types run analogously. Because

$$V_{ab;q}W^{ab} = 4k_{a;q}m^a, \tag{32.19}$$

we have from (32.18) and (32.6)

$$0 = \overset{*}{R}{}^{arsq}{}_{;q}V_{ar} = \left(\overset{*}{R}{}^{arsq}V_{ar}\right)_{;q} - \overset{*}{R}{}^{arsq}V_{ar;q} \tag{32.20}$$
$$= 4(\Psi_2 V^{sq})_{;q} - 8\Psi_3 V^{sq}k_{a;q}m^a - 2\Psi_2 U^{ar}V^{sq}V_{ar;q} - 8\Psi_2 W^{sq}k_{a;q}m^a.$$

Contraction with k_s yields

$$V^{sq}{}_{;q}k_s + 2k_{a;q}m^a k^q = k_{a;q}m^a k^q = 0; \tag{32.21}$$

that is, the vector field k^a is geodesic. Contraction of (32.20) with m_s leads to

$$V^{sq}{}_{;q}m_s + 2k_{s;q}m^s m^q = 3k_{s;q}m^s m^q = 0; \tag{32.22}$$

that is, the vector field is also shear-free ($\sigma = 0$).

The (multiple) eigenvectors of degenerate vacuum solutions thus form a geodesic, shear-free congruence of world lines (as also do the eigenvectors of degenerate electromagnetic fields). The converse of this statement holds as well (Goldberg–Sachs theorem): if in a vacuum solution a shear-free, geodesic null congruence exists, then this solution is degenerate and the congruence is a (multiple) eigencongruence. This law is often used for determining the Petrov type, since a conclusion can be drawn from knowledge of first derivatives ($k_{a;n}$) alone, while the Petrov type is generally only determinable from the curvature tensor (second derivatives).

The example of the Schwarzschild solution (Exercise 32.1) shows that degenerate solutions can in general not be interpreted as radiation fields.

Exercises

32.1   Show that for the Schwarzschild metric (23.23) the vectors

$$\sqrt{2}k_a = \left(e^{\lambda/2}, 0, 0, e^{-\lambda/2}\right), \quad -\sqrt{2}l_a = -\left(e^{\lambda/2}, 0, 0, -e^{-\lambda/2}\right),$$
$$\sqrt{2}m_a = (0, r, -ir\sin\vartheta, 0), \quad e^{-\lambda} = 1 - 2M/r$$

form a null tetrad satisfying the relations (17.44) and (17.46). Show that both vectors k_a and l_a are geodesic and shearfree. Which Petrov type occurs?

32.2   Which is the Petrov type of the plane waves (29.39)–(29.40)?

33
Killing vectors and groups of motion

33.1 The problem

When we are handling physical problems, symmetric systems have not only the advantage of a certain simplicity, or even beauty, but also special physical effects frequently occur then. One can therefore expect in General Relativity, too, that when a high degree of symmetry is present the field equations are easier to solve and that the resulting solutions possess special properties.

Our first problem is to define what we mean by a symmetry of a Riemannian space. The mere impression of simplicity which a metric might give is not of course on its own sufficient; thus, for example, the relatively complicated metric (31.1) in fact has more symmetries than the 'simple' plane wave (29.39). Rather, we must define a symmetry in a manner independent of the coordinate system. Here we shall restrict ourselves to continuous symmetries, ignoring discrete symmetry operations (for example, space reflections).

33.2 Killing vectors

The symmetry of a system in Minkowski space or in three-dimensional (Euclidean) space is expressed through the fact that under translation

along certain lines or over certain surfaces (spherical surfaces, for example, in the case of spherical symmetry) the physical variables do not
change. One can carry over this intuitive idea to Riemannian spaces
and ascribe a symmetry to the space if there exists an s-dimensional
$(1 \leq s \leq 4)$ manifold of points which are physically equivalent: under a
symmetry operation, that is, a motion which takes these points into one
another, the metric does not change.

These ideas are made more precise by imagining a vector $\xi^i(x^a)$ at
every point x^a of the space and asking for the conditions under which
the metric does not change when proceeding in the direction ξ^i. Since
every finite motion can be constructed from infinitesimal motions, it is
sufficient to ensure the invariance of the metric under the infinitesimal
motion

$$\bar{x}^a = x^a + \xi^a(x^n)\,\mathrm{d}\lambda = x^a + \delta x^a. \tag{33.1}$$

For such a transformation we have

$$\delta g_{ab} = g_{ab,n}\xi^n\,\mathrm{d}\lambda, \quad \delta(\mathrm{d}x^a) = \mathrm{d}(\delta x^a) = \xi^a{}_{,n}\,\mathrm{d}x^n\,\mathrm{d}\lambda, \tag{33.2}$$

so that the line elements at the point x^a and at the neighbouring point
$\bar{x}^a$ are identical only if

$$\begin{aligned}
\delta(\mathrm{d}s^2) &= \delta(g_{ab}\,\mathrm{d}x^a\,\mathrm{d}x^b) \\
&= (g_{ab,n}\xi^n + g_{nb}\xi^n{}_{,a} + g_{an}\xi^n{}_{,b})\,\mathrm{d}x^a\,\mathrm{d}x^b\,\mathrm{d}\lambda = 0.
\end{aligned} \tag{33.3}$$

A symmetry is present if and only if (33.3) is satisfied independently of
the orientation of $\mathrm{d}x^a$, that is, for

$$g_{ab,n}\xi^n + g_{nb}\xi^n{}_{,a} + g_{an}\xi^n{}_{,b} = 0. \tag{33.4}$$

For a given metric, (33.4) is a system of differential equations determining the vector field $\xi^i(x^n)$; if it has no solution, then the space has
no symmetry. In spite of the fact that it contains partial derivatives,
(33.4) is a covariant characterization of the symmetries present. One
can see this by substituting covariant for partial derivatives or formulating (33.4) with the help of the Lie derivative; (33.4) is equivalent to
the equation

$$\xi_{a;b} + \xi_{b;a} = \mathcal{L}_\xi g_{ab} = 0, \tag{33.5}$$

which is clearly covariant.

Vectors ξ^i which are solutions of the equations (33.4) or (33.5) are
called *Killing vectors*. They characterize the symmetry properties of
Riemannian spaces in an invariant fashion (Killing 1892).

If one chooses the coordinate system so that ξ^n has the normal form

$$\xi^n = (0,0,0,1), \tag{33.6}$$

then (33.4) reduces to

$$\partial g_{ab}/\partial x^4 = 0; \tag{33.7}$$

the metric does not depend upon x^4. This shows clearly that in (33.4) the alternative definition of symmetry as 'independence of a coordinate' has been covariantly generalized.

The world line congruence associated with the Killing vector field, that is, the family of those curves which link points which can be carried into one another by symmetry operations, is obtained by integration of the equations

$$\mathrm{d}x^n/\mathrm{d}\lambda = \xi^n(x^i). \tag{33.8}$$

33.3 Killing vectors of some simple spaces

The Killing equations (33.5) constitute a system of first-order linear differential equations for the Killing vectors $\xi^i(x^n)$; the number and type of solutions of these ten equations are dependent upon the metric and hence vary from space to space. Here we shall first of all determine the Killing vectors explicitly for two simple metrics, and only in the next section deduce some general statements about the diversity of solutions to the Killing equations.

The Killing vectors of Minkowski space can without doubt be obtained most simply in Cartesian coordinates. Since all Christoffel symbols vanish, in these coordinates the Killing equations read

$$\xi_{a,b} + \xi_{b,a} = 0. \tag{33.9}$$

If one combines the equations

$$\xi_{a,bc} + \xi_{b,ac} = 0, \qquad \xi_{b,ca} + \xi_{c,ba} = 0, \qquad \xi_{c,ab} + \xi_{a,cb} = 0, \tag{33.10}$$

which result from (33.9) by differentiation, then one obtains

$$\xi_{a,bc} = 0, \tag{33.11}$$

with the general solution

$$\xi_a = c_a + \varepsilon_{ab}x^b. \tag{33.12}$$

The Killing equations (33.9) are satisfied by (33.12), however, only if

$$\varepsilon_{ab} = -\varepsilon_{ba}. \tag{33.13}$$

Flat space thus processes ten linearly independent Killing vectors; the four constants c_a correspond to four translations and the six constants

ε_{ab} to six generalized rotations (three spatial rotations and three special Lorentz transformations).

One can also obtain relatively quickly the Killing vectors associated with the spherical surface

$$ds^2 = d\vartheta^2 + \sin^2\vartheta\, d\varphi^2 = (dx^1)^2 + \sin^2 x^1 (dx^2)^2. \tag{33.14}$$

Written out in full, equations (33.4) read

$$\xi^1{}_{,1} = 0, \qquad \xi^1{}_{,2} + \sin^2\vartheta\,\xi^2{}_{,1} = 0, \qquad \xi^1\cos\vartheta + \sin\vartheta\,\xi^2{}_{,2} = 0. \tag{33.15}$$

The general solution

$$\xi^1 = A\sin(\varphi + a), \qquad \xi^2 = A\cos(\varphi + a)\cot\vartheta + b \tag{33.16}$$

shows that there are three linearly independent Killing vectors, for example the vectors

$$\underset{1}{\xi^a} = (\sin\varphi, \cos\varphi\,\cot\vartheta), \quad \underset{2}{\xi^a} = (\cos\varphi, -\sin\varphi\,\cot\vartheta), \quad \underset{3}{\xi^a} = (0,1). \tag{33.17}$$

The sphere thus possesses exactly the same number of Killing vectors as the plane, which of course permits as symmetry operations two translations and one rotation.

33.4 Relations between the curvature tensor and Killing vectors

From the Killing equation (33.5)

$$\xi_{a;b} + \xi_{b;a} = 0 \tag{33.18}$$

and the relation

$$\xi_{a;b;n} - \xi_{a;n;b} = R^m{}_{abn}\xi_m \tag{33.19}$$

valid for every vector, a series of relations can be derived which enable one to make statements about the possible number of Killing vectors in a given space.

Because of the symmetry properties of the curvature tensor, the identity

$$(\xi_{a;b} - \xi_{b;a})_{;n} - (\xi_{n;a} - \xi_{a;n})_{;b} + (\xi_{b;n} - \xi_{n;b})_{;a} = 0 \tag{33.20}$$

follows from (33.19) for every vector. For Killing vectors it yields

$$\xi_{a;b;n} + \xi_{n;a;b} + \xi_{b;n;a} = 0 \tag{33.21}$$

and together with (33.19) and (33.5) leads to

$$\xi_{n;b;a} = R^m{}_{abn}\xi_m. \tag{33.22}$$

This equation shows that from the Killing vector ξ_n and its first derivatives $\xi_{n;a}$ all higher derivatives can be calculated in a given Riemannian space. To determine a Killing vector field uniquely it therefore suffices to specify the values of ξ_n and $\xi_{n;a}$ at one point. Since one must of course at the same time ensure that $\xi_{a;n} = -\xi_{n;a}$, then in an N-dimensional Riemannian space there are precisely $N + \binom{N}{2} = N(N+1)/2$ such initial values and, accordingly, a maximum of $N(N+1)/2$ linearly independent Killing vector fields. The physical space ($N = 4$) thus has at most ten Killing vectors and, as we shall show, it has ten only if the space is one of constant curvature.

The maximum number cannot always be realized in a given space, since the Killing equations are not necessarily integrable for every combination of the initial values, and there even exist spaces without any symmetry. Thus, for example, from the combination of the equation

$$\xi_{n;b;a;i} - \xi_{n;b;i;a} = R^m{}_{nai}\xi_{m;b} + R^m{}_{bai}\xi_{n;m}, \qquad (33.23)$$

which holds for every tensor $\xi_{a;b}$, with (33.22) and the Killing equation, we obtain the relation

$$(R^m{}_{abn;i} - R^m{}_{ibn;a})\xi_m$$
$$+ (R^m{}_{abn}g_i^k - R^m{}_{ibn}g_a^k + R^m{}_{bai}g_n^k - R^m{}_{nai}g_b^k)\xi_{m;k} = 0, \qquad (33.24)$$

which further restricts the freedom in specifying ξ_m and $\xi_{m;k}$. From the equations mentioned one can derive an algorithm for determining the number of possible Killing vector fields in a given space. We will not go into the details here, but rather refer the reader to the specialist literature, e.g. Eisenhart (1961).

It is relatively easy to answer the question of which spaces possess precisely the maximum number $N(N + 1)/2$ of Killing vectors. Clearly for such spaces (33.24) must imply no restrictions on the values of ξ_m and $\xi_{m;k}$, and therefore in this case we must have (remember that $\xi_{m;k} + \xi_{k;m} = 0$!)

$$R^m{}_{abp;i} = R^m{}_{ibp;a}, \qquad (33.25)$$

$$R^m{}_{abp}g_i^k - R^k{}_{abp}g_i^m - R^m{}_{ibp}g_a^k + R^k{}_{ibp}g_a^m$$
$$+ R^m{}_{bai}g_p^k - R^k{}_{bai}g_p^m - R^m{}_{pai}g_b^k + R^k{}_{pai}g_b^m = 0. \qquad (33.26)$$

By contraction first just over i and k, and then both over i and k and over a and b, one obtains from (33.26) the equations

$$(N - 1)R^m{}_{abp} = R_{ap}g_b^m - R_{ab}g_p^m, \qquad NR_p^m = Rg_p^m, \qquad (33.27)$$

and hence the curvature tensor of such spaces of maximal symmetry has the form

$$R_{mabp} = R(g_{ap}g_{mb} - g_{ab}g_{mp})/N(N-1). \tag{33.28}$$

The curvature scalar R must be constant, see Exercise 33.4. Spaces with these properties are called *spaces of constant curvature* (cp. Section 19.5). The curvature $R/N(N-1)$ can be positive, zero or negative. The quantity

$$K = \sqrt{N(N-1)/|R|} \tag{33.29}$$

is known as the radius of curvature. In these spaces no point and no direction is preferred. They are isotropic and homogeneous. Flat spaces with vanishing curvature tensor are special instances of these spaces.

As one can easily show, a four-dimensional space of constant curvature is not a solution of the vacuum field equations, apart from the trivial case of a Minkowski space. The question of the maximum number of Killing vectors in spaces which correspond to vacuum gravitational fields can be answered in the following way. Vacuum solutions of type *I* or *D* have at most four Killing vectors (to this group belongs the Schwarzschild metric, for example, with one timelike Killing vector $\xi^i = (0,0,0,1)$ and the three Killing vectors of the spherical symmetry); solutions of type *N* have at most six, and solutions of types *II* and *III* at most three Killing vectors.

33.5 Groups of motion

Translation in the direction of a Killing vector field can also be interpreted as a mapping of the space onto itself, or as a motion (for example, a rotation) of the space. Since we designate as motions precisely those transformations which do not alter the metric (for which the metric is the same, in a suitable coordinate system, at the initial point and the end point of the motion), these transformations form a group.

Groups of motion (Lie groups) are continuous groups whose elements are differentiable functions of a finite number of parameters r. One can imagine the entire group to be generated by repeated application of infinitesimal transformations (33.1) in the direction of the r Killing vectors of the space. These (linearly independent) Killing vectors thus serve as a basis for generating the group. Since every linear combination of Killing vectors is also a Killing vector, this basis is not uniquely determined.

One can characterize a group (and hence a space) by the number of linearly independent Killing vectors and their properties. An intuitive picture of the way in which the group acts is provided by the regions of transitivity, which are those regions of the space whose points can be carried into one another by the symmetry operations of the group. The surfaces of transitivity of the rotation group are spherical surfaces, for example, and the group is multiply transitive on them; that is, there exists more than one transformation which transforms one point into another.

The structure constants of a group of motion The structure of a group which is generated by r Killing vectors can be most clearly recognized if one examines the commutability of infinitesimal motions.

Two infinitesimal motions

$$\bar{x}^a = x^a + \underset{A}{\xi^a}(x^i)\,\mathrm{d}\lambda_A + \underset{A}{F^a}(\mathrm{d}\lambda_A)^2 + \cdots, \tag{33.30}$$

$$\tilde{x}^a = x^a + \underset{B}{\xi^a}(x^i)\,\mathrm{d}\lambda_B + \underset{B}{F^a}(\mathrm{d}\lambda_B)^2 + \cdots, \tag{33.31}$$

in the direction of the Killing vectors $\underset{A}{\xi^a}$ and $\underset{B}{\xi^a}$, respectively, give

$$
\begin{aligned}
\bar{\tilde{x}}^a &= \bar{x}^a + \underset{B}{\xi^a}(\bar{x}^i)\,\mathrm{d}\lambda_B + \underset{B}{F^a}(\mathrm{d}\lambda_B)^2 + \cdots \\
&= x^a + \underset{A}{\xi^a}(x^i)\,\mathrm{d}\lambda_A + \underset{A}{F^a}(\mathrm{d}\lambda_A)^2 + \underset{B}{\xi^a}(x^i)\,\mathrm{d}\lambda_B \\
&\quad + \underset{B}{\xi^a}_{,n}(x^i)\,\underset{A}{\xi^n}(x^i)\,\mathrm{d}\lambda_A\,\mathrm{d}\lambda_B + \underset{B}{F^a}(\mathrm{d}\lambda_B)^2 + \cdots
\end{aligned} \tag{33.32}
$$

when performed one after the other. If one performs the transformations in reverse order and then takes the difference of the two results, then only that part of (33.32) antisymmetric in A and B remains:

$$\bar{\tilde{x}}^a - \tilde{\bar{x}}^a = \left(\underset{B}{\xi^a}_{,n}\,\underset{A}{\xi^n} - \underset{A}{\xi^a}_{,n}\,\underset{B}{\xi^n} \right)\mathrm{d}\lambda_A\,\mathrm{d}\lambda_B + \cdots. \tag{33.33}$$

Infinitesimal motions thus commute only to first order; in second order a difference term is left over, according to (33.33). We know, however, that just as the point $\bar{P}$ (coordinates $\bar{x}^a$) is equivalent to the initial

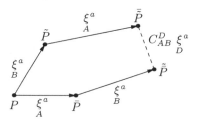

Fig. 33.1. Commuting of infinitesimal motions.

point P, so also is the point $\tilde{P}$ (coordinates $\tilde{x}^a$), see Fig. 33.1; because of the group property of the symmetry transformations it is thus also possible to construct a linear transformation from the Killing vectors which describes the transition from $\bar{P}$ to $\tilde{P}$. Because of (33.33) we then have for this transformation

$$\left(\underset{A}{\xi^n} \frac{\partial}{\partial x^n} \underset{B}{\xi^a} - \underset{B}{\xi^n} \frac{\partial}{\partial x^n} \underset{A}{\xi^a} \right) = C_{AB}^D \underset{D}{\xi^a}, \quad A, B, D = 1, \ldots, r. \quad (33.34)$$

The quantities C_{AB}^D are called the *structure constants* of the group; they are independent of the choice of coordinate system, but do depend upon the choice of basis $\underset{A}{\xi^a}$ and can be simplified (brought to certain normal forms) by suitable basis transformations.

Using the operators

$$\underset{A}{X} = \underset{A}{\xi^n} \frac{\partial}{\partial x^n}, \quad (33.35)$$

(33.34) can also be written in the form of a commutator relation

$$\underset{A}{X}\underset{B}{X} - \underset{B}{X}\underset{A}{X} = \left[\underset{A}{X}, \underset{B}{X} \right] = C_{AB}^D \underset{D}{X}. \quad (33.36)$$

One can show that for arbitrary specified structure constants a group always exists, if these constants have the antisymmetry property

$$C_{AB}^D = -C_{BA}^D \quad (33.37)$$

discernible in (33.34), and satisfy the Jacobi–Lie identity

$$C_{AD}^E C_{BC}^D + C_{BD}^E C_{CA}^D + C_{CD}^E C_{AB}^D = 0, \quad (33.38)$$

which follows from the Jacobi identity

$$\left[\underset{A}{X}, \left[\underset{B}{X}, \underset{C}{X} \right] \right] + \left[\underset{B}{X}, \left[\underset{C}{X}, \underset{A}{X} \right] \right] + \left[\underset{C}{X}, \left[\underset{A}{X}, \underset{B}{X} \right] \right] = 0. \quad (33.39)$$

Examples The group of translations

$$\underset{A}{\xi^a} = \delta_A^a, \quad \underset{A}{X} = \delta_A^a \, \partial/\partial x^a, \quad a = 1, \ldots, n, \quad A = 1, \ldots, n, \quad (33.40)$$

of an n-dimensional flat space is an Abelian group. All its transformations commute, all the structure constants vanish.

The group of rotations of a three-dimensional flat space

$$
\begin{aligned}
\underset{1}{\xi^a} &= (y, -x, 0), & \underset{1}{X} &= y \, \partial/\partial x - x \, \partial/\partial y, \\
\underset{2}{\xi^a} &= (z, 0, -x), & \underset{2}{X} &= z \, \partial/\partial x - x \, \partial/\partial z, \\
\underset{3}{\xi^a} &= (0, z, -y), & \underset{3}{X} &= z \, \partial/\partial y - y \, \partial/\partial z
\end{aligned} \quad (33.41)
$$

has the commutators

$$\left[\underset{1}{X},\underset{2}{X}\right]=\underset{3}{X}, \qquad \left[\underset{2}{X},\underset{3}{X}\right]=\underset{1}{X}, \qquad \left[\underset{3}{X},\underset{1}{X}\right]=\underset{2}{X}. \tag{33.42}$$

Since not all the structure constants vanish, but $C_{12}^3 = C_{23}^1 = C_{31}^2 = 1$, rotations do not commute. The operators $\underset{A}{X}$ and their commutators correspond to the angular-momentum operators of Quantum Mechanics and their commutation rules.

Classification of spaces according to their groups of motion One characterizes the group of motion of a space by the number of its Killing vectors, the structure of the group and the regions of transitivity. Establishing all the non-isomorphic groups G_r of r Killing vectors, that is, of groups whose structure constants cannot be converted into one another by linear transformations of the basis, is a purely mathematical problem of group theory. It is in principle solved: in the literature one can find tables of all such possible groups for the cases of interest in relativity theory.

Thus every group with two elements is either an Abelian group

$$\left[\underset{1}{X},\underset{2}{X}\right]=0, \tag{33.43}$$

or else we have

$$\left[\underset{1}{X},\underset{2}{X}\right]=c_1 \underset{1}{X} + c_2 \underset{2}{X}, \tag{33.44}$$

with $c_1 \neq 0$. In the second case, however, we can always arrive at the normal form

$$\left[\underset{1}{\bar{X}},\underset{2}{\bar{X}}\right]=\underset{1}{\bar{X}} \tag{33.45}$$

by means of a basis transformation

$$\underset{1}{\bar{X}}=c_1 \underset{1}{X} + c_2 \underset{2}{X}, \qquad \underset{2}{\bar{X}}=\underset{2}{X}/c_1. \tag{33.46}$$

The relations (33.43) and (33.45) characterize the two non-isomorphic groups G_2.

Of special interest in cosmology are those groups whose regions of transitivity are three-dimensional spaces (homogeneous models of the universe; all points of the three-dimensional universe are equivalent). All simply transitive groups G_3 lead to such models. A list of all non-isomorphic groups G_3 can be obtained by using the relation

$$\tfrac{1}{2}\varepsilon^{ABE}C_{AB}^D = N^{DE}, \quad A, B = 1, 2, 3, \tag{33.47}$$

which, because of the antisymmetry condition (33.37), maps the nine possible components of the matrix of the structure constants C_{AB}^D onto the 3×3 matrix N^{DE}, and then splitting this matrix further into its symmetrical part n^{DE} and the antisymmetric part, which in turn can be mapped onto a 'vector' a^A:

$$N^{DE} = n^{DE} + \varepsilon^{DEA} a_A. \tag{33.48}$$

If one substitutes the representation of the structure constants resulting from this decomposition,

$$C_{AB}^D = \varepsilon_{EAB}\, n^{DE} + \delta_B^D\, a_A - \delta_A^D\, a_B, \tag{33.49}$$

into the Jacobi–Lie identities (33.38), then these reduce to

$$n^{AB} a_A = 0. \tag{33.50}$$

One can always set $a_A = (a, 0, 0)$ by real linear transformations of the basis operators $\underset{A}{X}$ and moreover transform n_{AB} to principal axes, so that the diagonal elements have only the values $0, \pm 1$. One hence obtains the following normal form for the commutators and the structure constants of a group G_3:

$$
\begin{aligned}
\left[\, \underset{1}{X}, \underset{2}{X} \,\right] &= n_3\, \underset{3}{X} + a\, \underset{2}{X}, \\
\left[\, \underset{2}{X}, \underset{3}{X} \,\right] &= n_1\, \underset{1}{X}, \qquad\quad a n_1 = 0, \\
\left[\, \underset{3}{X}, \underset{1}{X} \,\right] &= n_2\, \underset{2}{X} - a\, \underset{3}{X}, \qquad n_i = 0, \pm 1.
\end{aligned}
\tag{33.51}
$$

As Table 33.1 shows, there are eleven types of groups G_3 altogether, which are distributed amongst the nine so-called Bianchi types I to IX. Notice that in the types VII and VI for $a \neq 0, 1$ one is always dealing with a whole family of non-isomorphic groups.

Table 33.1. *The structure constants of the groups* G_3

Bianchi type	I	II	VII	VI	IX	VIII	V	IV	III	VII	VI
a	0	0	0	0	0	0	1	1	1	a	a
n_1	0	0	0	0	1	1	0	0	0	0	0
n_2	0	0	1	1	1	1	0	0	1	1	1
n_3	0	1	1	−1	1	−1	0	1	−1	1	−1

We should mention further that there are also groups G_4 and G_6 which are (multiply) transitive in three-dimensional space and can therefore likewise correspond to homogeneous models of the Universe.

33.6 Killing vectors and conservation laws

The conservation laws of physics are closely connected with the symmetry properties of physical systems. In the theory of gravitation, the properties of the four-dimensional space also have a physical significance. In this section we shall show how symmetry properties (that is, the existence of Killing vector fields) lead to conservation laws or other simple statements.

Mechanics of a point mass The motion of a point mass on a surface or in Minkowski space in the absence of forces or the pure inertial motion in a Riemannian space (motion in the gravitational field) takes place along a geodesic:

$$\frac{\mathrm{D}^2}{\mathrm{D}\tau^2} x^a = \frac{\mathrm{D}}{\mathrm{D}\tau} u^a = u^a{}_{;b} u^b = 0. \tag{33.52}$$

Contraction of this equation with a Killing vector field $\underset{A}{\xi^a}$ leads to

$$\underset{A}{\xi_a} \mathrm{D}u^a/\mathrm{D}\tau = \mathrm{d}\big(\underset{A}{\xi_a} u^a \big)/\mathrm{d}\tau - u^a \underset{A}{\xi_{a;b}} u^b = 0, \tag{33.53}$$

and taking into account the Killing equation (33.5), that is, the anti-symmetry of $\underset{A}{\xi_{a;b}}$, we have

$$\underset{A}{\xi_a} u^a = \text{const.} \tag{33.54}$$

The quantities $\underset{A}{\xi_a} u^a$ do not change during the motion of the point mass; they are conserved quantities. Thus in mechanics a conservation law is associated with every Killing vector field. In Minkowski space with its ten Killing vectors (33.12) and (33.13) there are accordingly ten conservation laws: the four translational Killing vectors lead to the conservation law for the four-momentum, the three spatial rotations to the angular-momentum law and the three special Lorentz transformations to the centre-of-gravity law.

Interestingly enough, there exist Riemannian spaces in which there are more conservation laws than Killing vectors, that is to say, conservation laws which cannot be traced back to the presence of a symmetry. To see this consider the equation

$$\Xi_{ab;n} + \Xi_{bn;a} + \Xi_{na;n} = 0, \qquad \Xi_{na} = \Xi_{an}, \qquad (33.55)$$

which we shall take as the defining equations for a *Killing tensor* Ξ_{an}. If these equations possess a solution which is not a linear combination of products of Killing vectors, and which thus *cannot* be written in the form

$$\Xi_{ab} = c_0 g_{ab} + \sum_{A,B} c_{AB} \left(\underset{A}{\xi_a} \underset{B}{\xi_b} + \underset{A}{\xi_b} \underset{B}{\xi_a} \right), \qquad c_0, c_{AB} = \text{const.}, \qquad (33.56)$$

then the conservation laws

$$D(\Xi_{ab} u^a u^b)/D\tau = (\Xi_{ab} u^a u^b)_{;i} u^i = 0 \qquad (33.57)$$

which follow from (33.52) and (33.55) are independent of the conservation laws (33.54). One can show that in Minkowski space there exist only the trivial Killing tensors (33.56). An example of a space with a non-trivial Killing tensor is the Kerr metric discussed in Section 37.1. Killing tensors reflect symmetries of the (geodesic) differential equations in a space spanned by the variables (x^a, u^a) rather than those of space-time.

If forces are present and these have a potential,

$$Du_a/D\tau = -\Phi_{,a}, \qquad (33.58)$$

then the conservation law (33.54) is still valid if the potential does not change under the symmetry operation of the space:

$$\underset{A}{\xi^a} \, \Phi_{,a} = \underset{A}{X} \, \Phi = 0. \qquad (33.59)$$

The symmetry group of a mechanical (or general physical) system is thus always a subgroup of the symmetry group of the space in which the system is situated.

Scalar potentials in electrodynamics As the Killing equations (33.5) show, a space possesses a Killing vector field if and only if the Lie derivative of the metric in the direction of this vector field vanishes:

$$\mathcal{L}_\xi g_{nm} = g_{nm,i}\xi^i + g_{im}\xi^i_{,n} + g_{ni}\xi^i_{,m} = 0. \qquad (33.60)$$

We call a physical system in this space invariant under motion in the direction of the Killing vector field if the Lie derivatives of the physical variables vanish. This definition guarantees that the components of the field variables do not change under the motion when one introduces the old coordinate system again at the point reached by the motion (cp. the remarks on the intuitive interpretation of the Lie derivative in Section 18.5).

Thus if in a Riemannian space there exists an electromagnetic field (a test field, or a field which acts gravitationally), then this field possesses a symmetry if and only if, in an appropriate gauge, the four-potential satisfies the condition

$$\mathcal{L}_{\xi} A_m = A_{m,i}\xi^i + A_i\xi^i_{,m} = 0. \tag{33.61}$$

If the associated field tensor is contracted with the Killing vector, then the resulting vector E_m can be written as

$$E_m = F_{mn}\xi^n = (A_{n,m} - A_{m,n})\xi^n = \xi^n A_{n,m} + A_n\xi^n_{,m}; \tag{33.62}$$

that is, E_m can be represented as the gradient of a scalar function Φ:

$$E_m = (\xi^n A_n)_{,m} = -\Phi_{,m}. \tag{33.63}$$

In the absence of charges and currents, or in a simply connected region outside the sources, or, when the current density vector j^a and the Killing vector ξ^a are parallel, then one can derive an analogous statement for the vector

$$H_m = \tilde{F}_{mn}\xi^n \tag{33.64}$$

as well. From the Maxwell equation

$$F^{mn}_{\ ;n} = \tfrac{1}{2}\varepsilon^{mn}_{\ \ ab}\tilde{F}^{ab}_{\ ;n} = j^m/c \tag{33.65}$$

we obtain, upon contracting with $\varepsilon_{mrst}\xi^t$, the equation

$$\left(\tilde{F}_{rs,t} + \tilde{F}_{st,r} + \tilde{F}_{tr,s}\right)\xi^t = 0, \tag{33.66}$$

and since because of (33.61) the Lie derivative of the dual field tensor vanishes,

$$\mathcal{L}_{\xi}\tilde{F}_{mn} = \tilde{F}_{mn,a}\xi^a + \tilde{F}_{an}\xi^a_{,m} + \tilde{F}_{ma}\xi^a_{,n} = 0, \tag{33.67}$$

H_m satisfies the condition $H_{m,a} = H_{a,m}$ and consequently can be written as the gradient of a potential Ψ:

$$H_m = -\Psi_{,m}. \tag{33.68}$$

The six quantities E^m and H^m, which (for $\xi_a\xi^a \neq 0$) completely describe the Maxwell field, can thus be represented as gradients of two scalar potentials, if ξ^a is a Killing vector. These potentials are generalizations of the electrostatic and magnetic scalar potentials which one usually introduces in Minkowski space when the fields are static, that is, admit a timelike hypersurface-orthogonal Killing vector.

Equilibrium condition in thermostatics As we have shown in Section 21.5, a system is in thermodynamic equilibrium only if the Lie derivative of the metric in the direction of u^a/T vanishes; that is, if this vector is a Killing vector. It is static if, further, the vector is hypersurface-orthogonal. In the rest system $u^a = (0,0,0,c/\sqrt{-g_{44}})$ of the matter, the components $g_{4\alpha}$ then vanish, and the metric does not change under time reversal.

Substituting the vector $\xi^a = u^a/T$ into the Killing equations (33.60), we have (when $g_{4\alpha} = 0$)

$$g_{\alpha\beta,4} = 0, \qquad T_{,a} = 0, \qquad \left(\sqrt{-g_{44}}\,T\right)_{,\alpha} = 0. \qquad (33.69)$$

By means of a transformation of time only, $\mathrm{d}t' = \sqrt{-g_{44}}\,T\,\mathrm{d}t$, $g_{4'4'} = -1/T^2$, one can convert these equations into

$$g_{mn,4} = 0, \qquad \left(\sqrt{-g_{44}}\,T\right)_{,i} = 0 \qquad (33.70)$$

(we have once again dropped the dash on the indices).

The equations (33.70) are the equilibrium conditions in the rest system of the matter. A system is thus in equilibrium not when the temperature gradient vanishes, but rather when the gradient of $\sqrt{-g_{44}}\,T$ is zero. This condition can be interpreted in the following way: in equilibrium, the change in temperature just compensates the energy which has to be fed in or carried away under (virtual) transport of a volume element in the gravitational field.

Observables in quantum mechanics In the usual coordinate representation of quantum mechanics the operators of momentum and angular momentum associated with the physical observables correspond to the operators $\underset{A}{X}$ of the translations (33.40) and rotations (33.41) of the three-dimensional Euclidean space. There thus exists a close connection between those quantities which remain constant for a more extensive physical system (for example, an atom) because of the symmetry of the space, and those which can meaningfully be used to describe part of a system (for example, an electron). This connection explains the difficulties involved in carrying over the quantum mechanics of Minkowski space to a general Riemannian space, which of course possesses no Killing vectors at all.

Conservation laws for general fields In a Riemannian space the local conservation law

$$T^{ik}{}_{;k} = 0 \qquad (33.71)$$

holds for the energy-momentum tensor of an arbitrary field (an arbitrary matter distribution). But no genuine integral conservation law can be associated with it, because of the non-existence of a Gauss law for tensor fields of second or higher rank.

If, however, a Killing vector field $\underset{A}{\xi^a}$ exists in the space, then it follows from (33.60) and the Killing equation (33.5) that

$$\left(\underset{A}{\xi_i}T^{ik}\right)_{;k} = \underset{A}{\xi_{i;k}}T^{ik} + \underset{A}{\xi_i}T^{ik}{}_{;k} = 0, \tag{33.72}$$

and the Gauss law can be applied to this local conservation law for a vector field (see Section 20.5). Under certain mathematical assumptions a conservation law

$$\underset{A}{T} = \int\limits_{x^4=\text{const.}} T^{ia}\underset{A}{\xi_i}\,\mathrm{d}f_a = \int\limits_{x^4=\text{const.}} T^{i4}\underset{A}{\xi_i}\sqrt{-g}\,\mathrm{d}x^1\,\mathrm{d}x^2\,\mathrm{d}x^3 = \text{const.} \tag{33.73}$$

can then be associated with every Killing vector field of the space. If the Killing vector is timelike, the associated conserved quantity will be called energy. Whether for a spacelike Killing vector one uses the label 'momentum' or 'angular momentum' is sometimes only a matter of definition. In such a case one can be guided by the transitivity properties of the group of motion (the three translations in flat space yield a transitive group, the three spatial rotations are intransitive), by the commutators of the associated operators $\underset{A}{X}$ or by the structure of the Killing vectors in the asymptotically flat far-field zone.

Starting from the identity

$$(\xi^{a;b} - \xi^{b;a})_{;a;b} = 0, \tag{33.74}$$

valid for all vectors, one can recast the conservation law (33.72) in a different form (Komar 1959). If ξ^a is a Killing vector then this identity, (33.5) and (33.22) imply

$$(R^{mb}\xi_m)_{;b} = 0 = \left[(T^{mb} - \tfrac{1}{2}g^{mb}T)\xi_m\right]_{;b}, \tag{33.75}$$

which agrees with (33.72) since $T_{,m}\xi^m = 0$.

From the standpoint of the symmetry properties of a field and of the connection between symmetries and conservation laws, one would therefore answer the question, discussed in detail in Section 28.4, of the validity of an energy law for and in a gravitational field in the following way. The energy of a gravitating system can be defined if a timelike Killing vector exists, and then it is always conserved.

Exercises

33.1 Show that if $\mathcal{L}_\xi F_{ab} = 0$, then there is a gauge such that $\mathcal{L}_\xi A_a = 0$.

33.2   Assume there are two Killing vectors ξ and η. Is it always possible to gauge the four-potential A_n by $\mathcal{L}_\xi A_n = 0 = \mathcal{L}_\eta A_n$?

33.3   Find the Killing vectors of the metrics (34.1) and (19.41).

33.4   Show that the Killing vectors (33.17) satisfy (33.42).

33.5   In a Minkowski space, there is a rotationally symmetric Maxwell field. Use the two potentials ϕ and ψ to formulate Maxwell's equations in cylindrical coordinates.

34
A survey of some selected classes of exact solutions

A compendium of all currently known solutions of Einstein's equations fills a thick book, see for example Stephani *et al.* (2003). In spite of the complexity of the Einstein field equations many exact solutions are known, but most have little physical relevance, that is, it is most improbable that sources with that specific structure exist in our universe. On the other hand, exact solutions to many realistic problems, for example, the two-body problem, are unknown. Here we must restrict ourselves to a few brief references to, and remarks about, rather arbitrarily selected classes of solutions. In Chapters 37, 41 and 43 we shall discuss at greater length several solutions which can be used as models for stars or the universe.

Many of the known solutions have been found by assuming from the very beginning a high degeneracy (Petrov types D, N or 0) or a high symmetry. We shall follow this approach.

34.1 Degenerate vacuum solutions

Several classes of degenerate vacuum solutions, that is, solutions with (at least) one shear-free, geodesic null congruence, have been systematically investigated. These classes include the following.

Type D solutions They are all known; their most important representative is the Kerr solution, see Section 37.1. An example of a type D

solution which it has not (yet?) been possible to interpret physically is the metric

$$ds^2 = \frac{dz^2}{b/z - 1} + \left(\frac{b}{z} - 1\right)d\varphi^2 + z^2(dr^2 - \sinh^2 r\, c^2\, dt^2), \qquad (34.1)$$

which arises out of the Schwarzschild metric (23.23) via the transformation

$$\vartheta \to ir, \quad \varphi \to ict, \quad r \to z, \quad ct \to i\varphi \qquad (34.2)$$

and an overall change of sign. Note that the coordinate labels in (34.1) are completely arbitrary, for example, φ need not be an angular coordinate.

Degenerate solutions, whose eigenvector field is rotation-free and divergence-free In Section 31.3 it was explained that the most important physical and mathematical properties of a null vector field are contained in the three optical scalars σ (shear), ω (rotation) and θ (divergence). For degenerate solutions σ vanishes, in accordance with the definition, and the vanishing of the other scalars as well simplifies the field equations considerably. Besides some special type D solutions, which have this property, the structure of all types III and N which fall into this category is known. Their metrics have the form

$$ds^2 = |dx + i\, dy + W\, du|^2 + 2\, du\, dv + H\, du^2, \qquad (34.3)$$

where the functions W (complex) and H (real) must satisfy two partial differential equations. The most important representatives of this class are the plane fronted waves with parallel rays (29.39)–(29.40) found in Section 29.3.

Degenerate solutions, whose eigenvector field is rotation-free, but has a non-vanishing divergence These solutions are comprised by

$$ds^2 = r^2 P^{-2}(dx^2 + dy^2) + 2\, du\, dr + 2H\, du^2, \qquad (34.4)$$

$$2H = -\frac{2m}{r} - P^2\left(\frac{\partial^2}{\partial x^2} + \frac{\partial^2}{\partial y^2}\right)\ln P - 2r\frac{\partial \ln P}{\partial u}, \quad m = 0, 1,$$

where the function $P(x, y, u)$ is a solution of the equation (Robinson and Trautman 1962)

$$P^2\left(\frac{\partial^2}{\partial x^2} + \frac{\partial^2}{\partial y^2}\right)\left[P^2\left(\frac{\partial^2}{\partial x^2} + \frac{\partial^2}{\partial y^2}\right)\ln P\right] = -12m\frac{\partial \ln P}{\partial u}. \qquad (34.5)$$

It can be shown that some of these time-dependent solutions finally settle down to the Schwarzschild solution which is contained here as a special case. But the hope of finding metrics amongst the solutions (34.3) and (34.4) which describe the radiation field of a bounded physically meaningful matter distribution has not been realized.

34.2 Vacuum solutions with special symmetry properties

The Weyl class – axisymmetric, static vacuum fields In flat space, physical configurations which are static and also spherically or axisymmetric (in cylindrical coordinates: φ-independent) are particularly simple. The analogue of spherical symmetry leads immediately to the Schwarzschild solution (Birkhoff theorem, see Section 23.2). The relativistic generalization of axially symmetric, static vacuum fields is the Weyl class: in a suitable coordinate system the solution should not depend upon the time t nor the cyclic coordinate φ and should not change under the transformations $t \rightarrow -t$ and $\varphi \rightarrow -\varphi$ (should contain no terms $g_{\varphi r}$, $g_{\varphi t}$, $g_{\varphi \vartheta}$, g_{tr} or $g_{t\vartheta}$). The last condition means physically that a time-independent rotation of the source whose external field we are considering is forbidden.

To give an invariant definition: all vacuum solutions with two commuting, hypersurface-orthogonal Killing vectors, of which one is timelike, whilst the world line congruence associated with the spacelike vector consists of closed curves of finite length, belong to Weyl's class. (A metric is said to be *stationary* when it possesses a timelike Killing vector, and *static* if in addition that vector is hypersurface-orthogonal). One can show that for this symmetry the metric can be transformed to the normal form (Weyl 1917)

$$ds^2 = e^{-2U}[e^{2k}(d\rho^2 + dz^2) + \rho^2\, d\varphi^2] - e^{2U}c^2\, dt^2, \qquad (34.6)$$

where the functions $U(\rho, z)$ and $k(\rho, z)$ are to be determined from

$$U_{,\rho\rho} + \rho^{-1}U_{,\rho} + U_{,zz} = 0 \qquad (34.7)$$

and

$$\rho^{-1}k_{,\rho} = U_{,\rho}^{\,2} - U_{,z}^{\,2}, \quad \rho^{-1}k_{,z} = 2U_{,\rho}U_{,z}. \qquad (34.8)$$

Since (34.8) is always integrable when (34.7) holds, we have evidence which apparently suggests the astonishing fact that from every (φ-independent) solution U of the flat space potential equation (34.7), that is, for every static axisymmetric vacuum solution of the Newtonian gravitation theory, one can obtain a reasonable solution to the Einstein theory by simply performing two line integrals (34.8). This statement is, however, false in this oversimplified form. This is because we have not yet ensured that the singular line $\rho = 0$ of the coordinate system is not singular in the physical sense as well, with infinite mass density. To exclude the occurrence of such a singularity one has to demand that for every infinitesimal circle about the z-axis the ratio of the circumference to the radius is 2π (space-time is locally a Minkowski space); this is done by the condition

$$k = 0 \quad \text{for} \quad \rho = 0 \quad (z \text{ arbitrary}). \tag{34.9}$$

The differential equation (34.7) is of course linear, and solutions U can be superposed, but the sum of two solutions, which individually have a regular behaviour, will not in general satisfy the subsidiary condition (34.9). The simple superposition of fields of two sources does not yield a field whose sources are in gravitational equilibrium; to keep two attracting masses apart one needs a singular mass distribution on the axis.

In Newtonian theory, the spherically symmetric gravitational field is given by $U = 1/r = 1/\sqrt{\rho^2 + z^2}$. Surprisingly, this U does not lead to the Schwarzschild solution, see Exercise 34.1. Rather

$$e^{2U} = \frac{r_+ + r_- - 2m}{r_+ + r_- + 2m}, \quad e^{2k} = \frac{(r_+ + r_-)^2 - 4m^2}{4r_+ r_-},$$
$$r_\pm^2 = \rho^2 + (z \pm m)^2, \tag{34.10}$$

gives the Schwarzschild solution in Weyl coordinates, cp. Fig. 34.1. This is the analogue of the field of a massive rod (line) of length $2m$ in Newtonian physics. The singular surface $r = 2m$ is just this line. This example clearly shows that Weyl coordinates do not have an immediate physical meaning, and moreover, that in relativity it is dangerous to connect an intuitive meaning with coordinates suitably named.

Axisymmetric, stationary vacuum solutions In the Newtonian gravitation theory the gravitational field of an axisymmetric source distribution

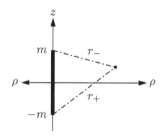

Fig. 34.1. Schwarzschild solution in Weyl coordinates.

does not depend upon a possible rotation of the source about the symmetry axis. In the Einstein theory, on the other hand, the metric will be altered by the corresponding matter current, which enters the energy-momentum tensor. For uniform rotation, the metric will, of course, be independent of t and φ, but the direction of time (the sense of rotation) will be important. Formulated invariantly: axisymmetric, stationary vacuum fields possess two commuting Killing vectors (an Abelian group of motion G_2), of which one is timelike (but not necessarily hypersurface-orthogonal). One can show that vacuum metrics of this class can be transformed into the canonical form

$$ds^2 = e^{-2U}[e^{2k}(d\rho^2 + dz^2) + \rho^2\,d\varphi^2] - e^{2U}[dt + A\,d\varphi]^2, \qquad (34.11)$$

where the functions U, k and A depend only upon ρ and z and have to satisfy the differential equations

$$\begin{aligned} U_{,\rho\rho} + \rho^{-1}U_{,\rho} + U_{,zz} &= -\tfrac{1}{2}e^{4U}\rho^{-2}(A_{,\rho}^2 + A_{,z}^2), \\ (e^{4U}\rho^{-1}A_{,z})_{,z} + (e^{4U}\rho^{-1}A_{,\rho})_{,\rho} &= 0, \end{aligned} \qquad (34.12)$$

$$\begin{aligned} \rho^{-1}k_{,\rho} &= U_{,\rho}^2 - U_{,z}^2 - \tfrac{1}{4}e^{4U}\rho^{-2}\left(A_{,\rho}^2 - A_{,z}^2\right), \\ \rho^{-1}k_{,z} &= 2U_{,\rho}U_{,z} - \tfrac{1}{2}e^{4U}\rho^{-2}A_{,\rho}A_{,z}. \end{aligned} \qquad (34.13)$$

The system (34.13) is always integrable when (34.12) holds, so that k can simply be calculated by quadrature. The system (34.12) has been extensively investigated with regard both to simple solutions and to the possibility of producing new solutions from those already known by, for example, Bäcklund transformations.

The best known and most important representative of the class of axisymmetric, stationary vacuum solutions is the Kerr metric, given by

$$e^{2U} = \frac{(r_+ + r_-)^2(2\cos^2\varphi - 1) + 4m(m - r_+ - r_-)\cos^2\varphi}{(r_+ + r_-)^2 + 4m(m + r_- + r_+)\cos^2\varphi}, \qquad (34.14)$$

$$r_\pm^2 = \rho^2 + (m\cos\Phi \pm z)^2$$

(the Weyl coordinates used here are related to the Boyer–Lindquist coordinates of (37.1) by $z = (r - m)\cos\vartheta$, $\rho = \sqrt{r^2 + a^2 - 2mr}\sin\vartheta$).

Cylindrically symmetric solutions The cylindrically symmetric solutions are the counterpart of the static axisymmetric Weyl solution; instead of a timelike and a spacelike Killing vector, one now has two spacelike ones. The metric is independent of z and φ and can be written as

$$ds^2 = e^{-2U}[e^{2k}(d\rho^2 - dt^2) + W^2 d\varphi^2] + e^{2U}dz^2. \qquad (34.15)$$

The function W has to obey the wave equation

$$W_{,\rho\rho} - W_{,tt} = 0, \tag{34.16}$$

which is solved by

$$W = f(t - \rho) + g(t + \rho). \tag{34.17}$$

Numerous solutions are known, which describe colliding plane waves.

For $W_{,\rho}^2 - W_{,t}^2 > 0$, one can adjust the coordinates by $W = \rho$, and the rest of the field equations then yield

$$U_{,\rho\rho} + \rho^{-1}U_{,\rho} - U_{,tt} = 0 \tag{34.18}$$

$$\rho^{-1}k_{,\rho} = U_{,\rho}^2 + U_{,t}^2, \quad \rho^{-1}k_{,t} = 2U_{,\rho}U_{,t}. \tag{34.19}$$

These are the *Einstein–Rosen waves*, which are the counterpart of the flat-space cylindrical waves which obey (34.18).

For $W = \sin\rho \sin t$, the equation for U reads

$$\sin t(U_{,\rho}\sin\rho)_{,\rho} - \sin\rho(U_{,t}\sin t)_{,t} = 0. \tag{34.20}$$

Standard separation leads to

$$U = \sum_n c_n h_n(\cos\rho)h_n(\cos t), \quad c_n = \text{const.}, \tag{34.21}$$

where the h_n are the Legendre functions (first and second kind). This form of the solution suggests that we should consider also ρ as a periodic coordinate. Indeed, the regularity conditions (e.g. at $\rho = 0 = \pi$) can be satisfied, and ρ, z and φ can be interpreted as generalized Euler angle coordinates. These solutions are known as *Gowdy universes*; they are closed universes containing only a gravitational radiation field, and they have final and initial collapse singularities at $t = 0$ and $t = \pi$.

34.3 Perfect fluid solutions with special symmetry properties

Perfect fluids are often used to model the interiors of stars, or the galaxy distribution in our Universe. Many exact perfect fluid solutions are known, but only very few of them can serve as realistic models. The most discussed ones are the (static or non-static) spherically symmetric solutions; contained here are also the Friedmann universes, see Section 41.2.

For a realistic model of a spherically symmetric star, one should prescribe an equation of state $f(\mu, p) = 0$. But then in most cases it turns out that the field equations cannot be solved analytically. So one rather likes to solve the field equations by making some assumptions on the metric functions, and only then does one calculate pressure p and mass density μ.

To give at least one example, we take the McVittie solution

$$ds^2 = (1+f)^4 e^{g(t)} [1 + r^2/4R^2]^{-2} [dr^2 + r^2(d\vartheta^2 + \sinh^2 \vartheta \, d\varphi^2)]$$
$$-(1-f)dt^2/(1+f), \tag{34.22}$$
$$2f = me^{-g(t)/2} [1 + r^2/4R^2]^{1/2}/r, \quad R = \text{const.}$$

For $m = 0$, this a Robertson–Walker metric, the space $t = \text{const.}$ is a space of constant (positive) curvature, cp. Section 19.5. For $g = 0$ and $r^2/R^2 \ll 1$, the solution approaches the exterior Schwarzschild solution (23.67). So the McVittie solution has been interpreted as a particle in a homogeneous universe.

Stars are usually rotating, and their interior should be modelled by a stationary axisymmetric perfect fluid solution. But here the situation is even worse than in the spherically symmetric case. A spherically symmetric perfect fluid with a surface $p = 0$ can nearly always be matched to the exterior Schwarzschild solution; there is only one spherically symmetric vacuum solution, and the surface of the star always is a sphere. For a rotating star, one cannot prescribe the shape of the surface, then solve the interior problem, and then find an exterior vacuum solution to be matched at the surface: such an exterior solution (asymptotically flat, with no singularities) need not exist! Rather one has to solve the interior and the exterior problem in one go, finding as a by-product the shape of the star's surface. No realistic model of a truly relativistic rotating star has been found so far. Only in the extreme limit of a rotating disc of dust is the complete solution known (Neugebauer and Meinel 1993).

Exercises

34.1   Determine the function k for the potential $U = r^{-1} = (\rho^2 + z^2)^{-1/2}$ and show that the resulting metric (the Chazy–Curzon particle) is not spherically symmetric.

34.2   Find the static cylindrically symmetric vacuum solutions ($U = U(\rho)$, $k = k(\rho)$ in (34.6) or in (34.15)).

34.3   Show that $ds^2 = K^2[(dx^1)^2 + \sin^2 x^1 (dx^2) + \sin^2 ct \, (dx^3)^2 - c^2 \, dt^2]$ is the gravitational field of a covariant constant electromagnetic field whose only non-vanishing omponent is $F_{34} = K^{-3} \sin ct$.

VI. Gravitational collapse and black holes

In the examples and applications considered up until now we have always correctly taken into account the non-linearity of the Einstein equations, but most of the properties and effects discussed do not differ qualitatively from those of other classical (linear) fields. Now, in the discussion of black holes and of cosmological models, we are going to encounter properties of the gravitational field which deviate clearly from those of a linear field. The structure of the space-time is essentially changed by comparison with that of Minkowski space, and essentially new types of questions arise.

35
The Schwarzschild singularity

35.1 How does one examine the singular points of a metric?

A quick glance at the Schwarzschild metric,

$$ds^2 = \frac{dr^2}{1 - 2M/r} + r^2 \left(d\vartheta^2 + \sin^2 \vartheta \, d\varphi^2 \right) - (1 - 2M/r)c^2 \, dt^2 \quad (35.1)$$

shows that a singularity of the metric tensor (of the component g_{rr}) is present at $r = 2M$. In our earlier discussion of the Schwarzschild metric in Chapter 23 we had set this problem aside with the remark that the radius $r = 2M$ lies far inside a celestial body, where the vacuum solution is of course no longer appropriate. Now, however, we shall turn to the question of whether and in what sense there is a singularity of the metric at $r = 2M$ and what the physical aspects of this are.

Places where a field is singular constitute a well-known phenomenon of classical physics. In electrostatics the spherically symmetric Coulomb field

$$U = \frac{e}{4\pi r} \tag{35.2}$$

is singular at $r = 0$, because an infinitely large charge density (point charge) is present there. In non-linear theories the situation is more complicated, because the singularity need not occur at the position of the source. Einstein hoped that the singularities of the gravitational field would represent elementary particles, that the general theory of relativity would thus to a certain degree automatically yield a (non-quantum-field-theoretical) theory of elementary particles. This hope has not been fulfilled. Meanwhile, however, much has been learnt about the nature of singularities of the gravitational field and about the physical effects which occur there. Here we shall have to limit ourselves to the description of a few basic ideas.

A singular coordinate system can evidently give a false indication of a singularity of the space. For example, in flat three-dimensional space spherical coordinates are singular at $r = 0$ in the sense that $\sqrt{g}$ is zero there and $g^{\vartheta\vartheta}$ and $g^{\varphi\varphi}$ become infinite, without the space showing any peculiar properties there. Therefore, if the metric is singular at a point, one investigates whether this singularity can be removed by introducing a new coordinate system. Or, appealing more to physical intuition, one asks whether a freely falling observer can reach this point and can use a local Minkowski system there. If both are possible, then the observer notices no peculiarities of the physical laws and phenomena locally, and hence there is no singularity present.

Singular points or lines can also arise if a hole has been cut in the universe by mistake, its edge appearing as a singularity. Of course one can repair such a defect by substituting a piece of universe back in, that is, one can complete the space by extension of a metric beyond its initially specified region of validity by unbounded extension of geodesics.

In distinction from these two local types of investigation, one can also examine the topological properties of the space in the neighbourhood of a singularity and, for example, ask what possibilities there are of interactions between the outside world and the neighbourhood of the singularity, that is, which points of the space can be linked to one another by test particles or by light rays.

We shall now elucidate some of these questions by reference to the simple example of the Schwarzschild metric. For a more exact discussion see for example Hawking and Ellis (1975).

35.2 Radial geodesics near $r = 2M$

Soon after the Schwarzschild metric had been obtained as a solution of the field equations, it was recognized that both the determinant of the metric

$$-g = r^4 \sin^2 \vartheta \qquad (35.3)$$

and also the invariant

$$R_{abcd} R^{abcd} = 48 M^2 / r^6 \qquad (35.4)$$

associated with the curvature tensor are regular on the 'singular' surface $r = 2M$. This suggests that no genuine singularity is present there, but rather that only the coordinate system becomes singular.

In order better to understand the physical conditions in the neighbourhood of $r = 2M$, we investigate the radial geodesics, information about which is provided by the line element

$$ds^2 = \frac{dr^2}{1 - 2M/r} - (1 - 2M/r)c^2 \, dt^2. \qquad (35.5)$$

From (35.5) or from (23.30) and (23.31) we obtain for the trajectories of test particles

$$\frac{dr}{d\tau} = \pm\sqrt{A^2 - c^2(1 - 2M/r)}, \quad \frac{dct}{d\tau} = \frac{A}{1 - 2M/r} \qquad (35.6)$$

($A = $ const.). For photons one has $ds^2 = 0$, that is

$$dr = \pm(1 - 2M/r)c \, dt. \qquad (35.7)$$

For a test particle (for a freely falling observer) passing from $r = r_0$ to $r = 2M$, equations (35.5) and (35.6) tell us that an infinitely long time

$$\int dt = \int_{r_0}^{2M} \frac{A}{c} \frac{dr}{(1 - 2M/r)\sqrt{A^2 - c^2(1 - 2M/r)}} \quad \rightarrow \quad \infty \qquad (35.8)$$

is required to traverse the finite distance

$$L_0 = \int_{r_0}^{2M} \frac{dr}{\sqrt{1 - 2M/r}}, \qquad (35.9)$$

but that the destination is reached in the finite proper time

$$\tau_0 = \int_{r_0}^{2M} \frac{dr}{\sqrt{A^2 - c^2(1 - 2M/r)}}. \qquad (35.10)$$

The freely falling observer would therefore probably not notice anything special at $r = 2M$; but the coordinates r and t are not really suitable for describing his motion.

A photon would likewise require an infinitely long time, namely,

$$T_0 = \frac{1}{c} \int_{r_0}^{2M} \frac{\mathrm{d}r}{1 - 2M/r} \qquad (35.11)$$

to cover the finite stretch L_0 (35.9) – and again the coordinate time t proves physically unsuitable for describing the process.

35.3 The Schwarzschild solution in other coordinate systems

We seek coordinate systems which are better adapted to the description of physical processes in the neighbourhood of $r = 2M$ than is the usual Schwarzschild metric, coordinate systems which may possibly even cover the space-time completely. Notice that an extension of the Schwarzschild metric from the exterior space across the surface $r = 2M$ does not necessarily have to lead from (35.1) to the metric

$$\mathrm{d}s^2 = (2M/r - 1)c^2\,\mathrm{d}t^2 + r^2\big(\mathrm{d}\vartheta^2 + \sin^2\vartheta\,\mathrm{d}\varphi^2\big) - \frac{\mathrm{d}r^2}{2M/r - 1}, \qquad (35.12)$$

which – with $r < 2m$ – one could of course regard as the metric 'inside' $r = 2M$ (where r is a timelike, and t a spacelike coordinate). For $r = 2M$ the metric (35.1) is completely undefined, and by extension of the metric of the exterior space into $r < 2M$ one could also arrive in a completely different region of the 'universal' Schwarzschild solution, just as by crossing a branch cut one can reach another branch of the Riemannian surface of an analytic function. We must therefore distinguish between the Schwarzschild *metric*, which is only valid for $r > 2M$, and the general Schwarzschild *solution*, which is the (yet to be revealed) maximal extension of the Schwarzschild metric, which contains (35.1) as one section, but which can also be described in completely different coordinates. We shall now meet three new coordinate systems (metrics) which describe various sections of the Schwarzschild solution.

One can adapt the coordinate system to a freely falling observer by the transformations

$$\mathrm{d}T = \mathrm{d}t + \sqrt{\frac{2M}{r}}\,\frac{\mathrm{d}r}{c(1 - 2M/r)}, \qquad (35.13)$$

$$cT(r,t) = ct + 2\sqrt{2Mr} + 2M\ln\big|(\sqrt{r} - \sqrt{2M})/(\sqrt{r} + \sqrt{2M})\big|,$$

$$dR = c\,dT + \frac{dr\sqrt{r}}{\sqrt{2M}} = c\,dt + \frac{\sqrt{r}}{\sqrt{2M}}\frac{dr}{1-2M/r},$$

$$r(R,cT) = \left[(R-cT)3\sqrt{M/2}\right]^{2/3}. \tag{35.14}$$

In this way we pass from the Schwarzschild metric to the Lemaître metric (Lemaître 1933)

$$ds^2 = \frac{2M}{r}\,dR^2 + r^2\left(d\vartheta^2 + \sin^2\vartheta\,d\varphi^2\right) - c^2\,dT^2, \quad r = r(R,cT). \tag{35.15}$$

T is clearly the proper time for particles which are at rest in the coordinate system (35.15); and because of (35.14) and (35.6) $dR = 0$ holds exactly for those particles which are initially at rest at infinity $(A = c)$ and then fall freely and radially. The line element (35.15) is regular at $r = 2M$, and a freely falling observer notices nothing peculiar there; only the point $r = 0$ is singular. A drawback of this metric is that the static Schwarzschild solution is described by time-dependent metric functions.

In another coordinate system null geodesics are preferred to the time-like geodesics used above. If one introduces a retarded time v by

$$dv = dt - dt^* = dt + dr/c(1 - 2M/r),$$

$$cv = ct + r + 2M\ln(r - 2M) - 2M\ln 2M \tag{35.16}$$

(dt^* is the time needed by a radially falling photon to complete the distance dr), then from the Schwarzschild metric (35.1) one arrives at the Eddington–Finkelstein metric (Eddington 1924, Finkelstein 1958)

$$ds^2 = 2c\,dr\,dv + r^2\left(d\vartheta^2 + \sin^2\vartheta\,d\varphi^2\right) - (1 - 2M/r)c^2\,dv^2, \tag{35.17}$$

in which light rays travelling inward radially are described by $dv = 0$. In these coordinates, too, the metric functions are only singular at $r = 0$ (the vanishing of g_{rr} at $r = 2M$ implies no loss of dimension, since the determinant of the metric (35.17) does not vanish there). The line element (35.17) is not invariant under time reversal $v \to -v$, which corresponds to a time reflection $t \to -t$ and a substitution of inward-travelling by outward-travelling photons. But from this time reversal we obtain another section of the universal Schwarzschild solution.

The maximal extension of the Schwarzschild metric is a metric which contains all the sections considered up until now as component spaces and which cannot be further extended. It is reached by introducing the advanced time u,

$$cu = ct - r - 2M\ln(r - 2M) + 2M\ln 2M,$$

$$c(v - u) = 2r + 4M\ln(r - 2M) - 4M\ln 2M, \tag{35.18}$$

into the metric (35.17), which hence (after elimination of dr) in '*null coordinates*' u, v takes the form

$$ds^2 = r^2(d\vartheta^2 + \sin^2\vartheta \, d\varphi^2) - c^2 \, du \, dv \, (1 - 2M/r), \qquad (35.19)$$

and by then making the coordinate transformations

$$v' = e^{cv/4M}, \quad u' = -e^{-uc/4M}, \quad z = \tfrac{1}{2}(v' - u'), \quad w = \tfrac{1}{2}(v' + u'). \quad (35.20)$$

The result is the Kruskal form of the metric (Kruskal 1960) representing the Schwarzschild solution

$$ds^2 = 32M^3 r^{-1} e^{-r/2M} (dz^2 - dw^2) + r^2(w, z)(d\vartheta^2 + \sin^2\vartheta \, d\varphi^2), \quad (35.21)$$

which is related to the original Schwarzschild metric by

$$z^2 - w^2 = \left(\frac{r}{2M} - 1\right) e^{r/2M}, \quad \frac{w}{z} = \frac{1 - e^{-2ct/4M}}{1 + e^{-2ct/4M}} = \tanh\frac{ct}{4M}. \quad (35.22)$$

In the Kruskal metric (35.21) ϑ and φ are spherical coordinates (coordinates on the subspaces with spherical symmetry). The coordinates z and w are spacelike and timelike, respectively; they can take positive and negative values, but are restricted so that r is positive.

We now want to describe briefly how the Schwarzschild metric and its singularity appear from the standpoint of the Kruskal metric. The exterior space of the Schwarzschild metric ($r > 2M$, t finite) corresponds to region I of Fig. 35.1, where $z > |w|$. The rays $w = \pm z$, $z \geq 0$, form the

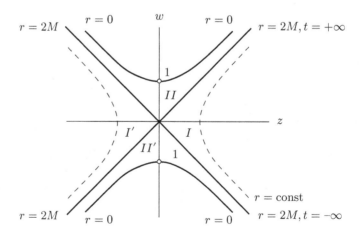

Fig. 35.1. The Kruskal diagram of the Schwarzschild solution (ϑ, φ suppressed).

boundary of this region, which is described in the r, t coordinates by $t = \pm\infty$ or $r = 2M$. If one crosses this boundary inwards into region II, that is, if one crosses $t = \pm\infty$, $r = 2M$, then one arrives in the 'interior' of the usual Schwarzschild metric; the Eddington–Finkelstein metric (35.17) covers precisely these two regions. The regions I' and II', which can be reached by further varying the coordinates w and z, are isometric to (metrically indistinguishable from) regions I and II: the maximal extension of the Schwarzschild solution contains both the exterior part $(r > 2M)$ and the interior part $(r < 2M)$ of the Schwarzschild metric twice. No boundaries or singularities occur, with the exception of (what can be shown to be) the genuine singularity $r = 0$, which cannot be removed by coordinate transformations, and which is represented in the Kruskal diagram by two hyperbolae.

If we are inside the gravitational field of a spherically symmetric star, that is, within region I, then because of the existence of the star, whose surface is described in Fig. 35.1 by the line $r = $ const., the regions I', II and II' are of course to be replaced by an interior solution. Before turning to the question of whether and in what manner the *complete* Schwarzschild solution (including I', II and II') can be realized in nature, we want to discuss more closely the physical consequences of the maximal extension of the Schwarzschild metric.

35.4 The Schwarzschild solution as a black hole

From the mathematical description given in the previous section of the different regions of the Schwarzschild solution, one might gain the impression that it would be possible for an observer to pass from our universe (region I) through the Schwarzschild singularity $r = 2M$ and the interior space (region II) into another universe (region I'), which is again the exterior space of a Schwarzschild metric. Since, however, this observer requires an infinite time (as measured in proper time by the people left behind) just to reach $r = 2M$, he would have vanished forever to those remaining behind. Or, alternatively, while they believe him to be still on the way to $r = 2M$, he has long since (as measured in his proper time) been exploring the new universe I'. To see whether such journeys are possible, we must examine more carefully the properties of the geodesics of the Schwarzschild solution. Our traveller need not necessarily fall freely (move on a geodesic), since he can of course use a rocket; but he can never travel faster than light. Hence, with regard to possible journeys and to the physical relations between the different regions of the Schwarzschild solution, the course of the null geodesics (light rays) is particularly important.

If we limit ourselves to purely radial motions ($\vartheta = $ const., $\varphi = $ const.), then the line element

$$ds^2 = 32M^3 r^{-1} e^{-r/2M} \left(dz^2 - dw^2 \right) \qquad (35.23)$$

determines the course of geodesics. This metric is conformally flat. Null geodesics ($ds^2 = 0$) have

$$dz = \pm \, dw, \qquad (35.24)$$

they are straight lines inclined at 45° (or 135°) in the zw-plane. This simple form for the null geodesics follows from the choice of our coordinates, which were deliberately adapted to light propagation.

If one inserts into the Kruskal diagram all the null geodesics which in region I run radially inwards (t increasing, r decreasing) or radially outwards (t increasing, r increasing) and extends them across $r = 2M$, then one obtains the result sketched in Fig. 35.2. All light rays going radially inwards intersect $r = 2M$ (for $t = \infty$), penetrate into region II and end up at the singularity $r = 0$; all light rays running outwards come from the region II' and the singularity there.

Thus one cannot send radially directed light rays from our world (I) into the regions I' or II'; only region II is within reach, and once the photon is there it cannot avoid the singularity $r = 0$. One might think

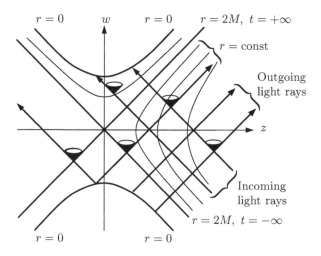

Fig. 35.2. Null geodesics and local light cones in the Schwarzschild solution.

of escaping the singularity $r = 0$ by using non-radially directed light rays or observers with suitable rockets. But in fact once a photon is in region II, it cannot avoid 'falling' to $r = 0$. Addition of new degrees of freedom to (35.23) implies for photons that because

$$\mathrm{d}s^2 = 0 = 32M^3 r^{-1} \mathrm{e}^{-r/2M} \left(\mathrm{d}z^2 - \mathrm{d}w^2\right) + \mathrm{d}\sigma^2, \quad \mathrm{d}\sigma^2 > 0, \quad (35.25)$$

$\mathrm{d}w^2$ must be larger in relation to $\mathrm{d}z^2$ than for radial photons since the term $\mathrm{d}\sigma^2$ must also be compensated. In Fig. 35.2 such light rays would bend up more steeply and reach $r = 0$ even earlier. The same conclusion can be drawn for observers with rockets. Their fate can also be described as follows. While in our part of the universe it is always possible with the aid of rockets to remain at a fixed point, in spite of the gravitational field, r cannot remain constant in region II; as one can see from (35.12), it plays the rôle of a time coordinate there, and the observer cannot prevent the passage of time. By the way, in region II the Schwarzschild solution is no longer static, since the Killing vector which is timelike in region I becomes spacelike here.

Let us return to the observer whom we wanted to send into region I'. While we believe him to be on the way to $r = 2M$, he has long ago been lost to the singularity $r = 0$ in region II; he can never reach regions I' or II'.

We can thus establish the following causal structure for the Schwarz-schild solution. From our world I we can influence region II, but not regions I' or II', and we can be influenced by region II', but not by II or I'. While region I' is therefore rather uninteresting for us, region II is a 'black hole' (everything can go in, nothing can come out), and region II' is a 'white hole' (nothing can go in, things can only come out).

Is all this just playing with mathematical constructions? The current opinion is that in our universe gravitational fields of this structure were not present from the beginning, but have possibly developed since. If they have thus only been in existence for a finite (coordinate) time, then region II' has no interest for us, since in a finite time nothing can reach us from there. Since, therefore, of the three regions I', II and II' on the other side of the Schwarzschild singularity (it is probable that) only II has relevance for us, we often speak of a 'black hole' when we mean the full Schwarzschild solution.

The story of how a black hole comes into being, of the evolution of the gravitational field of a normal star to a field whose Schwarzschild radius $r = 2M$ can be crossed from outside (no longer lies within the star) – this will be the subject of the next chapter.

Exercises

35.1 Are the curves $\{r, \varphi, \vartheta\} = $ const. geodesics for $r < 2M$?

35.2 Find all light rays which have $r = $ const.

35.3 Show that in the metric (35.19) the coordinates u and v are null coordinates in that $u^{,n} u_{,n} = 0 = v^{,n} v_{,n}$ holds.

35.4 Show that for small r the Schwarzschild solution behaves like $ds^2 = -d\tau^2 + (a\tau)^\alpha dt^2 + (a\tau)^\beta 4M^2 (d\vartheta^2 + \sin^2 \vartheta \, d\varphi^2)$. Determine a, α, and β and discuss the fate of a freely falling observer by using the results of the discussion of the Kasner-type solutions in Chapter 43 and Fig. 43.1.

36

Gravitational collapse – the possible life history of a spherically symmetric star

36.1 The evolutionary phases of a spherically symmetric star

In our universe a star whose temperature lies above that of its surroundings continuously loses energy, and hence mass, mainly in the form of radiation, but also in explosive outbursts of matter. Here we want to sketch roughly the evolution of such a star which is essentially characterized and determined by the star's innate properties (initial mass and density, ...) and its behaviour in the critical catastrophic phases of its life.

According to observation, stars exist for a very long time after they have formed from hydrogen and dust. Therefore they can almost always settle down to a relatively stable state in the interplay between attractive gravitational force, repulsive (temperature-dependent) pressure and outgoing radiation.

The first stable state is reached when the gravitational attraction has compressed and heated the stellar matter to such a degree that the conversion of hydrogen into helium is a long-term source of energy sufficient to prevent the star cooling and to maintain the pressure (a sufficiently large thermal velocity of the stellar matter) necessary to compensate the gravitational force. The average density of such a star is of the order of magnitude $1\ \mathrm{g\,cm}^{-3}$. A typical example of such a star is our Sun.

When the hydrogen of the star is used up, the star can switch over to other nuclear processes (possibly only after an unstable phase associated with explosions) and produce nuclei of higher atomic number. These processes will last a shorter time and follow one another more quickly. For sufficiently massive celestial bodies (the Earth would be too small!) the resulting pressure is then so great that the nuclei lose their electrons and a degenerate electron gas results. The density of this star is of order $10^7\ \mathrm{g\,cm}^{-3}$; stars of such density are known as white dwarfs. Using quantum mechanics for calculating the equilibrium states of such a degenerate electron gas, Chandrasekhar (1931) could show that white dwarfs with a mass above ~ 1.44 solar masses cannot be stable (their radius thus amounts at most to 5000 km). Stars with masses exceeding this *Chandrasekhar limit* must therefore either lose a part of their mass after hydrogen burning or else evolve towards a different final state.

If the pressure (the mass of the star) is large enough, this can happen by the electrons and the protons of the nuclei (starting, for example, from ^{56}Fe) turning into neutrons, so that the whole star finally consists of the most closely packed nuclear matter. The density of such neutron stars is about $10^{24}\ \mathrm{g\,cm}^{-3}$. Although the details of the nuclear interactions are not known exactly, one can nevertheless show that neutron stars are only stable (the pressure can only then support the gravitational force in equilibrium) if their mass does not appreciably exceed the mass of the Sun. Neutron stars hence have radii of about 10 km. We are now convinced that the so-called pulsars are neutron stars. Pulsars are stars which send out optical or radio signals at regular intervals of 10^{-3} to 1 s; the period is kept so exactly that it can only be caused by the rotation of the star, and rotation times of this order are only possible for exceptionally small stars. However, the fact that more massive stars cannot end their lives as neutron stars is crucial here to the question of the final state of a star.

Before turning to the possible fate of more massive stars, we shall bring in the Einstein field equations and ask what they have to say about stable, spherically symmetric accumulations of matter.

36.2 The critical mass of a star

As we have shown in the discussion of the interior Schwarzschild solution in Chapter 26, the gravitational field inside a static, spherically symmetric star is described by the metric

$$ds^1 = e^{\lambda(r)} \, dr^2 + r^2 \big(d\vartheta^2 + \sin^2 \vartheta \, d\varphi^2\big) - e^{\nu(r)} c^2 \, dt^2. \qquad (36.1)$$

A good approximation to stellar matter is given by the model of a perfect fluid medium with rest mass density $\mu(r)$ and pressure $p(r)$. Here the field equations lead to

$$e^{-\lambda(r)} = 1 - 2m(r)/r, \qquad (36.2)$$

where $m(r)$ is the mass function defined by

$$m(r) = \tfrac{1}{2}\kappa c^2 \int_0^r \mu(x) x^2 \, dx. \qquad (36.3)$$

The remaining field equations can be converted, upon using (36.2), into

$$\nu' = -\frac{2p'}{p + \mu c^2}, \quad \kappa p = \frac{\nu'}{r}\left(1 - \frac{2m}{r}\right) - \frac{2m}{r^3}. \qquad (36.4)$$

While we integrated these field equations earlier for the special case $\mu = \text{const.}$, we now want to derive a conclusion valid for arbitrary $\mu(r)$, following Weinberg (1972).

Suppose we have a star with finite (coordinate) radius r_0. The pressure p will vanish on the surface and will be greatest at the centre $r = 0$ of the star; it cannot, however, be infinitely great there. The density $\mu(r)$ should likewise remain finite for $r = 0$ and (on grounds of stability) decrease outwards:

$$\mu'(r) < 0. \qquad (36.5)$$

Since e^ν and its derivative must be continuous on the surface, $m(r)$ takes the value

$$m(r_0) = M \qquad (36.6)$$

there, where M is just the mass parameter occurring in the exterior, Schwarzschild solution; m/r^3 is finite at $r = 0$ because of (36.3).

Our aim is to derive a condition for the maximum possible mass M for given r_0 from the condition that μ and p are finite.

From (36.4) one can see at once that ν'/r must be finite at $r = 0$. If one introduces in place of e^ν the function $f(r)$, where

$$f(r) = e^{\nu(r)/2}, \qquad (36.7)$$

then this requirement becomes

$$f'/rf \quad \text{finite at } r = 0. \qquad (36.8)$$

By eliminating p from (36.4) one obtains, after some transformations,

$$\frac{\mathrm{d}}{\mathrm{d}r}\left[\frac{1}{r}\sqrt{1 - \frac{2m}{r}}\frac{\mathrm{d}f}{\mathrm{d}r}\right] = \frac{f}{\sqrt{1 - 2m/r}}\frac{\mathrm{d}}{\mathrm{d}r}\left(\frac{m}{r^3}\right). \qquad (36.9)$$

Since m/r^3 is the average mass density of the sphere of (coordinate) radius r, because of the definition (36.3), and since the average mass density cannot increase with r if μ decreases, then the right-hand side of (36.9) is negative or (and this only for $\mu = $ const.) zero:

$$\frac{\mathrm{d}}{\mathrm{d}r}\left[\frac{1}{r}\sqrt{1 - \frac{2m}{r}}\frac{\mathrm{d}f}{\mathrm{d}r}\right] \leq 0. \qquad (36.10)$$

On the surface of the star the metric must go over smoothly to the exterior Schwarzschild metric and the pressure must vanish, so that

$$f^2(r_0) = 1 - 2M/r_0, \qquad \mathrm{d}f/\mathrm{d}r|_{r=r_0} = M/(r_0^2\sqrt{1 - 2M/r_0}) \quad (36.11)$$

must hold, see (36.3) and (36.7). If one integrates (36.10) from r to r_0 using these relations, then one obtains

$$f'(r) \geq Mrr_0^{-3}(1 - 2m/r)^{-1/2}. \qquad (36.12)$$

Since for finite f and m the right-hand side of (36.9) is finite, then $f'(r)/r$ will be bounded. The finiteness condition (36.8) for the pressure then reduces to the requirement that $f(0) > 0$. Integration of (36.12) between 0 and r_0, using (36.11), gives, however,

$$f(0) \leq \left(1 - \frac{2M}{r_0}\right)^{1/2} - \frac{M}{r_0^3}\int_0^{r_0}\frac{r\,\mathrm{d}r}{(1 - 2m/r)^{1/2}}. \qquad (36.13)$$

If we now split $\mu(r)$ into a constant density $\mu_0 = 6M/\kappa c^2 r_0^3$ and a variable part $\rho(r)$,

$$\mu = \mu_0 + \rho, \qquad \int_0^{r_0}\rho(x)x^2\,\mathrm{d}x = 0, \qquad \rho' \leq 0, \ \rho(0) \geq 0, \qquad (36.14)$$

then we see that the integral in

$$m(r) = Mr^3 r_0^{-3} + \int_0^r \rho(x)x^2\,\mathrm{d}x \qquad (36.15)$$

is always positive. The right-hand side of (36.13) can therefore be increased in magnitude by substituting Mr^3/r_0^3 for $m(r)$. We then obtain the final result of the analysis, namely,

$$f(0) \leq \tfrac{3}{2}(1 - 2M/r_0)^{1/2} - \tfrac{1}{2}. \qquad (36.16)$$

As we have shown above, the central pressure $p(0)$ is only finite if $f(0)$ is greater than zero. Thus we can formulate the following important statement: a spherically symmetric star can only exist in a state of stable equilibrium (can only compensate its own gravitational attraction with a finite pressure) if its mass M and its radius r_0 satisfy the inequality

$$r_0 > \tfrac{9}{8} 2M. \tag{36.17}$$

For the special case of the interior Schwarzschild solution with the equation of state $\mu = $ const. we have already derived this inequality in Chapter 26. Now we know that it is valid for an arbitrary equation of state. In discussing this relation we must be careful about the definitions of M and r_0. M is (up to a factor) the integral of the mass density μ over the coordinate volume; it has the invariant significance of being the gravitating mass of the star as determined in the Newtonian far field. The stellar radius r_0 is defined so that the surface area of the star is $4\pi r_0^2$.

The inequality (36.17) expresses the fact that a star of fixed surface area is only stable as long as its mass lies below a critical mass. A star whose mass transgresses this limit must inevitably collapse into itself as a consequence of its now too strong gravitational attraction. While in the linear Newtonian gravitational theory a predominance of the gravitational force can be compensated by a contraction and the associated finite increase in pressure, or by additional forces, in the non-linear Einstein theory above the critical mass (36.17) a pressure increase or an extra force acts (via the energy-momentum tensor) to further increase the gravitational field.

An analysis of the maximum stable mass for given *constant* mass density μ, using the model of the interior Schwarzschild solution, leads from (36.3), (36.6) and (36.17) to the critical mass M_{crit}

$$M_{\text{crit}} = \frac{8}{9} \sqrt{\frac{2}{3\kappa c^2 \mu}}. \tag{36.18}$$

With $c^2 = 1.86 \times 10^{-27}$ cm g^{-1} one obtains for typical densities the following critical masses, which are compared with the mass of the Sun:

μ (in g cm^{-3})	1	10^6	10^{15}
M_{crit} (in cm)	1.685×10^{13}	1.685×10^{10}	0.532×10^6
$M_{\text{crit}}/M_\odot$	1.14×10^8	1.14×10^5	3.96

These very rough considerations already show that neutron stars can have only a few solar masses; more massive stars have no stable final state.

As an interesting side result we observe that, because of the general formula (23.57) and the inequality (36.17), the redshift of a light signal coming from the surface of a stable star has a maximum value of $z = 2$.

Intuitively it is obvious that a rotation of a star will diminish the central pressure and thus permit a larger critical mass. Indeed, Schöbel and Ansorg (2003) could show that a rigid rotation enlarges the critical mass (36.18) by (at most) a factor 1.3425.

36.3 Gravitational collapse of spherically symmetric dust

The considerations of the previous section have shown that if, during its evolution, a massive, spherically symmetric star does not succeed in ejecting or radiating away sufficient mass to become a neutron star, then there is no stable, final state available to it. At some time or other it will reach a state in which the pressure gradient can no longer balance the gravitational attraction. Consequently it will continue to contract further and its radius will pass the Schwarzschild radius $r = 2M$ and tend to $r = 0$: the star suffers a gravitational collapse.

Of course one would like to confirm these plausible intuitive ideas by making exact calculations on a stellar model with a physically reasonable equation of state (a reasonable relation between pressure and mass density). The only model for which this is possible without great mathematical complexity is that of dust ($p = 0$). Because the pressure vanishes it is to be expected here that once a star started to contract it would 'fall in' to a point. Nevertheless, this example is not trivial, since it yields an exact solution of the Einstein equations which is valid inside and outside the collapsing star, and which in a certain sense can serve as a model for all collapsing stars.

As the starting point for treating this collapsing stellar dust we do not take the canonical form (23.5) of the line element used earlier, but a system comoving with the dust (cp. Section 16.4). We obtain it by carrying out a transformation $r = r(\rho, c\tau)$, $t = t(\rho, c\tau)$ and hence bringing the metric (23.5) into the form

$$\mathrm{d}s^2 = \mathrm{e}^{\lambda(\rho, c\tau)}\,\mathrm{d}\rho^2 + r^2(\rho, c\tau)\big(\mathrm{d}\vartheta^2 + \sin^2\vartheta\,\mathrm{d}\varphi^2\big) - c^2\,\mathrm{d}\tau^2. \quad (36.19)$$

The coordinate τ is clearly the proper time of a particle at rest in the coordinate system (36.19), and the curves $\rho = \text{const.}$, $\vartheta = \text{const.}$, $\varphi = \text{const.}$ are geodesics (note that because of (21.87), dust always moves along geodesics). Since $u^n = (0, 0, 0, c)$, the energy-momentum tensor has as its only non-vanishing component

$$T_4^4 = -c^2\mu(\rho, c\tau). \quad (36.20)$$

The non-vanishing Christoffel symbols of the metric (36.19) are

$$\Gamma_{11}^1 = \lambda'/2, \quad \Gamma_{22}^1 = -\mathrm{e}^{-\lambda}rr', \quad \Gamma_{33}^1 = -\mathrm{e}^{-\lambda}rr'\sin^2\vartheta,$$

$$\Gamma_{12}^2 = r'/r, \quad \Gamma_{24}^2 = \dot{r}/r, \quad \Gamma_{33}^2 = -\sin\vartheta\cos\vartheta,$$

$$\Gamma_{13}^3 = r'/r, \quad \Gamma_{34}^3 = \dot{r}/r, \quad \Gamma_{23}^3 = \cot\vartheta, \quad \Gamma_{14}^1 = \dot{\lambda}/2,$$ (36.21)

$$\Gamma_{11}^4 = \dot{\lambda}\mathrm{e}^{\lambda}/2, \quad \Gamma_{22}^4 = r\dot{r}, \quad \Gamma_{33}^4 = r\dot{r}\sin^2\vartheta,$$

with $x^1 = \rho$, $x^2 = \vartheta$, $x^3 = \varphi$ and $x^4 = c\tau$, and denoting partial derivatives with respect to ρ and $c\tau$ by $'$ and $\cdot$ respectively. The field equations finally take the form

$$R_1^1 - \frac{R}{2} = \frac{r'^2}{r^2}\mathrm{e}^{-\lambda} - \frac{2\ddot{r}}{r} - \frac{\dot{r}^2}{r^2} - \frac{1}{r^2} = 0,$$ (36.22)

$$R_2^2 - \frac{R}{2} = R_3^3 - \frac{R}{2} = \left(\frac{r''}{r} - \frac{r'\lambda'}{2r}\right)\mathrm{e}^{-\lambda} - \frac{\dot{r}\dot{\lambda}}{2r} - \frac{\ddot{\lambda}}{2} - \frac{\dot{\lambda}^2}{4} - \frac{\ddot{r}}{r} = 0,$$ (36.23)

$$R_4^4 - \frac{R}{2} = \left(\frac{2r''}{r} - \frac{\lambda'r'}{r} + \frac{r'^2}{r^2}\right)\mathrm{e}^{-\lambda} - \frac{\dot{r}\dot{\lambda}}{r} - \frac{\dot{r}^2}{r^2} - \frac{1}{r^2} = -\kappa\mu c^2,$$ (36.24)

$$R_{14} = \dot{\lambda}r'/r - 2\dot{r}'/r = 0.$$ (36.25)

First integrals of these equations can be obtained very easily. The first step is to write (36.25) as

$$\dot{\lambda} = \left(r'^2\right)^{\cdot}/r'^2$$ (36.26)

and then integrate it to give

$$\mathrm{e}^{\lambda} = \frac{r'^2}{1 - \varepsilon f^2(\rho)}, \quad \varepsilon = 0, \pm 1,$$ (36.27)

with $f(\rho)$ as an arbitrary function. Substitution into (36.22) leads to

$$2\ddot{r}r + \dot{r}^2 = -\varepsilon f^2(\rho).$$ (36.28)

If one now chooses r as the independent variable and $u = (\dot{r})^2$ as the new dependent variable, then one obtains the linear differential equation

$$\mathrm{d}(ru)/\mathrm{d}r = -\varepsilon f^2(\rho),$$ (36.29)

whose solution (with a still to be fixed function of integration $F(\rho)$) is

$$\dot{r}^2 = -\varepsilon f^2(\rho) + F(\rho)/r.$$ (36.30)

If one next eliminates f^2 in (36.27) with the aid of (36.30), then one finds that (36.23) is satisfied identically and that (36.24) leads to

$$\kappa\mu c^2 = \frac{F'}{r'r^2}.$$ (36.31)

The partial differential equation (36.30) can be integrated completely, since ρ only plays the part of a parameter. For $\varepsilon \neq 0$ one can, through introducing

$$\mathrm{d}\eta = f\,\mathrm{d}c\tau/r, \tag{36.32}$$

bring the differential equation into the form

$$\left(\frac{\partial r}{\partial \eta}\right)^2 = \frac{F}{f^2}r - \varepsilon r^2 \tag{36.33}$$

and solve it by

$$r = \frac{F(\rho)}{2f^2(\rho)}h'_\varepsilon(\eta),$$

$$h_\varepsilon = \begin{cases} \eta - \sin\eta & \text{for } \varepsilon = +1, \\ \sinh\eta - \eta & \text{for } \varepsilon = -1, \end{cases} \tag{36.34}$$

$$c\tau - c\tau_0(\rho) = \pm\frac{F(\rho)}{2f^3(\rho)}h_\varepsilon(\eta),$$

while for $\varepsilon = 0$ one immediately has from (36.30) that

$$c\tau - c\tau_0(\rho) = \pm\tfrac{2}{3}F^{-1/2}(\rho)r^{3/2}, \quad \varepsilon = 0. \tag{36.35}$$

The general spherically symmetric dust solution, the *Tolman (1934) solution*, thus has in comoving coordinates the form

$$\mathrm{d}s^2 = \left(\frac{\partial r}{\partial \rho}\right)^2 \frac{\mathrm{d}\rho^2}{1 - \varepsilon f^2(\rho)} + r^2(\rho, c\tau)\big(\mathrm{d}\vartheta^2 + \sin^2\vartheta\,\mathrm{d}\varphi^2\big) - c^2\,\mathrm{d}\tau^2,$$

$$\kappa c^2\mu(\rho, c\tau) = \frac{F'(\rho)}{r^2\,\partial r/\partial\rho}, \tag{36.36}$$

where $r(\rho, c\tau)$ is to be taken from (36.34) and (36.35). Of the three free functions $F(\rho)$, $f(\rho)$ and $\tau_0(\rho)$, at most two have a physical significance, since the coordinate ρ is defined only up to scale transformations $\bar\rho = \bar\rho(\rho)$. Unfortunately one cannot simply specify the matter distribution $\mu(\rho, c\tau)$ and then determine the metric, but rather through a suitable specification of f, F and τ_0 one can produce meaningful matter distributions. Since layers of matter which move radially with different velocities can overtake and cross one another, one must expect the occurrence of coordinate singularities in the comoving coordinates used here.

We now want to apply the Tolman solution to the problem of a star of finite dimensions. To do this we have to obtain an interior ($\mu \neq 0$) solution and an exterior ($\mu = 0$) solution and join these two solutions smoothly at the surface $\rho = \rho_0$ of the star.

We obtain the simplest *interior solution* when μ does not depend upon position (upon ρ) and r has (for a suitable scale) the form $r = K(c\tau)\rho$.

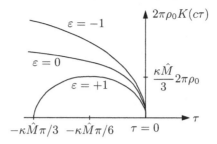

Fig. 36.1. Radius of a collapsing star as a function of time.

These restrictions lead to

$$f = \rho, \quad F = \tfrac{1}{3}\kappa\hat{M}\rho^3, \quad \mu c^2 K^3(c\tau) = \hat{M} = \text{const.}, \quad \tau_0 = 0, \quad (36.37)$$

and the metric

$$ds^2 = K^2(c\tau)\left[\frac{d\rho^2}{1 - \varepsilon\rho^2} + \rho^2\big(d\vartheta^2 + \sin^2\vartheta\, d\varphi^2\big)\right] - c^2\, d\tau^2,$$

$$
\begin{aligned}
K(\eta) &= \tfrac{1}{6}\kappa\hat{M}h'_\varepsilon(\eta), \\
c\tau &= -\tfrac{1}{6}\kappa\hat{M}h_\varepsilon(\eta),
\end{aligned}
\qquad
h_\varepsilon(\eta) =
\begin{cases}
\eta - \sin\eta & \text{for } \varepsilon = +1, \\
\eta^3/6 & \text{for } \varepsilon = 0, \\
\sinh\eta - \eta & \text{for } \varepsilon = -1.
\end{cases}
\qquad (36.38)
$$

As comparison with (19.56) shows, the interior $\rho \le \rho_0$ of the star is a three-dimensional space of constant curvature, whose radius K depends on time (in the language of cosmological models, it is a section of a Friedmann universe, see Section 41.2). A great circle on the surface of the star has the radius $\rho_0 K(c\tau)$, and because of the time-dependence of K the star either expands or contracts.

As Fig. 36.1 shows, models with $\varepsilon = 0$ or -1 correspond to stars whose radius decreases continuously from arbitrarily large values until at the time $\tau = 0$ a collapse occurs, while models with $\varepsilon = +1$ represent stars which first expand to a maximum radius and then contract.

The solution in the exterior space to the star is clearly a spherically symmetric vacuum solution, and because of the Birkhoff theorem it can only be the Schwarzschild solution (see Fig. 36.2). Since the Tolman solution (36.36) holds for arbitrary mass density μ, it must contain the exterior Schwarzschild solution $\mu = 0$ as a special case ($F = \text{const.}$). In the Tolman solution the coordinates in the exterior space are chosen so that the surface of the star is at rest. In the usual Schwarzschild metric, on the other hand, the stellar surface is in motion. But in both

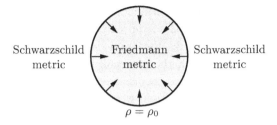

Fig. 36.2. Snapshot of a collapsing star.

cases the motion of a particle of the surface takes place on a geodesic.
The equation (35.6) of the radial geodesics of the Schwarzschild metric,
namely,

$$(\mathrm{d}r/\mathrm{d}\tau)^2 = A^2 - c^2 + 2Mc^2/r, \qquad (36.39)$$

must therefore coincide with (36.30) for all times τ at $\rho = \rho_0$; that is,
the relation

$$F = 2M \qquad (36.40)$$

must hold. Since scale transformations $\bar{\rho} = \bar{\rho}(\rho)$ are still possible, $f(\rho)$
cannot be uniquely determined here; in the following we shall not need
$f(\rho)$.

We must now ensure that the interior solution (36.38) and the exterior
solution (36.34), (36.35), (36.36) and (36.40) match smoothly at the
stellar surface $\rho = \rho_0$. The necessary condition for this is clearly

$$r(\rho_0, c\tau) = K(c\tau)\rho_0. \qquad (36.41)$$

If we choose the origin of time in the exterior metric so that $\tau_0(\rho_0) = 0$
then for $\varepsilon \neq 0$ the relation (36.41) can only be satisfied for all time τ
if both sides have the same functional dependence on τ, that is, only
if in (36.34) and (36.38) $h_\varepsilon(\eta)$ has the same factor. This leads to the
condition

$$6M/f^3(\rho_0) = \kappa\mu c^2 K^3. \qquad (36.42)$$

From this and the above equations we obtain $f(\rho_0) = \rho_0$, and hence for
condition (36.42) we have finally

$$\kappa\mu c^2 \rho_0^3 K^3 = 6M. \qquad (36.43)$$

For $\varepsilon = 0$ one immediately obtains the same condition from (36.35),
(36.38) and (36.41). It is easy to convince oneself that when (36.43) is

satisfied, then the metric is continuous on the surface of the star and the normal derivatives have the required continuity behaviour (30.41).

The condition (36.32) links the mass density μ of the star and its coordinate radius $K\rho_0$ with the externally acting Schwarzschild mass parameter M, in the same way as occurred in (26.32) when the interior and exterior Schwarzschild solutions were joined. If we recall the relation (23.25), that is, the relation $2M = \kappa mc^2/4\pi$ between the Schwarzschild radius $2M$ and the mass m which we would associate with the source of the Schwarzschild solution in the Newtonian gravitational theory, then we have

$$\tfrac{1}{3}4\pi\mu\rho_0^3 K^3(c\tau) = \tfrac{1}{3}4\pi\mu r_0^3 = m. \tag{36.44}$$

Notice that only for $\varepsilon = 0$ is m the same as the integral over the mass density μ, calculated in the interior metric (36.38).

The solution found here for the gravitational field of a collapsing star clearly shows that in the interior of the star no peculiarities occur even when the stellar surface $\rho = \rho_0$ lies *inside* the Schwarzschild radius $r = K(c\tau)\rho_0 = 2M$; only at $K(c\tau) = 0$ does the interior field become singular.

To end this discussion, we shall follow the fate of a collapsing star in the Kruskal space-time diagram. To do this we draw a radial geodesic on which the points at the surface of the dust star move (Fig. 36.3). On its left is the stellar interior with a metric which is regular up to the point $\tau = 0$ (from outside: $r = 0$). During the collapse a part of the region II is revealed to an observer in the exterior space, but the regions I' and II' (see Fig. 35.1) are not realized.

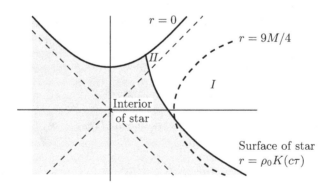

Fig. 36.3. Collapse of a star in the Kruskal diagram.

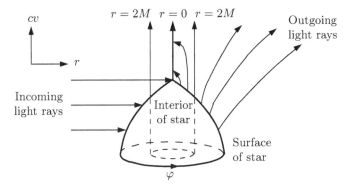

Fig. 36.4. Spherically symmetric collapse of a star in Eddington–Finkelstein coordinates.

For many purposes it is more convenient to describe the collapse in Eddington–Finkelstein coordinates (35.17), since one of the two rotational degrees of freedom can be more easily represented here (remember that the Eddington–Finkelstein coordinates describe just the regions I and II of the Kruskal diagram; that is, they include just those parts of the Schwarzschild solution essential to the collapse). One can see from Fig. 36.4 how the star contracts until it vanishes behind the Schwarzschild radius $r = 2M$, that is, until the emitted light rays

$$c\,\mathrm{d}v = \frac{2\,\mathrm{d}r}{1 - 2M/r} \qquad (36.45)$$

no longer succeed in reaching the exterior space ($\mathrm{d}v < 0$ for $r < 2M$!). The radially ingoing light rays

$$\mathrm{d}v = 0, \qquad (36.46)$$

however, can always reach the surface of the star (or the singularity $r = 0$). In order to interpret Fig. 36.4 or similar diagrams one has to remember that the metrical relations of a two-dimensional Riemannian surface are not correctly included in the plane of the paper. It is the possibilities of interaction represented by light rays or by geodesics that are essential.

Further reading for Chapter 36

Hawking and Ellis (1975), Shapiro and Teukolsky (1983).

37

Rotating black holes

37.1 The Kerr solution

Most known stars are rotating relative to their local inertial system (relative to the stars) and are therefore not spherically symmetric; their gravitational field is not described by the Schwarzschild solution. In Newtonian gravitational theory, although the field certainly changes because of the rotational flattening of the star, it still remains static, while in the Einstein theory, on the other hand, the flow of matter acts to produce fields. The metric will still be time-independent (for a time-independent rotation of the star), but not invariant under time reversal. We therefore expect that the gravitational field of a rotating star will be described by an axisymmetric stationary vacuum solution which goes over to a flat space at great distance from the source. Depending on the distribution of matter within the star there will be different types of vacuum fields which, in the language of the Newtonian gravitational theory, differ, for example, in the multipole moments of the matter distribution. One of these solutions is the Kerr (1963) solution, found almost fifty years after the discovery of the Schwarzschild metric. It proves to be especially important for understanding the gravitational collapse of a rotating star. To avoid misunderstanding we emphasize that the Kerr solution is not the gravitational field of an arbitrary axisymmetric rotating star, but rather only the exterior field of a very special source.

We shall now discuss the Kerr solution and its properties. Since its mathematical structure is rather complicated, we shall not construct a derivation from the Einstein field equations.

The line element of the Kerr solution has the form, in the so-called *Boyer–Lindquist coordinates*,

$$
\begin{aligned}
\mathrm{d}s^2 &= \Sigma \left(\mathrm{d}r^2 / \Delta + \mathrm{d}\vartheta^2 \right) + \left(r^2 + a^2 \right) \sin^2 \vartheta \, \mathrm{d}\varphi^2 - c^2 \, \mathrm{d}t^2 \\
&\quad + 2Mr \left(a \sin^2 \vartheta \, \mathrm{d}\varphi - c \, \mathrm{d}t \right)^2 / \Sigma, \\
\Sigma &\equiv r^2 + a^2 \cos^2 \vartheta, \qquad \Delta \equiv r^2 - 2Mr + a^2.
\end{aligned}
\tag{37.1}
$$

For very large r it goes over to the line element of a flat space. To disclose the meaning of the two parameters M and a, we take the far field (large r, $\Sigma \approx \Delta \approx r^2$) and transform the metric (37.1) there to

'Cartesian coordinates' by $r^2 = x^2 + y^2 + z^2$, $\vartheta = \arctan\left(\sqrt{x^2 + y^2}/z\right)$, and $\varphi = \arctan(y/x)$. We obtain

$$g_{4x} = 2May/r^3, \quad g_{4y} = -2Max/r^3, \quad g_{4z} = 0, \quad g_{44} = 1 - 2M/r. \quad (37.2)$$

By comparison with the representation (27.32), which is valid for every far field, we can deduce that M is the mass and Ma the z-component (the magnitude) of the angular momentum of the source of the Kerr field. This physical interpretation of the two constants of the Kerr metric is further consolidated by the facts that for $a = 0$ (absence of rotation) (37.1) reduces to the Schwarzschild metric and that the Kerr metric is invariant under the transformation $t \rightarrow -t$, $a \rightarrow -a$ (time reversal and simultaneous reversal of the sense of rotation).

The Boyer–Lindquist coordinates are generalized Schwarzschild coordinates and like these are not suitable for describing the solution over its full mathematical realm of validity. Provided that $0 < a^2 < M^2$, the coordinates (37.1) are clearly singular for $\Delta = 0$, that is, for the two values

$$r_+ = M + \sqrt{M^2 - a^2} \qquad r_- = M - \sqrt{M^2 - a^2}. \qquad (37.3)$$

For $a = 0$, r_+ goes over to the Schwarzschild radius $r = 2M$, while r_- goes over to $r = 0$. From now on we shall ignore the parameter region $M^2 < a^2$, which would correspond to very rapidly rotating bodies and does not lead to black holes.

In analogy to the transition from Schwarzschild coordinates to Eddington–Finkelstein coordinates, one can also transform the Kerr solution into a form which has no singularities at $r_\pm$. One introduces a new coordinate v adapted to light propagation by

$$c\,\mathrm{d}v = c\,\mathrm{d}t + (r^2 + a^2)\,\mathrm{d}r/\Delta \qquad (37.4)$$

and a new 'angular coordinate' Φ by

$$\mathrm{d}\Phi = \mathrm{d}\varphi + a\,\mathrm{d}r/\Delta, \qquad (37.5)$$

which takes into account the corotation of the local inertial system (cp. our discussion of the action of an angular momentum in Section 27.5). The result of these transformations is the Kerr solution in *Kerr coordinates*,

$$
\begin{aligned}
\mathrm{d}s^2 = {}& \Sigma\,\mathrm{d}\vartheta^2 - 2a\sin^2\vartheta\,\mathrm{d}r\,\mathrm{d}\Phi + 2c\,\mathrm{d}r\,\mathrm{d}v \\
& + \Sigma^{-1}\sin^2\vartheta\left[\left(r^2 + a^2\right)^2 - \Delta a^2 \sin^2\vartheta\right]\mathrm{d}\Phi^2 \qquad (37.6) \\
& - 4Mar\Sigma^{-1}\sin^2\vartheta\,\mathrm{d}\Phi\,c\,\mathrm{d}v - (1 - 2Mr/\Sigma)\,c^2\,\mathrm{d}v^2.
\end{aligned}
$$

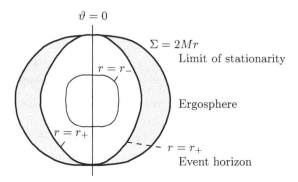

Fig. 37.1. The limiting surfaces of the Kerr solution.

The Kerr solution possesses (like every axially-symmetric stationary metric) two commuting Killing vectors, namely – in the coordinates $(r, \vartheta, \varphi, ct)$ or (r, ϑ, Φ, cv) – the vectors

$$\eta^i = (0,0,1,0), \qquad \xi^i = (0,0,0,1). \qquad (37.7)$$

The Killing vector ξ^i, which in the far field is associated with the stationarity (time independence), has an interesting property. Its magnitude

$$\xi_i \xi^i = -(1 - 2Mr/\Sigma) \qquad (37.8)$$

changes sign when one crosses the surface

$$\Sigma - 2Mr = r^2 - 2Mr + a^2 \cos^2 \vartheta = 0. \qquad (37.9)$$

Inside this surface the Killing vector ξ^i is spacelike. The surface is therefore called the *limiting surface of stationarity* (*stationary limit*). Observers (particles) on it cannot be at rest with respect to infinity, but must corotate.

The physical properties of the Kerr space-time are best brought out (as for the Schwarzschild solution) by studying the possible trajectories of test particles or photons. The details which one assembles in this way are, however, so complicated and confusing that we shall eschew an exhaustive description with proofs and merely give a qualitative discussion of the most important results.

When we approach (Fig. 37.1) the singularity of the Kerr solution, coming from the far field, we encounter first the stationary limit (37.9). Between it and the surface $r = r_+$ lies the so-called *ergosphere*. Particles and light rays can penetrate this region from outside and leave it again.

Even the following physical process is in principle possible. A particle of positive energy E_0,

$$-m_0 u_i \xi^i = E_0 > 0, \tag{37.10}$$

falls from outside along a geodesic into the ergosphere, E_0 remaining conserved ((37.10) is a special case of (33.54)). Under conservation of four-momentum, the particle then is split into two parts:

$$m_0 u^i = m_1 u_1{}^i + m_2 u_2{}^i. \tag{37.11}$$

Since the Killing vector ξ^i is spacelike inside the ergosphere, because of (37.8), then the timelike vector $u_1{}^i$ can be chosen so that $E_1 = -m_1 u_1{}^i \xi_i$ is negative. In the exterior space such a choice is impossible, because ξ^i is timelike there and the product of two timelike vectors is always negative. We then have

$$-m_2 u_2{}^i \xi_i = E_2 = E_0 - E_1 > E_0, \tag{37.12}$$

and the second particle leaves the ergosphere with a greater energy than that of the particle shot in. The rotation of the source is what yields the energy for this process. A similar enhancement may happen to radiation (*'superradiance'*).

The surface $r = r_+$ can of course be crossed by particles or photons from outside, but it is impossible for photons or particles ever to leave the interior space: like the Schwarzschild solution, the Kerr metric describes the gravitational field of a black hole. Since all events which occur inside the radius $r = r_+$, such as the disintegration or radiation of test particles or real particles, are never recorded by an external observer (no photon can reach him from there), this surface is called the *event horizon*.

Further inside is the surface $r = r_-$, which has no particular physical significance, and finally one reaches at $r = 0$ a ring singularity, and not, as one might at first suppose, a point singularity.

The regions of the Kerr solution discussed up until now correspond to the regions I and II of the Kruskal diagram (Fig. 35.1) of the Schwarzschild solution. It is also possible to extend maximally the Kerr metric; that is, the points corresponding to the regions I' and II' can be made mathematically accessible, see for example Hawking and Ellis (1975) for details.

37.2 Gravitational collapse – the possible life history of a rotating star

The life history of a rotating star differs from that of a spherically symmetric star (considered in Section 36.1) not in the phases and the sequence of processes which yield energy and the possible final stages of these processes, but rather by the influence of the rotation in the contraction phase. If a rotating star contracts very strongly, then because of the conservation of angular momentum it will rotate more and more rapidly and possibly break up into separate fragments (only the nucleus carries on contracting). Or, put another way, a star can in general become extremely contracted (e.g. to a neutron star) only if it gives up angular momentum to its surroundings. This can occur through ejection of matter or through gravitational interaction with other masses. If, however, at the end it still possesses sufficient mass at very high density, then the gravitational forces become so strong that gravitational collapse takes place, and then the Kerr solution remains as the external gravitational field. Although these ideas seem very plausible and are supported by a large number of facts and calculations, there are two gaps in the theory of the gravitational collapse of a rotating star which to date have not been closed.

The first gap is the lacking 'internal' Kerr solution. We do not have a (stable or unstable) interior solution with a reasonable equation of state which can be joined to the Kerr metric at the surface of the star, and probably such a solution does not exist. Nor does one know any timevariable (interior and exterior) solution whose exterior part changes into a Kerr metric under collapse of the star. For these reasons one cannot say in detail exactly how the collapse proceeds, although numerical relativity is making progress here.

The second gap in our present knowledge is our ignorance as to whether, under a gravitational collapse, a Kerr metric always results, or whether there are other, differently constituted, (singular) vacuum solutions, which describe the end stage of the gravitational field of a collapsed star. It is presumed that the star either does not collapse at all or just tends to a Kerr metric; but the proof of this has so far eluded us. The (supposed) uniqueness of the Kerr solution would be a typical property of the Einstein theory: the gravitational field of a collapsed star is characterized by only two parameters, namely, the mass M and the angular momentum Ma, in contrast to the infinitely many parameters (multipole moments) of a non-collapsed star. For the collapse itself this restriction to two parameters signifies that the star must lose all the higher moments not appropriate in this scheme by ejection or radiation of mass before it disappears behind the event horizon.

37.3 Some properties of black holes

In this section we want to collect some properties of black holes, taking particular account of those which are important for an observer in the exterior space. To some extent we shall repeat things said in Chapters 35–37, but we shall also bring out some new aspects and in particular take into regard the fact that a black hole can be electrically charged (the Kerr–Newman solution which is then appropriate contains the Kerr solution as a special case).

(a) Black holes are solutions of the field equations which describe the gravitational field of collapsed masses. This field is characterized by three parameters: by the mass M, the angular momentum Ma and the electrical charge Q. (An external observer can determine these three parameters by observing the trajectories of uncharged and charged particles.) Other possible physical properties the source of the field had before collapse (baryon number, electrical dipole moment,...) are lost during the collapse. The relation between angular momentum and the magnetic moment produced by the rotation is, moreover, the usual 'anomalous' one for the electron.

(b) Black holes contain a closed event horizon. Within this surface the gravitational field is so strong that particles, light rays and time-dependent fields produced inside can no longer leave this region. Particles and light rays from outside can penetrate the horizon; for this they need (as seen by a distant observer) an infinite time. An observer can reach and pass the horizon in a finite time and inside can, it is to be hoped, convince himself of the correctness of the theory described; but he can never report back to the outside.

(c) Inside the event horizon there is a genuine singularity of the gravitational field, which forms during the collapse. Fortunately the universe is so constituted that (because of the event horizon) we cannot see this singularity ('cosmic censorship') and so it is without meaning for physics in the outside universe.

(d) Once it has formed, a black hole is (probably) stable and cannot be destroyed. Matter (mass, radiation) which reaches the black hole from outside can, however, change the charge Q, the mass M and the angular momentum $P = aM$ (the 'indigestible remains' of physical properties of the matter fed in will be emitted in the form of radiation, from outside the horizon of course). But during all these processes the quantity

$$A = 4\pi\left[2M^2 - Q^2 + 2\left(M^4 - M^2Q^2 - P^2\right)^{1/2}\right] \qquad (37.13)$$

can only increase. In the Kerr metric (37.6) A can be visualized as the surface area of $r = $ const. $= r_+$, $v = $ const. (the event horizon). This law is also called (because of certain analogies to thermodynamics) the second law of black-hole dynamics. The rearrangement

$$M^2 = \frac{A}{16\pi} + \frac{4\pi P^2}{A} + \frac{\pi Q^4}{A} + \frac{Q^2}{2} \qquad (37.14)$$

of (37.13) clearly shows that it is indeed at the cost of charge and angular momentum that one obtains energy (mass) from a black hole, but that one cannot go below $M = (A/16\pi)^{1/2}$. These statements also hold for the possible union of two black holes into one. If, for example, two spherically symmetric black holes (of masses M_1 and M_2) coalesce to form one black hole, again spherically symmetric, then we must have

$$16\pi M^2 = A \geq A_1 + A_2 = 16\pi\left(M_1^2 + M_2^2\right), \qquad (37.15)$$

and so at most the fraction

$$\eta = \frac{M_1 + M_2 - M}{M_1 + M_2} \leq 1 - \frac{\sqrt{M_1^2 + M_2^2}}{M_1 + M_2} \leq 1 - \frac{1}{\sqrt{2}} \qquad (37.16)$$

of the mass can be given off in the form of gravitational radiation.

(e) The inclusion of quantum effects could alter this picture radically. We shall return to this point in Section 38.4.

37.4 Are there black holes?

The question as to whether these black holes with their remarkable properties really exist and are necessarily formed in the final stages of collapsing matter (stars, galaxies, . . .) has various aspects.

There seems no doubt that there are no stable configurations when mass becomes extremely concentrated. In those cases there must be a collapse, and according to General Relativity a horizon is formed. Stated otherwise, the gravitational field of very massive objects is so strong that light cannot escape.

One can get rid of the sense of uneasiness which the existence of an event horizon may imply. The lack of a genuine interaction with the matter behind the horizon is only apparent; if the matter has disappeared behind the horizon, then it is left *only* with the properties of mass, angular momentum and charge, and these act outward, are determinable and (within limits) can even be changed from outside.

Doubts may arise whether the physics near to the horizon is adequately described by Einstein's theory. In the next chapter we shall see that quantum physics certainly will change the classical picture, although the details of the interplay of quantum physics and relativity are not yet understood.

Even more doubts may arise whether the fate of a collapsing star which contracts to a point is really well understood. Perhaps this deep extrapolation of the realm of validity of the Einstein theory – which was developed mainly in weakly curved spaces – is wrong. But how, where and whether at all the Einstein theory must be modified can only be established by exploration of this theory and by comparison with astrophysical observation.

Do black holes exist somewhere or other in our Universe?

Gravitational collapse is itself very probably associated with an explosive outburst of matter, so that the star would suddenly flare up, rather like a supernova. But this flaring up is not on its own very conclusive, because it could also indicate the formation of a neutron star. Since the far field of a black hole in no way differs from that of an ordinary star, only processes close to the horizon can provide reliable evidence.

This evidence for the existence of a black hole becomes more convincing when one finds very compact mass distributions which cannot be stable if extended. This could be a partner of a double star, but the most promising candidates are the centres of galaxies with masses exceeding 10^6 $M_\odot$. This evidence can be supported by observing X-rays originating therefrom: matter falling into a rotating black hole, via an accretion disc in the equatorial plane, may during the sharply accelerated terminal stages emit X-rays or gravitational waves of high intensity, and thus provide evidence for the existence of black holes.

In all these cases General Relativity is used to conclude the existence of a black hole from the observation of a very massive object. By its very definition, the horizon itself (and what is beyond it) cannot be seen. So it is not yet possible to assert with absolute certainty whether or not the black holes (in the form predicted by Einstein's theory) exist. Whatever the final answer turns out to be will improve our understanding of spacetime, that is, gravity, significantly.

Further reading for Chapter 37

Hawking and Ellis (1975), Chandrasekhar (1998), Frolov and Novikov (1998), Celotti *et al.* (1999).

38

Black holes are not black – Relativity Theory and Quantum Theory

The picture of black holes we have drawn so far changes drastically if quantum effects are taken into account. Before we go into the details of this in Section 5 of this chapter, we want to make a few general remarks on the interplay of Relativity Theory and Quantum Theory. For a more detailed discussion we refer the reader to the literature given at the end of the chapter.

38.1 The problem

The General Theory of Relativity is completely compatible with all other classical theories. Even if the details of the coupling of a classical field (Maxwell, Dirac, neutrino or Klein–Gordon field) to the metric field are not always free of arbitrariness and cannot yet be experimentally tested with sufficient accuracy, no doubt exists as to the inner consistency of the procedure.

This optimistic picture becomes somewhat clouded when one appreciates that besides the gravitational field the only observable *classical* field in our universe is the Maxwell field, while the many other interactions between the building blocks of matter can only be described with the aid of Quantum Theory. A unification of Relativity Theory and Quantum theory has not yet been achieved, however.

One of the main postulates of relativity theory is that a locally geodesic coordinate system can be introduced at every point of space-time, so that the action of the gravitational force becomes locally ineffective and the space is approximately a Minkowski space. Hence it is easily understandable why in our neighbourhood, with its relatively small space curvature, space is, to very good approximation, as it is assumed to be in quantum theory. But it also shows us the limits of this more or less undisturbed coexistence of quantum theory and relativity theory: in regions of strong curvature (close to singularities) and in questions which concern the behaviour of far-extended physical systems, the two theories are no longer compatible, since they start out from different space structures. Quantum theory presupposes a Minkowski space *of infinite extent* both in its fundamental commutation rules, which are formulated explicitly using the group of motions of the space (the Lorentz group), and in more technical issues like expansion in plane waves, asymptotic behaviour at infinity or the formulation of conservation laws. Relativity theory shows, however, that the space is a Riemannian space.

On the other hand, the idea of relativity theory, that the properties of space are properties of the interaction of the matter and can be measured out by material test bodies, leads to contradictions when defining or measuring very small distances (the metric in very small regions of space); if the dimensions are so small that atoms or elementary particles should be taken as test objects, then quantum theory shows their location is no longer so precisely defined that one can really speak of a measurement, even be it only in a gedanken experiment.

In nature, however, there exist stars, which consist of elementary particles and whose motion obeys the gravitational laws, and therefore a self-consistent synthesis of relativity theory and quantum theory must be possible, where it is to be expected that at least one of these two theories must be modified. Theoretical physicists are certainly in a difficult situation: in contrast to the physics of elementary particles, which provides large amounts of experimental data seeking interpretation, here there are no experimental findings (or at least none recognized as such) which could give an indication of the course to be followed.

There are some likely candidates for the unification of gravity and quantum theory; three of them will be outlined in the following sections.

38.2 Unified quantum field theory and quantization of the gravitational field

Recently, after the successful unification of the weak, electromagnetic and strong interactions in a unified quantum field theory (see e.g. Weinberg 1996), theoretical physicists have set themselves the ambitious task of describing all four known interactions in a unified quantum field theory, for example, in a supergravity or superstring theory. Complicated theories, sometimes with mathematical beauty and elegance, have been proposed, but as yet there have been no resounding successes. But perhaps this 'theory of everything' sought after both by relativists and elementary particle physicists is an illusion; the unity of our world need not be reflected, even at the most basic physical level, by a simple comprehensive set of formulae.

A less demanding approach is to quantize only the gravitational field. Various physical and formal mathematical grounds suggest that all fields and interactions should be handled in a uniform manner; thus the gravitational field also should be quantized. Many attempts in this direction have already been made. So one may start from the analysis of the Cauchy problem for the vacuum Einstein field equations made in Chapter 30, use it to isolate the true dynamical degrees of freedom of the gravitational field, that is, those which do not arise from pure coordinate transformations, and quantize them. This is the starting point for canonical quantization. Although a great deal of effort has been invested not only to construct a formal theory, but also to understand and interpret it physically, the task is still in its infancy. The picture of the (four-dimensional) world would look quite different in a quantized theory of gravity. At each event the world is a mixture of states, each with a certain probability. Each of these states corresponds to a possible three-geometry, including its topological properties, and can be described by a point in *superspace*. How one couples in non-metric fields, how man is to interpret the wave function of the universe and how measurement processes and observers are to be described is unclear.

There is no problem in quantizing the *linearized* Einstein field equations, that is, the classical field $\bar{f}_{mn}$ described by the equations

$$\Box \bar{f}_{mn} = 0, \qquad \bar{f}^{mn}{}_{,n} = 0 \qquad (38.1)$$

(see Section 27.2). It shows that the massless particles of this field, analogous to the photons of the electromagnetic field, have spin 2. Of course by restricting consideration to source-free weak fields the real problems have been swept under the carpet.

38.3 Semiclassical gravity

A possible resolution of the problems caused by attempting to quantize gravity is to treat the gravitational field classically, but quantize all other fields. This school of thought is supported especially by those who regard the gravitational field as playing a privileged rôle, which should not and cannot be quantized. An extreme standpoint of this conservative-relativistic view was taken by Einstein himself. For a time he believed that quantum theory could be encompassed in a (possibly generalized) theory of relativity that would link space-time singularities to elementary particles. This hope has not been realized.

In a semiclassical theory the coupling of gravity to the quantized fields depends, on the one hand, on the fact that the field equations of the latter can be formulated covariantly, and thus can be made to depend on the gravitational field. On the other hand the gravitational field is generated by the quantum fields; these occur, however, in the 'source' of the Einstein field equations, the energy-momentum tensor, not as operators but as expectation values:

$$R_{mn} - \tfrac{1}{2}Rg_{mn} = \kappa\langle T_{mn}\rangle. \tag{38.2}$$

In order for the field equations (38.2) to be integrable the expectation values for the components of the energy-momentum tensor must be divergence free,

$$\langle T^{mn}\rangle_{;n} = 0. \tag{38.3}$$

However, as a deeper analysis reveals, (38.3) is not a simple consequence of the equations governing the quantum fields (which have not been given explicitly), but rather a constraint on those quantities, for example, the states, which are used to form the expectation values. One sees immediately that the main problem in this form of unification of quantum and relativity theory is the choice, meaning and interpretation of states, even the 'vacuum state'. In addition there are the difficulties in carrying over to a non-linear theory, in which the superposition principle is invalid, the usual interpretation of measurement processes. It cannot be said with certainty whether such a semiclassical theory is self-consistent, and to what extent it is a good approximation or even consistent with observations.

38.4 Quantization in a given classical gravitational field

One can obtain an insight into the problems and consequences of the as yet unknown unified theory by considering the influence of a given gravitational field on the quantum field and ignoring the back-reaction, that is, the inertia field produced by the quantum field. As an example we outline the typical procedure, some results and some problems by considering a real massless scalar field $\phi(x^i)$,

$$\Box\phi = \phi^{,n}{}_{;n} = 0. \tag{38.4}$$

In order to quantize classical fields ϕ satisfying the wave equation (38.4) in *Minkowski space* one can proceed as follows. One first represents the general (classical) solution of (38.4) by its Fourier transform with respect to time t and splits the inversion integral into waves of positive $(e^{-i\omega t})$ and negative $(e^{+i\omega t})$ frequency:

$$\phi(x^a) = \int_0^\infty \left[\phi_\omega(x^\alpha)\,e^{-i\omega t} + \bar\phi_\omega(x^\alpha)\,e^{i\omega t}\right] d\omega, \quad \alpha = 1, 2, 3. \quad (38.5)$$

On the surfaces $t = $ const. one constructs a complete orthonormal system, $f_p(x^\alpha)$ and $\bar f_p(x^\alpha)$, of solutions of the time-independent wave equation which can be used to represent ϕ_ω and $\bar\phi_\omega$. The norm used thereby is defined by

$$(\Psi_1, \Psi_2) = -i \int \left(\Psi_1 \dot{\bar\Psi}_2 - \dot\Psi_1 \bar\Psi_2\right) d^3x = \overline{(\Psi_2, \Psi_1)}. \quad (38.6)$$

Every solution of the wave equation can be represented as a superposition of partial waves g_n of the form $f_p(x^\alpha)\,e^{-i\omega t}$ and their complex conjugates $\bar g_n$; the index n represents symbolically the possible values of p, which are often discrete, and the continuous frequency parameter ω. Because of the structure of the norm (38.6) g_n and $\bar g_n$ satisfy the equations

$$(g_n, \bar g_m) = 0, \qquad (g_n, g_m) = -(\bar g_n, \bar g_m). \quad (38.7)$$

The general Hermitian field operator $\phi(x^\alpha, t)$ can then be written in the form

$$\phi(x^\alpha, t) = \sum_n \left\{\mathbf{a}_n g_n(x^\alpha, t) + \mathbf{a}_n^\dagger \bar g_n(x^\alpha, t)\right\}, \quad (38.8)$$

where the operators $\mathbf{a}_n$ and $\mathbf{a}_n^\dagger$ satisfy the commutator rules

$$[\mathbf{a}_n, \mathbf{a}_{n'}] = 0 = [\mathbf{a}_n^\dagger, \mathbf{a}_{n'}^\dagger], \qquad [\mathbf{a}_n, \mathbf{a}_{n'}^\dagger] = \delta_{nn'}. \quad (38.9)$$

The set of states which can be constructed by single or multiple application of the creation operator $\mathbf{a}_n^\dagger$ to the vacuum state $|0\rangle$ forms the Hilbert space of the system. Here the vacuum state is defined as that state in which no particles can be annihilated,

$$\mathbf{a}_n|0\rangle = 0. \quad (38.10)$$

A single particle state (of type n) $|1_n\rangle$ is then constructed via

$$\mathbf{a}_n^\dagger|0\rangle = |1_n\rangle. \quad (38.11)$$

The total number of particles in a given state can be found by using the number operator

$$N = \sum_n N_n = \sum_n \mathbf{a}_n^\dagger \mathbf{a}_n. \quad (38.12)$$

It can be shown that this quantization procedure is Lorentz invariant. In particular the vacuum state is independent of the (arbitrary) choice of surfaces $t = $ const.

However, the attempt to carry over the procedure sketched above to a *curved space-time* leads to a series of difficulties, which occur essentially because of the non-existence of a preferred foliation of space-time by three-dimensional surfaces and because the topology of space-time may differ from the Minkowski one. Two different foliations of space-time lead in general to different systems g_n and $\hat{g}_n$ of partial waves, that is, to different definitions of particles.

Consider the representation of a general field operator with respect to two such systems

$$\phi = \sum_n (\mathbf{a}_n g_n + \mathbf{a}_n^\dagger \bar{g}_n) = \sum_m (\hat{\mathbf{a}}_m \hat{g}_m + \hat{\mathbf{a}}_m^\dagger \bar{\hat{g}}_m), \tag{38.13}$$

with corresponding vacuum states $|0\rangle$ and $|\hat{0}\rangle$

$$\mathbf{a}_n|0\rangle = 0, \qquad \hat{\mathbf{a}}_n|\hat{0}\rangle = 0. \tag{38.14}$$

Because of the completeness of both systems, the functions and operators of each system can be represented in terms of the other. In particular, there exist relations of the form

$$\hat{g}_n = \sum_m (\alpha_{nm} g_m + \beta_{nm} \bar{g}_m), \quad g_n = \sum_m (\bar{\alpha}_{nm} \hat{g}_m - \beta_{nm} \bar{\hat{g}}_m), \tag{38.15}$$

with constant (complex) coefficients α_{nm} and β_{nm} (the relations (38.15) describe a 'Bogoliubov transformation'). On inserting (38.15) in (38.13) one obtains the transformation law for the operators

$$\mathbf{a}_m = \sum_n (\alpha_{nm} \hat{\mathbf{a}}_n + \bar{\beta}_{nm} \hat{\mathbf{a}}_n^\dagger). \tag{38.16}$$

Thus not only are the particle (partial wave) definitions in the two systems different, but also, if $\beta_{nm} \neq 0$, what one observer regards as a vacuum state $|\hat{0}\rangle$ is seen by the other to be a mixture of particles

$$\mathbf{a}_m|\hat{0}\rangle = \bar{\beta}_{nm} \hat{\mathbf{a}}_n^\dagger|\hat{0}\rangle = \bar{\beta}_{nm}|\hat{1}_n\rangle. \tag{38.17}$$

This surprising result shows clearly that within General Relativity the concept of particles is more problematical than one might have expected. Proper Lorentz transformations in Minkowski space-time have $\beta_{nm} = 0$ and so do not alter the vacuum state. However, an accelerated observer in the 'usual' Minkowski space-time vacuum state would detect particles (with a thermal spectrum).

An immediate consequence of this property of a quantum field is the possibility that a gravitational field can create particles. Suppose, for example, that initially (as $t \to -\infty$) there is a flat space with vacuum state $|\hat{0}\rangle$, then a gravitational field is switched on and off, and finally (as $t \to +\infty$), the space is again Minkowski. However, the final vacuum state $|0\rangle$ will not always agree with the initial one $|\hat{0}\rangle$; particles have been produced.

38.5 Black holes are not black – the thermodynamics of black holes

The most spectacular example for the creation of particles by a gravitational field is produced by the gravitational field of a collapsing star, that is, the creation of a black hole. We shall sketch the basic ideas in the case of spherically symmetric collapse. We shall use the Eddington–Finkelstein coordinates introduced in Chapter 35, in which the Schwarzschild line element has the form

$$ds^2 = 2c\,dr\,dv + r^2\left(d\vartheta^2 + \sin^2\vartheta\,d\varphi^2\right) - (1 - 2M/r)c^2\,dv^2,$$
$$cv = ct + r + 2M\ln(r/2M - 1) = ct + r^*. \tag{38.18}$$

Of course this metric represents only the exterior of the star, whose boundary is given by

$$f(r,v) = 0, \tag{38.19}$$

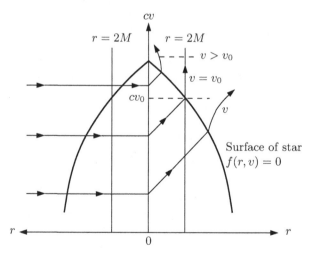

Fig. 38.1. A collapsing star and light rays in Eddington–Finkelstein coordinates.

see Figs. 36.4 and 38.1. The metric in the interior could be, for example, a part of the Friedmann universe (36.38). What matters is that it is regular and shows no peculiarities even when the surface of the star disappears behind the horizon.

We consider a state of the system in which incoming waves do not occur, especially as $t \to -\infty$. This corresponds to the choice of partial waves

$$\hat{g}_\omega = \hat{h}_\omega(r, \vartheta, \varphi)\, e^{-i\omega v} = \hat{h}_\omega(r, \vartheta, \varphi)\, e^{-i\omega(t+r^*/c)} \qquad (38.20)$$

as waves of positive frequency (with respect to v). Thus we write the solution of the wave equation $\Box \phi = 0$ as

$$\phi = \int_0^\infty (\hat{a}_\omega \hat{g}_\omega + \hat{a}_\omega^\dagger \hat{\bar{g}}_\omega)\, d\omega, \qquad (38.21)$$

and require the system to be in the corresponding vacuum state $|\hat{0}\rangle$.

If the gravitational field creates particles, outgoing particles should be present, although there are no ingoing ones. However, outgoing waves are best described in terms of retarded time u given by

$$cu = ct - r^* = cv - 2r^* = cv - 2r - 4M \ln(r/2M - 1), \qquad (38.22)$$

and the corresponding preferred system of partial waves with positive frequency with respect to u is

$$g_{\omega'} = h_{\omega'}(r, \vartheta, \varphi)\, e^{-i\omega' u} = h_{\omega'}(r, \vartheta, \varphi)\, e^{-i\omega'(v-2r^*/c)}. \qquad (38.23)$$

Unlike (38.20), this system is not complete, for in a general state some incoming waves will be absorbed by the black hole and not propagate to infinity as those in (38.23). However, in spite of this one can represent the functions $g_{\omega'}$ in terms of the *complete* system $\hat{g}_\omega$ and $\hat{\bar{g}}_\omega$,

$$g_{\omega'} = \int_0^\infty \left(\bar{\alpha}_{\omega\omega'} \hat{h}_\omega\, e^{-i\omega v} - \beta_{\omega\omega'} \hat{\bar{h}}_\omega\, e^{i\omega v} \right) dv. \qquad (38.24)$$

Outgoing particles occur if and only if the Bogoliubov coefficients $\beta_{\omega\omega'}$ do not vanish.

It is not possible to carry out exactly the transformation between the two systems of functions $g_{\omega'}$ and $\hat{g}_\omega$; in particular, the radial dependence of the functions $h_{\omega'}$ and $\hat{h}_\omega$ cannot be given as simple analytic expressions. However, as often in the discussion of wave propagation, a geometrical optics (eikonal) approximation (see Section 21.4) allows further progress. Both systems of functions have the form

$$\phi = A(x^i)\, e^{-iW(r,v)}, \qquad (38.25)$$

where, because of the wave equation (38.4), the eikonal $W(r,v)$ is to satisfy

$$0 = cW_{,a}W^{,a} = W_{,r}\left[2W_{,v} + c(1 - 2M/r)W_{,r}\right] = W_{,r}[2W_{,v} + cW_{,r^*}]. \tag{38.26}$$

Clearly the solutions of this equation form two classes: ingoing waves having $W = W(v)$, and outgoing waves having $W = W(v - 2r^*/c) = W(u)$.

The approximation that we shall now make is to convert only the eikonal function W of the outgoing wave to the ingoing form, thus neglecting the factors $\hat{h}_\omega$ and $h_{\omega'}$, and to include only those terms in the eikonal whose derivative is especially large. (In the geometrical optics ansatz (38.25) it is always implicitly assumed that derivatives of W are large in comparison with those of A.)

The conversion of the eikonal

$$W = \omega'u = \omega'\left[cv - 2r - 4M\ln(r/2M - 1)\right]/c, \tag{38.27}$$

of the outgoing waves (38.23), that is, the determination of the eikonal $\hat{W} = \hat{W}(v)$ of the corresponding ingoing wave can be done as follows. Since $\hat{W}$ does not depend on r it is obviously sufficient to know $\hat{W}$ on the surface of the star. In order to extract from W, the eikonal of the outgoing wave, the eikonal of the ingoing wave on the surface of the star, we have to trace the outgoing wave back to the surface of the star, thence to the centre and further back to the surface again, and sum up all changes in phase occurring along this path. Although we do not know the eikonal or phase change within the star, we can at least estimate the required quantities.

If the surface of the star is at rest or almost so (e.g. at the start of the collapse, $t \to -\infty$), the forward and backward directions within the star are equivalent and in particular are v-independent. Therefore the eikonal $\hat{W}(v)$ coincides with $W(v)$ (up to an additive constant) on the surface of the star,

$$\begin{aligned}\hat{W}(v) &= W(u)|_{r=r_0} + \text{const.} = W(v,r)|_{r=r_0} + \text{const.} \\ &= \omega'\left[v - 2r^*(r_0)/c\right] + \text{const.} = \omega'v + \text{const.}\end{aligned} \tag{38.28}$$

A constant contribution to the eikonal is inessential and can be incorporated in the amplitude A (see (38.25)). Thus an outgoing wave $\mathrm{e}^{-\mathrm{i}W} = \mathrm{e}^{-\mathrm{i}\omega'u}$ is associated with an ingoing wave $\mathrm{e}^{-\mathrm{i}\hat{W}} = \mathrm{e}^{-\mathrm{i}\omega'v}$ of the same frequency; if there is no ingoing wave then outgoing waves will not exist and particles are not produced.

If the surface of the star moves, and especially when it approaches the horizon $r = 2M$, the forward and backward directions in the stellar interior are no longer equivalent; the eikonal $\hat{W}(v)$ on the surface of the star (and hence for all values of v) differs from $W(u)$ on the surface $f(r, v) = 0$ by an additive v-dependent function $F(v)$,

$$\hat{W}(v) = W(u)|_{f(r,v)=0} + F(v). \tag{38.29}$$

Near the horizon the equation $f(r, v) = 0$ for the surface of the star has the approximate form

$$r = 2M + Bc(v_0 - v) + \cdots, \quad B = \text{const.}, \quad v < v_0. \tag{38.30}$$

From (38.27) and (38.30) follows

$$W(u)|_{f=0} = \omega'c^{-1}\big(cv - 4M - 2Bc(v_0 - v) \tag{38.31}$$
$$- 4M \ln\big[Bc(v_0 - v)/2M\big] + \cdots\big).$$

If we are only interested in the dominant part of $\hat{W}(v)$ near $v = v_0$ we need only retain the ln term in (38.32), and we can also discard $F(v)$ in (38.29) which has a finite derivative at $v = v_0$ because of the regularity of the metric in the stellar interior. Thus for small positive $v_0 - v$

$$\hat{W}(v) \approx -\omega'c^{-1}4M \ln c(v_0 - v). \tag{38.32}$$

Ingoing waves cannot produce outgoing waves with $v > v_0$ because the former must pass within the horizon $r = 2M$ and can never return.

Let us summarize. The outgoing partial wave (38.23) has the following representation in terms of incoming waves:

$$g_{\omega'} = 0, \qquad\qquad v > v_0,$$
$$g_{\omega'} \sim e^{+i\omega'4M[\ln(v_0-v)]/c}, \quad c(v_0 - v)/2M \ll 1, \tag{38.33}$$
$$g_{\omega'} \sim e^{-i\omega'v}, \qquad\qquad v \to -\infty.$$

Because the Fourier transform of the middle line contains $\Gamma(1 - i\omega4m/c)$, $g_{\omega'}$ contains all frequencies ω (and not just the positive ones), and the $\beta_{\omega\omega'}$ of (38.24) are non-zero; particles are produced! The same result (38.33) would have been obtained when instead of a collapsing star we had considered a shrinking reflective spherical surface. Here too the essential condition is that on the surface the phase (eikonal) of the incoming wave coincides with that of the outgoing one.

The important parameter for particle production and the frequency spectrum is the quantity $4M/c$. A more precise analysis (Hawking 1975)

shows that the particles have a thermal spectrum, that is, a black hole
with (Newtonian) mass m radiates like a black body of temperature

$$T = \frac{\hbar c}{8\pi M k} = \frac{\hbar}{k m \kappa c^2} \approx \frac{1.2 \times 10^{26} \text{ K}}{m[\text{g}]}, \quad (38.34)$$

where in the last equation the mass m is to be given in grams and the
temperature is obtained in degrees Kelvin. Black holes are therefore
not black but emit radiation continuously ('*Hawking effect*'); smaller
(lighter) ones are hotter.

Even before Hawking had found this astonishing relationship between
thermodynamics, quantum theory and gravitation, Bekenstein (1973)
had suggested that a temperature and entropy could be associated with
a black hole. Just as the total entropy of a process involving several thermodynamical systems can never decrease, so the sum of all surface areas
A_i of a system of (rotating or non-rotating) black holes cannot decrease
(see Section 37.3). In fact Hawking's discussion can be generalized to
rotating black holes, whose temperature T and entropy S are given by

$$T = \frac{2(r_+ - M)\hbar c}{Ak}, \qquad S = \frac{2\pi k}{\kappa c \hbar} A. \quad (38.35)$$

For spherically symmetric black holes the first law of thermodynamics
then reads

$$T \, dS = d(8\pi M/\kappa) = dmc^2. \quad (38.36)$$

Let us return to the derivation and discussion of particle production
in the gravitational field of a spherically symmetric collapsing star. The
derivation of the effect outlined above may appear to include somewhat arbitrary approximation procedures. However, the main equation
(38.32) furnishes all the important details about outgoing waves (particles) that would be observed by a distant observer at late times, since all
those come from a neighbourhood of the horizon. One sees immediately
that the basic idea can easily be carried over to other massless fields (e.g.
the Maxwell fields) because the eikonal equation (38.26) is the same for
all such fields. It can also be shown that massive particles are produced.

Where precisely do these particles originate? The analytic structure
of the eikonal suggests that the creation can be localized in a close
neighbourhood of the horizon; however, the global nature of the particle
concept in quantum field theory suggests caution before accepting so
simple an interpretation.

If a collapsing star emits particles continuously, its energy (mass) must
of course decrease. Because a solar mass black hole has a temperature

$T = 6 \times 10^{-8}$ K this mass loss for conventional celestial objects undergoing collapse is totally negligible. However, very low mass black holes can have only a short life; because of the energy loss the temperature rises rapidly, more is radiated, and in a self-accelerating process the black hole disappears.

In order to decide whether these considerations are correct one needs a theory which correctly describes the back-reaction of the quantum field on the gravitational field, and we do not yet have one. Therefore it is not clear whether in gravitational collapse a black hole must occur, or whether particle production (which will have started before the star disappears within the horizon) decreases the mass so quickly, and so forces the horizon $r = 2M$ to shrink so rapidly, that the outer surface of the star always remains outside the horizon. It is highly plausible that a horizon is created, but as yet we have no detailed ideas or theory as to how it might subsequently disappear.

Further reading for Chapter 38
Bekenstein (1973), Hawking (1975), Birrell and Davies (1984), Green *et al.* (1988), Ashtekar (1991), Wald (1994), Rovelli (1998).

39
The conformal structure of infinity

39.1 The problem and methods to answer it

Where does one end if one moves unwaveringly straight on? The naive answer will be: at infinity. But is there only one infinity? From mathematics one knows that the *complex* plane is closed by a single point, whereas infinity of the *projective* plane is a straight line. How is it in a general space or space-time: does infinity depend on direction and/or on velocity?

On the other hand one knows from the theory of complex functions that the complex plane can be (stereographically) mapped onto the surface of a sphere, so that infinity of that plane corresponds to a point of the sphere. So, to study infinity, one should perhaps always per-

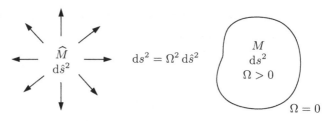

Fig. 39.1. Conformal transformation mapping infinity on $\Omega = 0$.

form a transformation which carries the infinitely extended manifold over into a finite one, where the former infinity now is a point or in general the boundary of a hopefully finite region. This transformation can in general not be a coordinate transformation, and so we have to decide which transformations should be admitted; the structure of infinity may depend on that decision.

In space-time, the causal structure mediated by light cones is most important. If we want that causal structure to be conserved, we are led to use conformal transformations. A conformal transformation is a transformation between two spaces $\widehat{M}$ (with metric $d\hat{s}^2$) and M (with metric ds^2) such that

$$ds^2 = \Omega^2 \, d\hat{s}^2 \quad \leftrightarrow \quad g_{ab} = \Omega^2 \widehat{g}_{ab}, \tag{39.1}$$

all distances are (locally) scaled by the same factor, independent of their directions. These transformations leave the source-free Maxwell equations invariant, as was shown in Section 21.3.

To study the structure of infinity for a given physical space $\widehat{M}$ we have to find a conformal transformation (39.1) which relates $\widehat{M}$ to a conformally equivalent mathematical world M such that the infinity of $\widehat{M}$ is mapped onto the 'boundary' of M which is given by $\Omega = 0$. This boundary may be a point in an infinite region or – if we are lucky – it looks like that in Fig. 39.1.

There is no algorithmic way of finding such a conformal transformation. One usually looks at the geodesics, extends them to arbitrarily large values of their affine parameter to see where infinity is, and then one tries some coordinate transformations in $\widehat{M}$ and hopes that one can identify an Ω. The boundary $\Omega = 0$ may a point, a line, or a hypersurface.

39.2 Infinity of the three-dimensional Euclidean space (E_3)

In three-dimensional Euclidean space

$$d\widehat{s}^2 = d\xi^2 + d\eta^2 + d\zeta^2, \tag{39.2}$$

the potential equation

$$\widehat{\Delta V} = (\sqrt{-\widehat{g}}\,\widehat{g}^{mn}\widehat{V}_{,m})_{,n}/\sqrt{-\widehat{g}} \tag{39.3}$$

can be considered to be the physically most important equation.

As a first trial to discuss infinity, one may take the straight lines in the ξ-direction as starting point. These geodesics extend to infinity. We introduce a new coordinate $x = 1/\xi$ (which brings infinity to $x = 0$). The metric (39.2) then reads

$$d\widehat{s}^2 = x^{-4}[dx^2 + x^4(d\eta^2 + d\zeta^2)]. \tag{39.4}$$

One can easily read off Ω and ds^2 as

$$\Omega = x^2, \quad ds^2 = dx^2 + x^4(d\eta^2 + d\zeta^2). \tag{39.5}$$

For $\Omega = 0$ we have $ds^2 = 0$; infinity is a point, all parallel straight lines end in a single point – which may be a different point for each direction of the lines. The potential equation is no help in deciding this – it is not invariant.

In a second trial we write the metric as

$$d\widehat{s}^2 = dr^2 + r^2(d\vartheta^2 + \sin^2\vartheta\,d\varphi^2) \tag{39.6}$$

and take the radial geodesics as starting point. Again we choose an inversion $x = 1/r$ to map infinity to $x = 0$. We now get

$$d\widehat{s}^2 = x^{-4}[dx^2 + x^2(d\vartheta^2 + \sin^2\vartheta\,d\varphi^2)] \tag{39.7}$$

and can read off

$$\Omega = x^2, \quad ds^2 = dx^2 + x^2(d\vartheta^2 + \sin^2\vartheta\,d\varphi^2). \tag{39.8}$$

We see that $d\widehat{s}^2$ and ds^2 give the same metric – which reflects the fact that the inversion is a conformal transformation in an E_3. The boundary $\Omega = 0$ is a point, independent of the direction of the radial geodesics. Moreover, the potential equation is invariant under this special conformal transformation if $\widehat{V}$ is suitably transformed, see Exercise 39.1. We conclude that infinity of E_3 is a point.

39.3 The conformal structure of Minkowski space

The study of geodesics Since we want to use 'radial' geodesics, we start with the form

$$\mathrm{d}\widehat{s}^2 = \mathrm{d}r^2 + r^2(\mathrm{d}\vartheta^2 + \sin^2\vartheta\,\mathrm{d}\varphi^2) - c^2\,\mathrm{d}t^2 \qquad (39.9)$$

of the line element.

To deal with the *spacelike geodesics*, we parametrize them by $r = \rho\cosh\chi, ct = \rho\sinh\chi$ (χ, ϑ and φ label the different geodesics, and infinity is approached for $\rho \to \infty$). Taking ρ and χ as coordinates, we get

$$\begin{aligned}\mathrm{d}\widehat{s}^2 &= \mathrm{d}\rho^2 + \rho^2\cosh^2\chi(\mathrm{d}\vartheta^2 + \sin^2\vartheta\,\mathrm{d}\varphi^2) - \mathrm{d}\chi^2 \\ &= x^{-4}[\mathrm{d}x^2 + x^2\cosh^2\chi(\mathrm{d}\vartheta^2 + \sin^2\vartheta\,\mathrm{d}\varphi^2) - x^4\mathrm{d}\chi^2], \quad x = 1/\rho > 0.\end{aligned} \qquad (39.10)$$

We read off $\mathrm{d}s^2 = \mathrm{d}x^2 + x^2\cosh^2\chi(\mathrm{d}\vartheta^2 + \sin^2\vartheta\,\mathrm{d}\varphi^2) - x^4\mathrm{d}\chi^2$, $\Omega = x^2$. We see that $\Omega = 0$ is the point $x = 0$ which we call I^0; it represents spacelike infinity.

Similarly, for *timelike geodesics* we take $r = \rho\sinh\chi$, $ct = \rho\cosh\chi$, and arrive at

$$\mathrm{d}\widehat{s}^2 = x^{-4}[-\mathrm{d}x^2 + x^2\sinh^2\chi(\mathrm{d}\vartheta^2 + \sin^2\vartheta\,\mathrm{d}\varphi^2) + x^4\mathrm{d}\chi^2], \quad x = 1/\rho. \qquad (39.11)$$

But now ρ can have either sign, depending on the geodesics going into the future or into the past. Timelike infinity consists of two points, called I^+ and I^-.

For *null geodesics* we best use null coordinates $u = (ct - r)/\sqrt{2}$, $v = (ct + r)/\sqrt{2}$, and label the null geodesics pointing into the future by (v, ϑ, φ). We obtain

$$\begin{aligned}\mathrm{d}\widehat{s}^2 &= -2\,\mathrm{d}u\,\mathrm{d}v + \tfrac{1}{2}(v - u)^2(\mathrm{d}\vartheta^2 + \sin^2\vartheta\,\mathrm{d}\varphi^2), \\ &= x^{-2}[2\,\mathrm{d}x\,\mathrm{d}u + \tfrac{1}{2}(1 - ux)^2(\mathrm{d}\vartheta^2 + \sin^2\vartheta\,\mathrm{d}\varphi^2)], \quad x = 1/v. \end{aligned} \qquad (39.12)$$

We read off $\mathrm{d}s^2 = 2\,\mathrm{d}x\,\mathrm{d}u + (1 - ux)^2(\mathrm{d}\vartheta^2 + \sin^2\vartheta\,\mathrm{d}\varphi^2)/2$, $\Omega = x$ and see that future null infinity $\mathcal{I}^+$ (scri-plus, from scri = script i) is characterized by

$$\Omega = x = 0, \quad \mathrm{d}s^2 = \tfrac{1}{2}(\mathrm{d}\vartheta^2 + \sin^2\vartheta\,\mathrm{d}\varphi^2). \qquad (39.13)$$

It is not only a sphere, since the null coordinate u can also vary, but a three-dimensional null surface. For the null geodesics pointing into the past, we have to interchange u and v, and get past null infinity $\mathcal{I}^-$. Null infinity comprises $\mathcal{I}^+$ and $\mathcal{I}^-$. The results are summarized in Fig. 39.2.

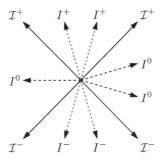

Fig. 39.2. Infinity of Minkowski space.

Conformal mapping to the Einstein universe A clearer picture of the
structure of infinity than that given by Fig. 39.2 can be obtained by the
conformal mapping of Minkowski space in the Einstein universe

$$ds_E^2 = -c^2 dt^2 + d\chi^2 + \sin^2\chi(d\vartheta^2 + \sin^2\vartheta\, d\varphi^2),$$
$$0 \le \chi \le \pi,\ 0 \le \vartheta \le \pi,\ 0 \le \varphi \le 2\pi,$$

(39.14)

which we will discuss in detail in Section 41.2.

Since both Minkowski space and Einstein universe are conformally
flat, there is at least a conformal relation of the type (39.1) between the
two. To make that explicit, one first transforms the Einstein universe

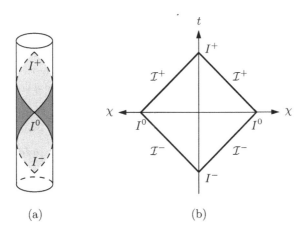

(a) (b)

Fig. 39.3. Conformal Minkowski space (a) as part of the Einstein universe
(b) developed.

by introducing null coordinates p, q to

$$\mathrm{d}s_E^2 = -4\,\mathrm{d}p\,\mathrm{d}q + \sin^2(p-q)[\mathrm{d}\vartheta^2 + \sin^2\vartheta\,\mathrm{d}\varphi^2], \quad \chi = p-q, \quad ct = p+q.$$
$$(39.15)$$

Starting with (39.12), one then gives Minkowski space a similar structure by

$$\mathrm{d}\hat{s}^2 = \cos^{-2}p\cos^{-2}q\left\{-4\,\mathrm{d}p\,\mathrm{d}q + \sin^2(p-q)[\mathrm{d}\vartheta^2 + \sin^2\vartheta\,\mathrm{d}\varphi^2]\right\},$$
$$\tan p = v/\sqrt{2}, \quad \tan q = u/\sqrt{2}, \quad p \geq q, \tag{39.16}$$

(the condition $p \geq q$ originates in $r \geq 0$ in (39.9)). Comparing the two line elements, one reads off

$$\Omega = \cos p \cos q = \cos \tfrac{1}{2}(\chi + ct) \cos \tfrac{1}{2}(\chi - ct). \tag{39.17}$$

In the following discussion we shall suppress the two coordinates ϑ, φ.

The mapping between Minkowski space and Einstein universe is not one-to-one. Rather, because of $\tan p = v/\sqrt{2}$, $\tan q = u/\sqrt{2}$, the whole of Minkowski space $(-\infty \leq u, v \leq +\infty)$ is mapped onto the part $(-\tfrac{1}{2}\pi \leq p, q \leq +\tfrac{1}{2}\pi)$ of the Einstein universe, infinity of Minkowski space being the boundary $\Omega = 0$. We shall not discuss in detail the different parts of that boundary analytically, but just refer to Fig. 39.3.

The boundary $\Omega = 0$, where p and q are constant, is thus built from null geodesics on the Einstein universe. The corresponding picture of Minkowski space is that of a square handkerchief wound around a cylinder; if developed, we get the square of Fig. 39.3(b). This is also called the *Penrose diagram* of the Minkowski space. It shows that (conformal) infinity is a closed light cone.

Qualitative questions can often be discussed in terms of this diagram. The world lines of particles all originate in I^- and end in I^+. The point I^0 is the 'usual' spacelike infinity. Radiation going to infinity ends up in $\mathcal{I}^+$. Null geodesics are given by $p = \text{const.}$ or $q = \text{const.}$, both coordinates are null coordinates $(p_{,a}p^{,a} = 0 = q_{,a}q^{,a})$.

39.4 Asymptotically flat gravitational fields

Suppose one has a gravitational field which is isolated, all material sources being within a closed region of space. Only incoming and outgoing radiation extends to spatial infinity. Albeit the material sources will in general extend to timelike infinity, one may expect that far away from the sources space-time is nearly Minkowskian or 'asymptotically flat'. How can one specify this 'nearly Minkowskian'? One possibility is to assume that there is a conformal mapping as discussed above, and to impose conditions on the conformal factor Ω.

The essential condition arises from the inspection of the field equations. If two metrics are related by (39.1), i.e. by $g_{ab} = \Omega^2 \widehat{g}_{ab}$, then one gets for the traces $\widehat{R}$ and R of the Ricci tensors the relation

$$\widehat{R} = \Omega^2 R + 6\Omega\Omega_{,c}{}^{\|c} - 12\Omega_{,a}\Omega^{,a}, \tag{39.18}$$

where $\|$ denotes the covariant derivative with respect to the (unphysical) metric g_{ab}.

If at infinity, i.e. at $\Omega = 0$, one has vacuum ($\widehat{T}_{ab} = 0$) or electrovacuum ($\widehat{T} = 0$), then the field equations imply $\widehat{R} = 0$, and because of (39.18) we have

$$\Omega_{,a}\Omega^{,a} = 0 \quad \text{at} \quad \Omega = 0, \tag{39.19}$$

$\Omega = 0$ is a null surface (except for $\Omega_{,a} = 0$). A further condition is that $\Omega = 0$ has the same structure as in Minkowski space, any null geodesic should begin and end there.

Space-times $(\widehat{M}, \widehat{g}_{ab})$ which have the three properties that

(I) a relation $g_{ab} = \Omega^2 \widehat{g}_{ab}$ exists, with $\Omega \geq 0$, where g_{ab} and Ω are C^3 on M and its boundary,

(II) on the boundary one has $\Omega = 0$, $\Omega_{,a} \neq 0$, and

(III) every null geodesic intersects the boundary in two points

are called asymptotically simple. Together with the field equations these three conditions should lead to asymptotically flat space-times. Note that there may be singular points like $I^\pm$ and I^0, which have to be studied separately.

It is not known whether all 'reasonable' solutions for isolated sources are asymptotically flat in this sense.

39.5 Examples of Penrose diagrams

The Schwarzschild solution Starting with the metric (35.19), i.e.

$$\mathrm{d}\widehat{s}^2 = -c^2 \mathrm{d}u\,\mathrm{d}v(1-2M/r) + r^2[\mathrm{d}\vartheta^2 + \sin^2\vartheta\,\mathrm{d}\varphi^2], \quad r = r(u,v), \tag{39.20}$$

which is already adapted to light rays, we map the two null coordinates u and v to finite ranges by

$$\tan U = cu, \quad \tan V = cv. \tag{39.21}$$

This leads to

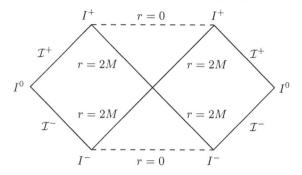

Fig. 39.4. Penrose diagram for the Schwarzschild solution.

$$d\hat{s}^2 = \cos^{-2} U \cos^{-2} V (1 - 2M/r) dU \, dV + r^2 [d\vartheta^2 + \sin^2 \vartheta \, d\varphi^2]$$
$$= \Omega^{-2} \left[(1 - 2M/r) dU \, dV + r^2 \Omega^2 (d\vartheta^2 + \sin^2 \vartheta \, d\varphi^2) \right], \quad (39.22)$$
$$\Omega = \cos U \cos V.$$

The boundary $\Omega = 0$ is where $U = \pm\pi/2$ or $V = \pm\pi/2$ (or both), i.e. $u, v = \pm\infty$. Since $r\Omega \neq 0$ at $\Omega = 0$, the boundary consists of (pieces of) a light cone. We will not present a detailed discussion, but rather give the results as Fig. 39.4; see also the diagrams Fig. 35.1 and 35.2.

Robertson–Walker metrics Cosmological models of the Robertson–Walker type, which will be discussed in the next chapters, are homogeneous in space and, therefore, not asymptotically flat. But one still may ask where null, or timelike, geodesics begin and end, and use conformal factors to do this.

As can be shown by calculation of the Weyl tensor (32.1) or by explicitly carrying out a coordinate transformation, all Robertson–Walker metrics are conformally flat. If we restrict ourselves to the closed spaces

$$d\hat{s}^2 = K^2(ct)[d\chi^2 + \sin^2 \chi (d\vartheta^2 + \sin^2 \vartheta \, d\varphi^2)] - c^2 dt^2, \quad (39.23)$$

then after the transformations

$$T = \int dt/K(ct), \quad r = \frac{2 \sin \chi}{\cos \chi + \cos cT}, \quad c\eta = \frac{2 \sin cT}{\cos \chi + \cos cT} \quad (39.24)$$

the line element takes the form

$$d\hat{s}^2 = \tfrac{1}{4} K^2(ct)[\cos \chi + \cos cT]^2 [dr^2 + r^2 (d\vartheta^2 + \sin^2 \vartheta \, d\varphi^2) - c^2 d\eta^2], \quad (39.25)$$

which differs from that of a Minkowski space only by a conformal factor. More important for our present purpose is the fact that these metrics are all conformal to the Einstein universe,

$$d\hat{s}^2 = \Omega^{-2}(cT)[d\chi^2 + \sin^2\chi(d\vartheta^2 + \sin^2\vartheta\,d\varphi^2) - c^2dT^2],$$
$$\Omega^{-1}(ct) = K[ct(cT)], \tag{39.26}$$

which – in contrast to (39.16) – describes a one-to-one mapping.

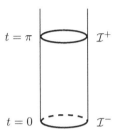

$t = \pi$ $\mathcal{I}^+$

$t = 0$ $\mathcal{I}^-$

Fig. 39.5. Conformal structure of the Friedmann dust universe.

For example, for the Friedmann dust universe (41.34) we have $\Omega = $ const.$/(1 - \cos cT)$: at the 'boundary' – beginning and end of the universe – the conformal factor Ω is not zero, but tends to infinity. But the three-spaces $cT = 0$, 2π (circles in Fig. 39.5) still have the property that all past and future null geodesics end there: they represent $\mathcal{I}^-$ and $\mathcal{I}^+$, which are spacelike for this universe. This is closely related to the occurrence of horizons, see Section 40.3 and the figures given there.

Exercises

39.1   Show that $\widehat{\Delta}\widehat{V} = 0$ in metric (39.6) leads to $\Delta V = 0$ in metric (39.8) if one sets $\widehat{V} = Vx$.

39.2   Identify the different parts of infinity of Minkowski space contained in $\Omega = \cos p \cos q = 0$.

VII. Cosmology

Gravitational forces are the only forces presently known which are long range (in contrast to the nuclear forces, for example) and which cannot be compensated (there are no negative masses). It is therefore to be expected that, for large quantities of matter distributed over wide regions of space, they will be the decisive forces, and hence the gravitational forces will determine the evolution and dynamics of the universe.

Physical laws get their importance from the fact that a single law describes many very different situations. Technically this comes out by writing the laws as differential equations (usually of second order), which admit a multitude of initial or boundary conditions. The law itself has often been found by extracting some common rules from the observed variety of effects. All these features are also present in the theory of gravitation.

In cosmology, however, we encounter a very different situation. There is only one realization of a cosmos, that which we are living in. And if there was an extra physical law for this cosmos, we could not find and prove it the usual way. That is to say, if we find a surprising new phenomenon, we cannot easily decide which of its properties are a new law, and which are due to initial conditions. Sometimes it is claimed that, in a proper theory, initial conditions should be excluded, the cosmos must not depend on them (this was one of the assumptions of the 'inflation' theory). Or one claims that the new law can be obtained from other principles, such us the anthropic principle. It seems that mankind does not like to be the outcome of an accidental initial value, but rather of an extra law.

In this book we shall set aside these more speculative ideas. Rather we shall take the conservative approach by asking whether our cosmos can be understood as a special case governed by the 'usual' laws. That is, we take the physical theories found and checked in our tiny neighbourhood, extending and stretching them to the outmost. The laws of gravitation, for example, have been checked only for a few hundred years, and mostly inside the Solar System, and for small matter densities; to apply them to the universe is an extreme extrapolation likely to fail.

For a recent review on many problems of cosmology see Ellis (1999).

40

Robertson–Walker metrics and their properties

40.1 The cosmological principle and Robertson–Walker metrics

Cosmology makes statements about the whole universe. Here as in many other areas of the natural sciences, every new discovery can revolution- ize the structure of our knowledge, our present picture of the universe being in no way complete and secure. But up until now this picture has always proved compatible with assuming initially the universal validity of natural laws, making calculations with strongly simplified models of reality, then comparing with the observations, and thus in a stepwise manner approximating models and formulations of the laws of nature to reality. Still General Relativity is able to explain, and helps us to understand, many features of the universe.

The simplest model of the universe is obtained from the cosmologi- cal principle, that is, from the assumption that in the rest system of matter there is no preferred point and no preferred direction, the three- dimensional universe being constituted in the same way everywhere. A

'The universe, my son,
is a large tank full of water.'

Fig. 40.1. The cosmological principle.

352

glance at the sky (which of course ought to be uniformly bright or dark) shows us that this model is a very great simplification and that the universe is uniform at best only on the average. We do not know how large are the spatial regions over which the average should be taken – at any rate the galaxies are not uniformly distributed, but tend to be clustered. Nevertheless, this most simple of cosmological models can explain observations surprisingly well – although we have to admit that we might be in the situation of the fish in Fig. 40.1.

Translated into the language of Riemannian geometry, this cosmological principle clearly asserts that three-dimensional position space is a space of maximal symmetry, that is, a space of constant curvature whose curvature can, however, depend upon time:

$$\overset{(3)}{\mathrm{d}s}{}^2 = g_{\alpha\beta}\,\mathrm{d}x^\alpha\,\mathrm{d}x^\beta = K^2(ct)\,\mathrm{d}\sigma^2, \qquad (40.1)$$

$$\mathrm{d}\sigma^2 = \frac{\mathrm{d}x^2 + \mathrm{d}y^2 + \mathrm{d}z^2}{(1 + \varepsilon r^2/4)^2} = \frac{\mathrm{d}\bar{r}^2}{1 - \varepsilon \bar{r}^2} + \bar{r}^2(\mathrm{d}\vartheta^2 + \sin^2\vartheta\,\mathrm{d}\varphi^2), \quad \varepsilon = 0, \pm 1.$$

cp. Section 19.5. Since the occurrence of terms $g_{4\alpha}$ in the full space-time metric picks out a spatial direction and $g_{44}(x^\alpha)$ signifies the dependence upon position of the proper time of a test particle at rest, then only the Robertson–Walker metrics (*R–W metrics*) (Robertson 1936, Walker 1936)

$$\mathrm{d}s^2 = K^2(ct)\,\mathrm{d}\sigma^2 - c^2\,\mathrm{d}t^2 \qquad (40.2)$$

are in accord with the cosmological principle. The metric of this model is thus already substantially determined by symmetry requirements; the Einstein field equations can (if they are satisfied at all) now fix only the time behaviour of the universe – the function $K(ct)$ – and the type of the local space – the choice of ε. For the reasons explained in Section 34.2, K is called the radius of the universe, although K can only be visualized in this way for closed (three-dimensional) spaces ($\varepsilon = +1$).

In the next section we shall discuss first of all some physical properties of the metrics (40.2)–(40.2); we shall not draw conclusions from the Einstein equations until Chapter 41.

40.2 The motion of particles and photons

We use the R–W metric in the form

$$\mathrm{d}s^2 = K^2(ct)\big[\mathrm{d}\chi^2 + f^2(\chi)\big(\mathrm{d}\vartheta^2 + \sin^2\vartheta\,\mathrm{d}\varphi^2\big)\big] - c^2\,\mathrm{d}t^2,$$

$$f(\chi) = \begin{cases} \sin\chi & \text{for} & \varepsilon = 1, \\ \chi & \text{for} & \varepsilon = 0, \\ \sinh\chi & \text{for} & \varepsilon = -1. \end{cases} \qquad (40.3)$$

It picks out the origin $\chi = 0$ as a preferred point, and the coordinate χ is directly related to the distance D of an arbitrary point (star at rest) from the origin by

$$D = K(ct)\chi. \qquad (40.4)$$

If the radius K of the universe changes with time, then the distances of the stars and galaxies between each other also change, just as the separations of fixed points (fixed coordinates) on a balloon change when the balloon is blown up or deflated. The velocity $\dot{D}$ which thereby results is proportional to the displacement D:

$$\dot{D} = \frac{\partial D}{\partial t} = \frac{\dot{K}}{K}cD. \qquad (40.5)$$

A test particle or a photon which moves in the absence of forces describes, under suitable choice of the coordinate system, a purely 'radial' trajectory $\chi(\tau)$, $\vartheta = \text{const.}$, $\varphi = \text{const.}$, that is, a geodesic of the metric

$$\mathrm{d}s^2 = K^2(ct)\,\mathrm{d}\chi^2 - c^2\,\mathrm{d}t^2 = -c^2\,\mathrm{d}\tau^2. \qquad (40.6)$$

For a test particle of mass m_0 we get, if we denote by v the speed $K\,\mathrm{d}\chi/\mathrm{d}t$ and by $p = mv = m_0 v/\sqrt{1 - v^2/c^2}$ the momentum of the particle, a conservation law in three-dimensional form as

$$pK = \text{const.}, \qquad (40.7)$$

(see Exercise 40.2). That is, the product of the radius of the universe and the magnitude of the momentum is constant for force-free motion.

For photons one expects a similar result, that is, a dependence of the wavelength and the frequency of an emitted photon upon the radius of the universe K. We want now to derive the formula for the more general case that the source and the observer move arbitrarily with respect to the coordinate system (which we shall later identify as the rest system of the matter), see Fig. 40.2.

The world line $x^a(v)$ (null geodesic) of a photon defines the null vector $k^a = \mathrm{d}x^a/\mathrm{d}v$ uniquely up to a factor which is constant along the world line if we use an affine parameter as the parameter v; that is, one which brings the geodesic equation to the form $k_{a;b}k^b = 0$ (see Section 21.4). The frequency ν which an observer moving with the four-velocity u^a associates with this photon is proportional to the timelike component

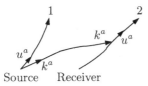

Fig. 40.2. The change in frequency of a photon in the gravitational field.

of k^a at the location and in the rest system of the observer, that is, proportional to $k^a u_a = g_{ab} u^a k^b$. The frequencies measured in the rest system of the source and by the observer are related according to

$$\frac{\nu_1}{\nu_2} = \frac{(k^a u_a)_1}{(k^a u_a)_2} = \frac{(g_{ab} k^a u^b)_1}{(g_{ab} k^a u^b)_2}. \tag{40.8}$$

This formula describes not only the change in frequency which is a consequence of the relative motion (dependence upon u_1^a and u_2^a), that is, of the Doppler effect, but also the shift in frequency in the gravitational field (dependence upon the metric g_{ab}), and shows that the two effects can only be separated in an artificial manner depending upon the coordinate system.

Applying the formula (40.8) to sources and receivers which are at rest in the coordinate system (40.6), we have to substitute

$$u^a = (0,0,0,c), \quad k_a = (1,0,0,-1/K), \tag{40.9}$$

and obtain

$$\frac{\nu_1}{\nu_2} = \frac{K(ct_2)}{K(ct_1)} \quad \Longrightarrow \quad \nu K = \text{const.}, \tag{40.10}$$

in complete analogy with (40.7).

One usually expresses the change in frequency of the light received at two points through the *red shift* (relative change in wavelength)

$$z = \frac{\lambda_2 - \lambda_1}{\lambda_1}. \tag{40.11}$$

The equation (40.10) thus yields the relation

$$z = \frac{K(ct_2)}{K(ct_1)} - 1 \tag{40.12}$$

between the redshift z of light received, for example, on the Earth at time t_2 and the radii of the universe at the times of emission (t_1) and reception (t_2).

If on the Earth at the present time $t = t_2$ one examines the light emitted by a star at the time $t = t_1$, then, if the radius of the universe does not change too quickly and the light travel time $t_2 - t_1$ is not too large, one can replace $K(ct_1)$ by the first few terms of the Taylor series

$$K(ct) = K(ct_2)\left[1 + Hc(t - t_2) - \tfrac{1}{2}qH^2c^2(t - t_2)^2 + \cdots\right]. \quad (40.13)$$

The parameters occurring here are the *Hubble parameter H*,

$$H(ct_2) = \dot{K}(ct_2)/K(ct_2), \quad (40.14)$$

and the *acceleration parameter* (retardation parameter) q,

$$q(ct_2) = -\ddot{K}(ct_2)K(ct_2)/\dot{K}^2(ct_2). \quad (40.15)$$

Substitution of the series (40.13) into (40.12) gives the relation

$$z = Hc(t_2 - t_1) + \left(1 + \tfrac{1}{2}q\right)H^2c^2(t_2 - t_1)^2 + \cdots \quad (40.16)$$

between the redshift z and the light travel time $t_2 - t_1$.

The validity or applicability to our universe of the model of a Robertson–Walker metric is usually tested in the relation between the redshift and the distance of the source. Since $ds^2 = 0$ for light, from (40.6) and (40.13) it follows that, to first approximation,

$$\chi = \int_{t_1}^{t_2} \frac{c\,dt}{K(ct)} \approx \frac{c(t_2 - t_1)}{K(ct_2)} + \frac{Hc^2(t_2 - t_1)^2}{2K(ct_2)} + \cdots, \quad (40.17)$$

and therefore, using (40.4) and (40.5),

$$z = HD + \tfrac{1}{2}(q+1)H^2D^2 + \cdots = \dot{D}/c + \tfrac{1}{2}\left(\dot{D}^2 - D\ddot{D}\right)/c^2 + \cdots. \quad (40.18)$$

The redshift is *to first approximation* proportional to the present distance D of the source or to the ratio of the (cosmological) escape velocity $\dot{D}$ of the source to the velocity of light.

40.3 Distance definitions and horizons

The determination of distance in astronomy is mostly done using the concepts and ideas of a three-dimensional Euclidean space. We therefore want to describe briefly how the laws of light propagation in R–W metrics influence the determination of distance.

One possible way of determining the distance of an object is to compare its absolute luminosity L, which is defined as the total radiated energy per unit time and is regarded as known, with the apparent luminosity l, which is the energy reaching the receiver per unit time and per unit surface area. The *luminosity distance D_L* is defined by

$$D_L = \sqrt{L/4\pi l}, \quad (40.19)$$

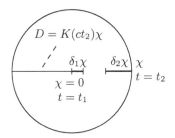

Fig. 40.3. The relation between apparent brightness and coordinate distance.

so that in the Euclidean space luminosity distance and geometrical distance coincide.

In a Robertson–Walker metric the relationship between true distance $D = K\chi$ and luminosity distance D_L is more complicated, see Fig. 40.3. The photons streaming out at $t = t_1$ from the source $\chi = 0$ are distributed, after a coordinate interval of χ, not of course over the surface $4\pi\chi^2$, but, in the metric (40.3), over the surface

$$F = 4\pi f^2(\chi)K^2(ct_2). \qquad (40.20)$$

Moreover, because $ds^2 = 0$, that is, because

$$K\,d\chi = c\,dt, \qquad (40.21)$$

near to the source the photons emitted during the time interval δt are distributed over the interval $\delta_1\chi = c\,\delta t/K(ct_1)$, while at the receiver in the time δt all those photons arrive which lie in an interval $\delta_2\chi = c\,\delta t/K(ct_2)$. And third, the energy of an individual photon has also changed during its passage through the gravitational field by the factor $K(ct_1)/K(ct_2)$. We therefore finally obtain for the apparent luminosity

$$l = \frac{L}{4\pi f^2(\chi)K^2(ct_2)}\cdot\frac{K^2(ct_1)}{K^2(ct_2)}, \qquad (40.22)$$

that is, the relation

$$D_L = \frac{f(\chi)K^2(ct_2)}{K(ct_1)} = (1+z)D\frac{f(\chi)}{\chi} \qquad (40.23)$$

between the luminosity distance D_L, the coordinate distance D (at time t_2) and the redshift z of a light source. Since one observes objects with $z > 5$, D and D_L can differ considerably.

A alternative way of determining distance is to compare the true

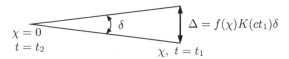

Fig. 40.4. Measurements of distance by determination of angle.

diameter Δ of a system with the angle δ which it subtends at the Earth. In Euclidean space we have of course

$$D_A \equiv D = \Delta/\delta \qquad (40.24)$$

for the distance D_A determined by measurement of angle.

In an R–W metric, however, (40.3) implies, according to Fig. 40.4,

$$D_A = \frac{\Delta}{\delta} = f(\chi)K(ct_1) = \frac{D}{1+z}\frac{f(\chi)}{\chi}. \qquad (40.25)$$

These two examples of how to determine distance show clearly how the space curvature comes into astronomical considerations via the laws of light propagation. Unexpected effects can occur. If, for example, the function $K(ct_1)$ decreases with t_1, for increasing $f(\chi)$, then the more distant of two objects of identical dimensions may have the greater angular diameter.

Of course optical methods can only be used to determine the distances of objects whose light reaches us. In flat space we can in principle see every flash of light, however distant, if we wait sufficiently long to allow for the finite velocity of light. In a curved space, however, the situation is more complicated. Imagine a fly (a photon) which is crawling at constant velocity away from the south pole of a balloon. By blowing up the balloon (increasing the radius of curvature) sufficiently rapidly, can one prevent the fly from reaching the north pole?

Light emitted at time t_1 at the origin $\chi = 0$ has, because of (40.3), reached the point

$$\chi = \int_{t_1}^{t_2} \frac{c\,dt}{K(ct)} \qquad (40.26)$$

by time t_2. If we want to know whether at the present time $t = t_2$ we can see all stars, then we must investigate whether the light sent out at the beginning of the universe t_B (the earliest possible time), from the furthest possible star, can reach us, or whether our signal sent out at the beginning of the universe and at the origin of the coordinate system has by now reached all stars. Depending on the cosmological model the

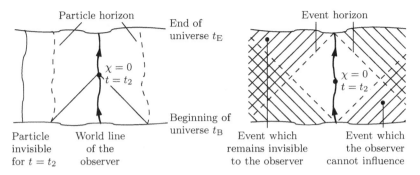

Fig. 40.5. Particle and event horizons.

beginning of the universe is here $t_B = -\infty$ or the first zero of $K(ct)$ (where the metric becomes singular) lying in the past.

At the present time $t = t_2$ we can see stars up to a coordinate distance

$$\chi_P = \int_{t_B}^{t_2} \frac{c\,dt}{K(ct)}. \tag{40.27}$$

If this value χ_P is smaller than the maximum coordinate distance (which is π in closed universes and ∞ in open ones), that is, if not all stars are visible, then χ_P defines the horizon up to which we can see. It is called the *particle horizon*.

If, for example, the radius of the universe changes according to the law $K(ct) = ct^2$ and we find ourselves in the contraction phase $t < 0$, then at the observer time $t_2 = -1$

$$\chi_P = \int_{-\infty}^{-1} \frac{dt}{t^2} = 1; \tag{40.28}$$

that is, in this cosmological model there is a particle horizon.

Another physically interesting question is whether (by means of the photons emitted there) we can learn about every event occurring in the universe, no matter when or where, or whether the end of the universe t_E ($t_E = \infty$ or the next zero of $K(ct)$ lying in the future) coming prematurely prevents this. An equivalent question is whether our light signal sent out at the present time $t = t_2$ reaches all points of the universe before its end t_E. Since this light signal traverses a maximum coordinate distance

$$\chi_E = \int_{t_2}^{t_E} \frac{dt}{K(ct)} \tag{40.29}$$

World line of the observer

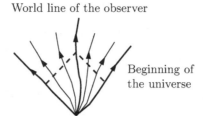

Beginning of
the universe

Fig. 40.6. A cosmological model without an event horizon: the past light cone going out from the observer intersects the world lines of *all* particles.

there exists an *event horizon* χ_E if χ_E is smaller than π or ∞: we shall never learn anything about events which at the present time $t = t_2$ are situated at distances greater than χ_E.

Figures 40.5 and 40.6 give a qualitative picture of how horizons work. The occurrence of horizons is closely related to the conformal structure of infinity discussed in Section 39.5.

A possible misinterpretation of the significance of the horizons should also be dealt with. Should there be an event horizon in our universe, that is, events about which we can never learn anything, then that would not imply the absence of an interaction with that part of the universe or the establishing of something which is in principle now knowable. Our (very poor) cosmological model presupposes from the beginning that the universe is everywhere the same, and we therefore know in advance that the same gravitational field and the same mass densities, and so forth, are present behind the horizon as close by, because without the presence of these masses as well the space in our neighbourhood would not be homogeneous and isotropic. The events which we cannot observe only affect test particles, that is, particles without a gravitational interaction, and it is only we who cannot detect these particles, which do act upon observers situated nearer to them.

40.4 Some remarks on physics in closed universes

There exists a multitude of cosmological models (up until now we have encountered in the Robertson–Walker metrics only the most primitive), which do not always differ significantly from one another, since they form a continuous sequence. There are, however, some characteristics of spaces which can be expressed by integers; in this class belongs the property of whether a universe is open or closed. It is to be expected that closed universes also differ from open ones in a clear physical way and that this difference may even possibly lead to statements which can be tested on the Earth. We therefore want to describe in more detail some properties of closed universes with R–W metrics.

As can be shown by calculation of the Weyl tensor (32.1) or by explicitly carrying out a coordinate transformation, all R–W metrics are conformally flat. If we restrict ourselves to closed spaces

$$ds^2 = K^2(ct)\left[d\chi^2 + \sin^2\chi\left(d\vartheta^2 + \sin^2\vartheta\,d\varphi^2\right)\right] - c^2\,dt^2, \qquad (40.30)$$

then after the transformations

$$T = \int \frac{dt}{K(ct)}, \qquad r = \frac{2\sin\chi}{\cos\chi + \cos cT}, \qquad c\eta = \frac{2\sin cT}{\cos\chi + \cos cT} \qquad (40.31)$$

the line element takes the form

$$ds^2 = \tfrac{1}{4}K^2(ct)\left[\cos\chi + \cos cT\right]^2\left[dr^2 + r^2\left(d\vartheta^2 + \sin^2\vartheta\,d\varphi^2\right) - c^2\,d\eta^2\right], \qquad (40.32)$$

which differs from that of a Minkowski space only by a conformal factor.

As one can immediately see from the transformation formula (39.24), this statement has only a local significance: the relations (40.31) map a section $(\cos\chi + \cos cT) \neq 0$ of the curved space-time onto the full Minkowski space, but a one-to-one mapping of the metrics (40.32) and (40.30) onto one another is impossible, see also the discussion in Section 39.2.

The source-free Maxwell equations are conformally invariant (see Section 21.3); if we know their general solution in the Einstein universe, then we also have at hand the general solutions in every closed R–W universe. Since all spatial coordinates are periodic in the Einstein universe, the source-free Maxwell equations have the character of eigenvalue equations for the frequency (Schrödinger 1940, see also Stephani 1974). Amongst the solutions one finds a generalized plane wave (eigenfunction with frequency c/λ), which in the neighbourhood of a point is practically a plane wave, but whose amplitude is noticeably different from zero only over a region

$$d \approx \sqrt{\lambda K}. \qquad (40.33)$$

The influence of the space curvature 'localizes' the plane wave and makes it similar to a particle trajectory. For visible light and a radius of the universe of 2×10^{10} light years, we obtain the value $d \approx 10^7$ km.

If in an Einstein universe one draws the field lines of the **D** field emanating from a point charge at rest, then they all intersect at the antipodal point, but they arrive there with the opposite sign: in a closed universe, to every charge there corresponds a charge of opposite sign (which, however, is not necessarily situated at the antipodal point). This intuitively obtained statement can be derived from Maxwell's equations. For the charge density j^4/c we have

$$\frac{1}{c}j^4 = F^{4n}{}_{;n} = \left[\sqrt{-\frac{4}{g}}F^{4n}\right]_{,n} \bigg/ \sqrt{-\frac{4}{g}} = \left[\sqrt{\frac{3}{g}}F^{4n}\right]_{,\alpha} \bigg/ \sqrt{\frac{3}{g}} \quad (40.34)$$

or, in three-dimensional form with $F^{4\alpha} = D^\alpha$,

$$D^\alpha{}_{;\alpha} = j^4/c. \quad (40.35)$$

Since the closed three-dimensional space has no surface (a spherical surface has no boundary), application of the Gauss law yields

$$\int j^4 \, \mathrm{d}^3 V = 0; \quad (40.36)$$

that is, Maxwell's equations can only be integrated if the total charge vanishes. Our universe is uncharged on the average, and so in this respect a closed cosmological model would not stand in contradiction with experiment.

The conclusion deduced from (40.35) evidently uses only the mathematical structure of this equation, not its physical interpretation: the volume integral of *any* quantity which can be written as a three-dimensional divergence must vanish. A Newtonian gravitational theory, for example, in which there are only positive mass densities μ,

$$\Delta U = U^{,\alpha}{}_{;\alpha} = -\mu, \mu > 0, \quad (40.37)$$

is not possible in a closed universe (40.30).

If there exists in a closed universe a Killing vector ξ^n proportional to a four velocity u^n,

$$\xi_{n;i} + \xi_{i;n} = 0, \quad \xi^n = \alpha u^n, \quad (40.38)$$

then because of the definition of the curvature tensor, the Einstein field equations and the general splitting (21.71) of the energy-momentum tensor of a fluid, we have

$$(\xi^{i;n} - \xi^{n;i})_{;n} = -2\xi^{n;i}{}_{;n} = -2\xi^m R_m{}^{ni}{}_n$$

$$= -2\kappa(\xi^m T_m{}^i - \tfrac{1}{2}\xi^i T^n{}_n) = \alpha\kappa(3p + \mu c^2)u^i + 2\alpha\kappa q^i. \quad (40.39)$$

Because of the formal similarity to Maxwell's equations one can conclude that for every closed universe the integral over the time component $(i = 4)$ of the right-hand side of (40.39) – over the analogue of the charge density – must vanish. But this is clearly not possible for perfect fluids ($q^i = 0$, $p > 0$, $\mu > 0$): there exists no static or stationary, spatially closed cosmological model with perfect fluid medium, whose Killing vector is parallel to the four-velocity. In the language of thermodynamics this can also be formulated as a cosmological model with closed three-dimensional space and perfect fluid ($q^i = 0$) cannot exist in complete thermodynamical equilibrium (the temperature vector u^a/T cannot be a Killing vector) (Neugebauer 1974).

In these last considerations we have already made use of the Einstein equations. We shall now turn to the problem of determining the evolution of the R–W metrics from these field equations.

Exercises

40.1 A balloon is inflated so that its circumference D grows as $D = Vt$. A fly starts at $t = 0$ at the south pole, creeping with velocity $v < V$. Will it ever reach the north pole?

40.2 Show that the conservation law (40.7) holds for the geodesics of the metric (40.6).

41

The dynamics of Robertson–Walker metrics and the Friedmann universes

41.1 The Einstein field equations for Robertson–Walker metrics

The Robertson–Walker metrics are completely determined by the temporal behaviour of the radius of the universe and by the sign of the curvature, that is, by $K(ct)$ and ε. We are thus confronted with the problem of calculating these parameters from the properties of the matter in our universe, and of seeing whether observational results and cosmological model can be brought into agreement.

The curvature tensor and the Ricci tensor of an R–W metric can be calculated relatively easily by applying the reduction formulae (30.25) to the line element

$$ds^2 = K^2(ct)\, d\sigma^2 - c^2\, dt^2 = g_{\alpha\beta}\, dx^\alpha\, dx^\beta - c^2\, dt^2. \qquad (41.1)$$

That is, we start from

$$R_{\alpha\beta\mu\nu} = \overset{3}{R}_{\alpha\beta\mu\nu} - \dot{K}^2 K^{-2}(g_{\beta\mu}g_{\alpha\nu} - g_{\beta\nu}g_{\alpha\mu}),$$
$$R^4{}_{\beta 4\nu} = \ddot{K} K^{-1} g_{\beta\nu}, \quad R^4{}_{\beta\mu\nu} = 0, \quad K_{\alpha\beta} = -\dot{K} K^{-1} g_{\alpha\beta}, \qquad (41.2)$$

and, in accordance with (19.40), substitute the relation

$$\overset{3}{R}_{\alpha\beta\mu\nu} = \varepsilon K^{-2}(g_{\alpha\mu}g_{\beta\nu} - g_{\beta\mu}g_{\alpha\nu}) \qquad (41.3)$$

for the three-dimensional curvature tensor of the R–W metric, whose space is of course a space of constant curvature. For the non-vanishing components of the Ricci tensor we obtain

$$R_{\beta\nu} = \left[\ddot{K}/K + 2(\dot{K}^2 + \varepsilon)K^{-2}\right]g_{\beta\nu}, \quad R_{44} = -3\ddot{K}/K. \qquad (41.4)$$

Together with the Einstein field equations

$$R_{mn} - \tfrac{1}{2} R g_{mn} = \kappa T_{mn}, \qquad (41.5)$$

the equations (41.4) show us that the energy-momentum tensor of the matter in the universe is spatially isotropic in the coordinates (41.1) and that no current of energy occurs ($T_{4\alpha} = 0$); in R–W metrics the energy-momentum tensor *must* be that of a perfect fluid,

$$T_{mn} = p g_{mn} + (\mu + p/c^2) u_m u_n, \qquad (41.6)$$

where the preferred coordinate system (41.1) is the rest system of the matter and μ and p depend only upon time.

As a consequence of (41.4) and (41.6) the field equations (41.5) reduce to

$$2\ddot{K}/K + (\dot{K}^2 + \varepsilon)/K^2 = -\kappa p, \qquad (41.7)$$
$$3(\dot{K}^2 + \varepsilon)/K^2 = -\kappa \mu c^2. \qquad (41.8)$$

These two equations are only mutually compatible if

$$\dot{\mu}/(\mu + p/c^2) = -3\dot{K}/K. \qquad (41.9)$$

Since for $\dot{K} \neq 0$ and $\mu c^2 + p \neq 0$ (41.7) also follows from (41.9) and (41.8), the field equation (41.7) can be replaced by (41.9).

Equation (41.8) is called the *Friedmann equation* (Friedmann 1922), and the special R–W metrics which satisfy it are called Friedmann universes. Occasionally, only the cosmological model arising from the special case $p = 0$ is designated the Friedmann universe. If one knows the equation of state $f(\mu, p) = 0$, then from (41.9) one can determine the radius K as a function of the mass density μ and hence calculate the behaviour of K and μ with respect to time from (41.8).

The Friedmann cosmological models can also be characterized invariantly by the fact that they are just those solutions of the Einstein field equations with a perfect fluid as source whose velocity fields $u^n(x^i)$ are free of rotation, shear and acceleration.

41.2 The most important Friedmann universes

The Einstein universe Soon after having set up his field equations, Einstein (1917) tried to apply them to cosmology. In accordance with the then state of knowledge, he started from a static cosmological model. Thus all the time derivatives in (41.7) and (41.8) vanish, so that we are left with the equations

$$\varepsilon/K^2 = -\kappa p, \qquad 3\varepsilon/K^2 = \kappa\mu c^2. \tag{41.10}$$

These can only be brought into agreement with the observed data, which require that $p \approx 0$, by rather artificial means, namely, by introduction of the cosmological constant Λ. According to this hypothesis, the energy-momentum tensor contains, in addition to the contribution due to the gravitating matter (here the dust), a contribution proportional to the metric tensor:

$$\kappa T_{mn} = -\lambda g_{mn} + \kappa\bar{\mu}u_m u_n, \quad \bar{\mu} > 0, \quad \Lambda = \text{const.} \tag{41.11}$$

Comparison of (41.11) with (41.10) gives us, upon use of

$$\kappa p = -\Lambda, \qquad \kappa\mu c^2 = \kappa\bar{\mu}c^2 + \Lambda, \tag{41.12}$$

the relations

$$\varepsilon = +1, \qquad \Lambda = +1/K^2, \qquad \kappa\bar{\mu}c^2 = 2/K^2. \tag{41.13}$$

The Einstein universe is a closed universe of constant curvature:

$$ds^2 = K^2\left[d\chi^2 + \sin^2\chi\left(d\vartheta^2 + \sin^2\vartheta\,d\varphi^2\right)\right] - c^2\,dt^2, \quad K = \text{const.} \tag{41.14}$$

As (41.12) shows, introducing a positive cosmological constant Λ is tantamount to admitting a negative pressure.

The de Sitter universes The introduction of the cosmological constant means that the space is curved even in the complete absence of matter. For if one substitutes

$$\kappa T_{mn} = -\Lambda g_{mn}, \quad \Lambda = \text{const.}, \tag{41.15}$$

into the field equations (41.7) and (41.8), then one obtains

$$K\ddot{K} - \dot{K}^2 = \varepsilon \tag{41.16}$$

$$3(\dot{K}^2 + \varepsilon)K^{-2} = \Lambda. \tag{41.17}$$

The best starting point for the integration of this system is equation (41.17) differentiated once, namely,

$$\ddot{K} - \tfrac{1}{3}\Lambda K = 0. \tag{41.18}$$

For positive Λ one obtains the proper de Sitter metrics (de Sitter 1917)

$$\varepsilon = +1: \quad K = B^{-1}\cosh Bct,$$

$$\varepsilon = -1: \quad K = B^{-1}\sinh Bct, \quad \Lambda = 3B^2, \tag{41.19}$$

$$\varepsilon = 0: \quad K = Ae^{Bct},$$

for negative Λ the anti-de Sitter metrics

$$\varepsilon = -1: \quad K = B^{-1}\cos cBt, \quad \Lambda = -3B^2, \tag{41.20}$$

and for $\Lambda = 0$ the flat space $\varepsilon = 0$, $K = $ const.

The de Sitter universes have a higher symmetry than might be supposed from their description by Robertson–Walker metrics. If from (41.2), (41.3), (41.16) and (41.17) one calculates the complete four-dimensional curvature tensor of these spaces, then one obtains

$$R_{abmn} = \tfrac{1}{3}\Lambda(g_{am}g_{bn} - g_{an}g_{bm}). \tag{41.21}$$

Thus we are dealing with four-dimensional spaces of constant curvature (of positive curvature for $\Lambda > 0$) in which neither any space-direction nor any time-direction is singled out, cp. Section 19.5. In particular, the three metrics (41.19) are only three different sections of the same four-dimensional space of constant positive curvature.

The radiation universe Incoherent, isotropic electromagnetic radiation can formally be described by the energy-momentum tensor (41.6) of a perfect fluid with

$$p = \tfrac{1}{3}\mu c^2. \tag{41.22}$$

With the aid of this equation of state we can at once integrate (41.9) and obtain

$$\mu c^2 K^4 = \text{const.} = 3A, \tag{41.23}$$

which says that when the universe expands or contracts the mass density (energy density) of the radiation is inversely proportional to the fourth power of the radius of the universe.

The behaviour of this universe with respect to time is determined by

$$\dot{K}^2 = \kappa A K^{-2} - \varepsilon; \tag{41.24}$$

upon introduction of $y = K^2$ the differential equation (41.24) becomes

$$\tfrac{1}{4}\dot{y}^2 = \kappa A - \varepsilon y, \tag{41.25}$$

which can easily be integrated. If we choose the constant of integration so that $y(t_0) = 0$, then we obtain the solutions

$$
\begin{aligned}
\varepsilon = 0: \quad & K^2 = 2c\sqrt{\kappa A}(t - t_0), \\
\varepsilon = -1: \quad & K^2 = c^2(t - t_0)^2 + 2c\sqrt{\kappa A}(t - t_0), \\
\varepsilon = +1: \quad & K^2 = -c^2(t - t_0)^2 + 2c\sqrt{\kappa A}(t - t_0).
\end{aligned}
\tag{41.26}
$$

Although we certainly do not live in a radiation universe now, several properties of these solutions are worth noting. One such is the occurrence of a singularity of the metric at $t = t_0$. There K goes to zero, the separation of two arbitrary points in the universe becomes arbitrarily small, and in the neighbourhood of this singularity the radius K becomes independent of ε; that is, the same for open and closed universes. Another interesting statement is that electromagnetic radiation (light) alone can, by virtue of its own gravitational interaction, produce a closed universe ($\varepsilon = 1$) whose radius K increases from zero to a maximum of $\sqrt{\kappa A}$ and then after the time $\Delta T = 2\sqrt{\kappa A/c^2}$ goes back to zero again.

The Friedmann universes By Friedmann universes in the strict sense one means cosmological models with dust:

$$T_{mn} = \mu u_m u_n. \tag{41.27}$$

For this special case one can immediately integrate (41.9) to give

$$\mu c^2 K^3 = \hat{M} = \text{const.} \tag{41.28}$$

The integration constant $\hat{M}$ is evidently proportional to the total mass for closed universes. Notice the changed power-dependence upon K in comparison with the radiation universe (41.23)!

The remaining field equation (41.8) simplifies to the 'Friedmann differential equation'

$$\dot{K}^2 = \kappa \hat{M}/3K - \varepsilon. \tag{41.29}$$

Introduction of the new variables

$$cT = \pm \int \frac{\mathrm{d}ct}{K(ct)} \tag{41.30}$$

brings it to the form

$$K'^2 = \kappa \hat{M} K/3 - \varepsilon K^2, \tag{41.31}$$

in which it can easily be solved by separation of variables. If we denote the time at which K vanishes by $t = 0$, then the solutions of (41.29) are parametrically (see Fig. 41.1)

$$\varepsilon = 0 : \ K = \kappa \hat{M} c^2 T^2 / 12, \ ct = \pm \kappa \hat{M} (cT)^3 / 36, \tag{41.32}$$

$$\varepsilon = -1 : \ K = \tfrac{1}{6}\kappa \hat{M}(\cosh cT - 1), \ ct = \pm \tfrac{1}{6}\kappa \hat{M}(\sinh cT - cT), \tag{41.33}$$

$$\varepsilon = +1 : \ K = \tfrac{1}{6}\kappa \hat{M}(1 - \cos cT), \ ct = \pm \tfrac{1}{6}\kappa \hat{M}(cT - \sin cT). \tag{41.34}$$

All three types have a singularity at the 'beginning of the universe' $t = 0$, where the radius K goes to zero. In the neighbourhood of this singularity the three types have the same dependence upon time, namely,

$$K(ct) \approx (3\kappa \hat{M}/4)^{1/3} t^{2/3}. \tag{41.35}$$

For the closed model ($\varepsilon = +1$), $K(ct)$ reaches a maximum and then goes back again to zero, describing a cycloid. In the two open models $K(ct)$ increases continuously (if we take as the positive direction of time that in which T increases).

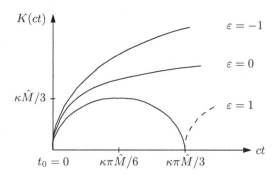

Fig. 41.1. The radius of the universe K as a function of time for the three Friedmann models.

41.3 Consequences of the field equations for models with arbitrary equation of state having positive pressure and positive rest mass density

From experience we know that the equation of state of ordinary matter lies between that of dust ($p = 0$, $\mu > 0$) and that of incoherent radiation ($\mu c^2 = 3p$) in the sense that $\mu c^2 - 3p \geq 0$ (in the microscopic picture the pressure is caused by collisions of particles with at most the velocity of light). Some notable properties of the Friedmann model follow, however, from the field equations alone and the assumptions that $\mu > 0$ and $p > 0$, independently of the particular kind of equation of state (temperature dependence!).

Thus from (41.7) and (41.8) one obtains the relation

$$6\ddot{K} = -\kappa(\mu c^2 + 3p)K, \qquad (41.36)$$

which can be interpreted in the following way. If $\dot{K}$ is positive at a time t, then, because of (41.36), $K(ct)$ is a curve concave downwards (like the curves of Fig. 41.1), which must have touched the axis $K = 0$ a finite time ago. If $\dot{K}$ is negative, this point $K = 0$ lies in the future. Since, as we shall describe in detail, we are currently observing a positive $\dot{K}$, the universes with Robertson–Walker metrics inevitably have a 'beginning of the universe' $t = 0$ at which the metric becomes singular (K becomes zero), independently of the equation of state and the choice of ε. By comparison of the function $K(ct)$ with its tangent at the time t, the age of the universe can be estimated in terms of the Hubble parameter H according to

$$ct < K(ct)/\dot{K}(ct) = 1/H(ct). \qquad (41.37)$$

If one writes (41.9) in the form

$$\partial(\mu c^2 K^3)/\partial K = -3pK^2, \qquad (41.38)$$

then one can see that $\mu c^2 K^3$ increases into the past, possibly even becoming infinite: for $K \to 0$, μ increases at least as fast as K^{-3}. Hence one can ignore the term proportional to ε in equation (41.8) and near the origin of the universe calculate with

$$3\dot{K}^2 = \kappa\mu c^2 K^2. \qquad (41.39)$$

The expansion behaviour of the early universe does not depend upon ε; it is the same for open and closed models.

The future behaviour of the universe, on the other hand, will essentially depend on ε. Because of (41.38), for increasing K the rest mass density μ decreases at least like K^{-3}, and hence the term $\kappa\mu c^2 K^2$ goes at least like K^{-1}. Thus for $\varepsilon = 1$, a maximum $\dot{K} = 0$ will be reached in a finite time, and since $K = $ const. is not a solution of (41.36), the radius function will decrease again and will necessarily reach $K = 0$ again: a closed universe with $\varepsilon = 1$ executes a cycle (or several cycles). For $\varepsilon = -1$, $\dot{K}^2$ can never become zero, and the universe expands continuously, $\dot{K}$ tending to the value 1 ($K(ct) \approx ct$ for $t \to \infty$). The universes with $\varepsilon = 0$ also expand continuously, only now $\dot{K}$ and $\ddot{K}$ go to zero.

In all these universes there are *particle horizons*; an observer cannot always see the whole universe at time t. Because of (40.26), the existence of such horizons obviously depends crucially upon the behaviour of $K(ct)$ at the origin $t_{\mathrm{B}} = 0$, and therefore we substitute the ansatz $K \sim (ct)^\alpha$, $\alpha > 0$, into (41.36). From the signs alone of both sides of the resulting equation it follows that $\alpha < 1$. For small times we have

$$\chi_{\mathrm{P}} \sim (ct)^{1-\alpha}, \quad 0 < \alpha < 1, \tag{41.40}$$

so that χ_{P} is finite near $t = 0$ and smaller than π, and also for arbitrary finite times χ_{P} is finite: in open models part of the universe is always invisible. In closed models, however, after a sufficient time χ_{P} can take the value π or even 2π. Thus, for example, in the Friedmann universe (41.34) $\chi_{\mathrm{P}} = cT = \pi$ for the time of maximum expansion (the whole space is visible) and $\chi_{\mathrm{P}} = 2\pi$ at the end of the universe (the observer sees his world line, that is, he himself, at the beginning of the universe).

The occurrence of *event horizons* depends upon the behaviour of $K(ct)$ at the end of the universe. Since for open models ($\varepsilon = 0, -1$) the radius function K goes at most like t^{-1} for large t, then the integral (40.4) diverges for $t_{\mathrm{E}} \to +\infty$: there is no event horizon; provided one waits long enough, one learns of every event. In closed models ($\varepsilon = +1$), however, there do exist event horizons (the proof runs as with the above considerations regarding particle horizons); an observer will not necessarily learn anything before the universe comes to an end about events which take place after the stage of maximal expansion.

Exercises

41.1   Show that for the de Sitter universes (41.21) holds.

41.2   Determine acceleration, rotation, shear and expansion (see Section 31.2) for the R–W solutions!

41.3   Are there R–W metrics which have a constant μ ($\dot{\mu} = 0$)?

42

Our universe as a Friedmann model

42.1 Redshift and mass density

It was one of the most important confirmations of the ideas of the theory of General Relativity and its application to cosmology when the cosmological redshift was found by Hubble in the year 1929, about thirteen years after the basic equations had been set up and seven years after the publication of the Friedmann model. In between had been the detour and error of Einstein, who believed he could arrive at a cosmological model only by the introduction of the cosmological constant, which led to the Einstein universe, a static model without redshift.

Not only does the redshift verify the cosmology of General Relativity, and in particular the concept of the expanding universe, but its exact evaluation also gives us data to determine which of the homogeneous, isotropic cosmological models our universe most closely resembles. From the redshift (as a function of distance) the Hubble parameter H and the acceleration parameter q can in principle be determined, see (40.18). If one substitutes them according to their definitions

$$H = \dot{K}/K, \qquad q = -\ddot{K}K/\dot{K}^2 \qquad (42.1)$$

into the field equations (41.7) and (41.8), then one obtains

$$6qH^2 = \kappa(\mu c^2 + 3p), \qquad 3H^2 = \kappa\mu c^2 - 3\varepsilon/K^2. \qquad (42.2)$$

In general these two equations are of course not sufficient to determine the four unknowns μ, p, ε and K from the redshift, that is, from a knowledge of q and H^2. But for our universe in its present state the predominant part of μ is contained in the masses of the galaxies and the pressure can consequently be ignored. For this dust we then have

$$6qH^2 = \kappa\mu c^2, \qquad H^2(2q - 1) = \varepsilon/K^2. \qquad (42.3)$$

Since ε can only take on the values $0, \pm1$, it can be determined from the value of q alone: $q > \frac{1}{2}$ gives a closed universe, $q \leq \frac{1}{2}$ the two open models. If ε is fixed, then from H and q one can determine the radius function K and the mass density μ, and compare them with observations. That mass density μ which corresponds precisely to the

371

critical value $q = \frac{1}{2}$ (the transition from an open to a closed model of the universe) is called the critical mass density:

$$\mu_{\text{crit}} = 2H^2/\kappa c^2. \tag{42.4}$$

Unfortunately present measurements and analyses of the redshift–distance relation (40.18) are still so incomplete and inexact that the relations (42.3) cannot yet be reliably evaluated. The following numerical values based on the redshift (H and q) and the analysis of galaxy counts (μ), are the most probable to date:

$$H = 6 \cdot 10^{-29}\ \text{cm}^{-1},\ cH = 55\ \text{km}\,\text{s}^{-1}\,\text{Mpc}^{-1},\ 1/cH = 18 \cdot 10^9\ \text{a}, \tag{42.5}$$

$$q = 1 \pm 1, \tag{42.6}$$

$$\mu = 3 \cdot 10^{-31}\ \text{g}\,\text{cm}^{-3}. \tag{42.7}$$

If one compares the three numerical values with the relations (42.3) then one establishes that:

(a) the presently observed mass density lies below the critical density

$$\mu_{\text{crit}} = 6 \cdot 10^{-30}\ \text{cm}^{-3} \tag{42.8}$$

which means that we ought to be living in an open universe;

(b) our universe has a radius of about $K \approx H^{-1} = 1.8 \cdot 10^{10}$ light years, is about $1 \cdot 10^{10}$ years old and is in an expansion phase;

(c) since q cannot yet be determined exactly enough from the redshift and also μ is not yet known with sufficient certainty, we cannot yet say whether our universe is open or closed.

Taking into consideration the surprisingly rapid change (oscillations) in the 'certain' numerical values of H, q and μ in the last few decades one can regard only the following as reliably established:

(a) the age of the universe, which follows from the Hubble parameter and from the age of rocks or of stellar systems, is of the order of magnitude 10^{10} years (the uncertainty is by a factor of 2);

(b) the average mass density is about $\mu = 10^{-31}$ g cm^{-3} (uncertain by a factor of 10);

(c) there is no doubt concerning the cosmological nature of the redshift and the applicability of relativity theory to cosmology.

As we have shown above, in the early epoch of the universe the parameter ε played no essential part. Our ignorance of the exact value of the acceleration parameter q, that is, of the value ε, thus does not put the value of the models for the earliest developmental stages of our universe in jeopardy. In the following we therefore want to sketch the ideas embodied in these models.

42.2 The earliest epochs of our universe and the cosmic background radiation

In direct optical observation of very distant objects we are looking a considerable way back into the past of our universe. But the origin of the universe, and times close to it, corresponds to an infinitely large redshift; it is therefore invisible in practice. Thus if from our observations we want to obtain statements about the constitution of the universe in its early phase, then we must look at physical objects closer to us, and judge from their present condition, and the laws governing their (local) evolution, the state of the universe when they were formed.

How then did our universe appear at the beginning? 'Beginning of the universe' does not mean that no matter was present before or that it was created at an instant; rather, this phrase should express the fact that on the basis of physical laws the state of the universe was *essentially* different from its present state (similarly at the 'end of the universe'). In the framework of the Friedmann universe, the beginning of the universe is that time in the past at which the radius of curvature K was zero, and the universe manifested a singular behaviour.

When speaking about 'time' in the early universe, one must take into account that the measurement of time should always be seen in relation to the properties of the matter. The time coordinate (universal time) t of the Friedmann universe is the proper time for the mass elements of the universe. The clocks which one uses for measuring proper time have zero dimensions in the abstract theory; in practice, this means that they are so small that the cosmological gravitational field does not change within the clocks and during the lapse of one period. While at the present time, therefore, the planetary system, for example, is a useful clock, in the early universe only elementary particles and their conversions are available. Measured by the number of characteristic individual physical processes going on, the beginning of the universe may be still very far away (possibly infinitely far away) even in the early phases (close to the singularity); the unit of measurement derived from the planetary system, the year and its subdivisions, does not correctly express this.

Let us return to the model of the early stages of the Friedmann universe. Today the major contribution to the energy-momentum tensor comes from stars (and possibly the mysterious dark matter hidden in galaxies); the contribution from radiation is negligibly small. But in the early stages of the universe a rather different balance must have occurred. On the one hand, as the radius of the universe K decreases, the energy density of radiation increases faster than that for matter because of equations (41.23) and (41.28). On the other hand, the energy density and temperature would rise so much that massive elementary particles and antiparticles, which would be unstable under terrestrial conditions, would be in thermal equilibrium with the high-energy photons. All observations and calculations point to the fact that about 10^{10} years ago the universe was probably in a state of very high density. Cosmological models therefore begin with conditions in which interactions of elementary particles are the decisive process. Thus a precise description of the earliest epoch is only possible if quantum physics (elementary particle physics) is taken into account; we can extrapolate into the past only as far as we know the laws of high-energy physics, *taking gravity into account*. This is a highly speculative area, but the following broad ideas are generally accepted.

The universe began in a state of extremely high temperature and density, which can only be described accurately through the not yet achieved unification of quantum theory and gravity. In the subsequent expansion an era may have occurred in which quantum effects produced an energy-momentum tensor proportional to the metric tensor, corresponding to a negative pressure. During this epoch the world is described by a de Sitter universe, in which the radius K increases exponentially (see equation (41.19)); this is known as an *inflationary universe*.

This rapid expansion of the universe reduced the temperature, so that equilibrium preferred the stable particles, namely, the electrons, protons, atomic nuclei, the lighter chemical elements, and the neutrinos and photons generated in particular by pair annihilation. All these are particles which are rather well understood, so that more confident predictions are possible from here on.

During further expansion and cooling the photons then decouple in the following sense. On the one hand, photons are not created to any great extent, they do not have sufficient energy for pair production, and they do not give their energy to the remaining matter (the universe is 'transparent'). On the other hand, the energy density of the photons decreases more rapidly than that of the rest of the matter, so that the subsequent behaviour of the expansion is not influenced by the photons.

From this time on the energy-momentum tensor of the photon gas *alone* thus obeys a conservation law

$$T^{ik}_{\text{Ph};k} = 0, \tag{42.9}$$

so that the energy density $\mu_{\text{Ph}}c^2$ obeys the relation

$$\mu_{\text{Ph}}c^2 K^4 = \text{const.}, \tag{42.10}$$

see (41.23), where now, however, in contrast to (41.24), the evolution of the radius function $K(ct)$ is dictated from outside (by the main component of the matter, that is, by the matter in atomic nuclei). Since we also have for the photon gas, according to the Planck radiation law,

$$\mu_{\text{Ph}}c^2 = \text{const.}\, T^4, \tag{42.11}$$

its temperature decreases with increasing radius function as

$$T \sim 1/K. \tag{42.12}$$

The experimental confirmation of these considerations (first made by Gamow as early as 1948), namely, the discovery of the incoherent *cosmic background radiation* by Penzias and Wilson (1965, Nobel Prize 1978), was certainly the greatest success of relativity theory in cosmology since the interpretation of the Hubble redshift. Observations show that the Earth is bathed by an (incoherent) electromagnetic radiation, whose frequency spectrum corresponds to the radiation of a black body at temperature

$$T_0 \approx 2.7 \text{ K} \tag{42.13}$$

with a maximum intensity near the wavelength $\lambda_0 \approx 0.2$ cm. (Since the earliest measurements could be fitted by a slightly higher temperature, this radiation is still also called the 3 K radiation.) The energy density of this radiation today corresponds to a mass density of about

$$\mu_{\text{Ph}} \approx 4.4 \times 10^{-34} \text{ g cm}^{-3}. \tag{42.14}$$

If one assumes that the photons uncouple from the rest of the matter at about 4000 K, then the cooling which has taken place in the meantime corresponds, because of (42.13), (42.12) and (40.12), to a red shift of

$$z = 4000 : 2,7 - 1 \approx 1480. \tag{42.15}$$

The cosmic background radiation thus gives us immediate optical access to the early epoch of the universe, back to much earlier times than are accessible to optical instruments by observation of distant objects (for which $z \leq 10$). The high degree of isotropy of this radiation shows that already at this time (if one assumes that initial anisotropies were

dissipated) or up to this time (if one thinks of the inhomogeneities resulting later from the formation of galaxies) the universe was Friedmann-like
and that the Earth moves at most with a small velocity relative to the
rest system of the total matter.

For people who believe that all properties of the universe must be due
to its evolution and not to its initial conditions (see the discussion at the
very beginning of this part), this high degree of isotropy was the main
reason to introduce the inflationary model: an inflation will smooth out
that part of the universe which is accessible to our observations and
make it Friedmann-like.

For all discussions about the early universe one should keep in mind
that if we use an R–W metric, then General Relativity permits any
behaviour of the world radius K: we can insert $K(ct)$ into (41.7)–(41.8),
read off p and μ, and invent a model sophisticated enough to give this
energy-momentum tensor.

To end this section we make some brief remarks about the evolution of
the universe *after* the formation of the electromagnetic background radiation. During the gradual cooling of the 'primeval fire-ball' the chemical
elements hydrogen and helium form in the preferred equilibrium ratio
of about $73:27$, almost no heavier elements being synthesized. Small
disturbances to the homogeneity of the universe then lead to galaxy formation, and there the subsequent compression and heating of matter in
the stars leads to nuclear processes, during which the heavier elements
are produced. All these things are still the subject of research.

42.3 A Schwarzschild cavity in the Friedmann universe

The assumption of a position-independent mass-density in the universe
leads, as we have seen, to useful cosmological models with properties
which approximate to the observations, but they stand in flat contradiction to the mass distribution to be found in our neighbourhood. Here
the mass is always concentrated into individual objects (planets, stars,
galaxies), and the practically matter-free space in between exceeds the
volume of these objects by several orders of magnitude.

This discrepancy can at least partially be removed, since the exact
solution for the gravitational field of a spherically symmetric star which
is surrounded by a matter-free space and situated in a special Friedmann
universe ($p = 0$) is known (Einstein and Strauss 1945).

The details of this model are as follows (Fig. 42.1). A spherically
symmetric star is surrounded by a space free of matter which again is

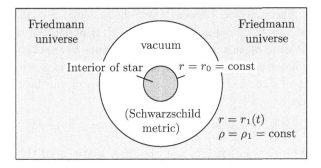

Fig. 42.1. The Schwarzschild vacuole in the Friedmann universe.

surrounded by a Friedmann dust universe. The gravitational field inside the star can be described, for example, by the interior Schwarzschild metric (for a static star) or by the section of a Friedmann universe (for a collapsing or exploding star). A Schwarzschild solution

$$ds^2 = \frac{dr^2}{1 - 2M/r} + r^2\big(d\vartheta^2 + \sin^2\vartheta\, d\varphi^2\big) - (1 - 2M/r)c^2\, dt^2,$$

$$r_0 \le r \le r_1,$$

(42.16)

can always be joined to this interior solution, and to this Schwarzschild solution a Friedmann universe. This last part has been discussed in detail and proved in Section 36.3, although always from the viewpoint of a connection 'inwards'. One can easily show, however, that all the calculations are equally valid for the connection 'outwards' which is used here. This connection can be most simply achieved by introducing a new coordinate system into the Schwarzschild metric via a coordinate transformation

$$r = r(\rho, c\tau), \qquad ct = ct(\rho, c\tau),$$

(42.17)

so that we have

$$ds^2 = \left(\frac{\partial r}{\partial \rho}\right)^2 \frac{d\rho^2}{1 - \varepsilon f^2(\rho)} + r^2(\rho, c\tau)\big(d\vartheta^2 + \sin^2\vartheta\, d\varphi^2\big) - c^2\, d\tau^2,$$

(42.18)

$$\left(\frac{\partial r}{\partial c\tau}\right)^2 + \varepsilon f^2(\rho) - 2M/r = 0,$$

where the boundary $\rho = \rho_1$ to the expanding or contracting universe

$$ds^2 = K^2(c\tau)\left[\frac{d\rho^2}{1 - \varepsilon\rho^2} + \rho^2\big(d\vartheta^2 + \sin^2\vartheta\, d\varphi^2\big)\right] - c^2\, d\tau^2$$

(42.19)

is at rest. The junction between the Schwarzschild metric (42.16) or
(42.18) and the Friedmann universe (42.19) is possible if and only if
between the mass density $\mu(c\tau)$ and the curvature $K(c\tau)$ of the universe,
on the one hand, and the gravitational radius $2M$, the Newtonian mass
$m = 8\pi M/\kappa c^2$ and the 'radius' ρ_1, or $r_1 = r_{(\rho_1, c\tau)}$, on the other hand,
the relations

$$\kappa\mu c^2 K^3 \rho_1^3 = 6M = 3\kappa m c^2/4\pi, \qquad r_1 = \rho_1 K(c\tau) \qquad (42.20)$$

hold. These relations ensure that the dust particles of the boundary
surface between the vacuum and the cosmic matter move on geodesics
both of the interior Schwarzschild metric and of the exterior universe;
that is, the system is in a dynamical equilibrium.

In a Friedmann universe one can thus construct one or several spherically symmetric cavities, gather up the originally dust matter within
each, and put it back into the middle of the cavities as stars. Notice,
however, that for $\varepsilon \neq 0$ the total extracted mass

$$m_{\mathrm{h}} = \int_0^{\rho_1} \sqrt{g}\, \mu \,\mathrm{d}\vartheta\, \mathrm{d}\varphi\, \mathrm{d}\rho = 4\pi K^3 \mu \int_0^{\rho_1} \frac{\rho^2 \,\mathrm{d}\rho}{\sqrt{1-\varepsilon\rho^2}} \qquad (42.21)$$

differs from the stellar mass m (only insignificantly for small radii ρ_1).

The most interesting property of the vacuole solution is that inside the
vacuole the field is static. The expansion or contraction of the universe
has no influence on the physical processes inside the vacuole, except that
the radius r of the vacuole is time-dependent. An observer inside is only
made aware of the cosmic expansion through the redshift of objects lying
beyond the boundary surface. Stars inside show no redshift.

How then is the vacuole in which we live constructed? The relation

$$r_1 = \sqrt[3]{6M/\kappa\mu c^2} \qquad (42.22)$$

between the coordinate radius r_1 of the vacuole, the Schwarzschild radius
$2M$ of the central body and the cosmic mass density μ is crucial for
the size of the vacuole. If we measure r_1 and $2M$ in centimetres and
substitute the value 3×10^{-31} g cm^{-3} for μ, then we have

$$r_1 \approx 1.75 \times 10^{19} \sqrt[3]{2M}. \qquad (42.23)$$

The vacuole radius of the Earth ($2M = 0.88$ cm) would thus extend
out far beyond the Sun, and many of the nearest fixed stars would in
fact be contained within the vacuole of the Sun ($2M = 2.95 \times 10^5$cm).
But we obtain a realistic model if we identify the central body with
the local group of galaxies ($2M \approx 5 \times 10^{17}$cm) to which belongs not

only, for example, our Galaxy, but also the Andromeda nebula; there is then no other galaxy in the associated vacuole. Inside this system the expansion of the universe is not effective, the gravitational field being determined exclusively by the masses contained within the vacuole (in so far as the model is applicable, that is, as the mass distribution is to good approximation spherically symmetric). The radius of the Earth or of the Earth's orbit thus does not change because of the cosmic expansion. But of course the central body, that is, our Galaxy, could also be in a state of general expansion which is independent of the cosmic expansion.

Of more theoretical interest is the fact that time runs differently inside and outside the vacuole. As a consequence of the method of joining, the universal time τ of the Friedmann universe is identical with the proper time τ of the particles on the boundary layer $r = r_1$, which move on radial geodesics. Because of (35.6) and (36.41)–(36.45), this proper time differs from the coordinate time t of the Schwarzschild metric by the factor

$$\frac{dt}{d\tau} = \frac{\sqrt{1 - \varepsilon\rho_1^2}}{1 - 2M/\rho_1 K(c\tau)}. \tag{42.24}$$

In general this factor is ignorably small.

Vacuoles cannot be arbitrarily large. Trivially, the vacuole radius $\rho_1 = \sin\chi_1$ may not exceed the maximum value π of the coordinate χ in a closed universe ($\varepsilon = +1$). (The coordinate system used here covers only the half universe with $0 \leq \rho \leq 1$.) A second, more important, condition follows from the requirement that the vacuole radius r_1 must be outside the Schwarzschild radius $r = 2M$. Because of the relation (42.20) between vacuole radius and mass and the special form

$$K(c\tau) = \tfrac{1}{6}\kappa\mu c^2 K^3 k(c\tau), \tag{42.25}$$

of the time dependence of the Friedmann universe (see (41.28) and (41.32)–(41.34)), the condition $r_1 > 2M$ just mentioned is only satisfied for

$$\rho_1^2 < \tfrac{1}{2}k(c\tau). \tag{42.26}$$

The maximum allowed vacuole radius $K\rho_1$ depends upon the age of the universe. In a closed Friedmann universe the vacuole is always smaller than the semi-universe ($\rho_1 < 1$). A galaxy (a group of galaxies) of mass M can thus only occur in a universe if the age of the universe τ is greater than that required by

$$M^2 = k(c\tau)K^2(c\tau)/8. \tag{42.27}$$

In a closed universe ($\varepsilon = +1$) with a period of about 6×10^9 years, that is, $\kappa M/6 \approx 9 \times 10^{27}$ s, a vacuole of mass $M \approx 2.5 \times 10^{17}$ cm, for example, can form at the earliest 128 days after the beginning of the universe.

43
General cosmological models

43.1 What is a cosmological model?

A cosmological model is a model of our universe which, taking into account and using all known physical laws, predicts (approximately) correctly the observed properties of the universe, and in particular explains in detail the phenomena in the early universe. Such a model must also explain *inter alia* why the universe was so homogeneous and isotropic at the epoch of last scattering of the cosmic microwave background, and how and when inhomogeneities (galaxies and stars) arose.

In a more restricted sense cosmological models are exact solutions of the Einstein field equations for a perfect fluid that reproduce the important features of our universe. Because there is only one actual universe the large number of known or possible cosmological models may at first seem surprising. There are, however, two reasons for this multiplicity.

Firstly, only a section of our universe is known, both in space and in time. All cosmological models which differ only near the origin of the universe must be accepted for competition. In fact solutions are known which are initially inhomogeneous or anisotropic to a high degree, and which then increasingly come to approximate a Friedmann universe. All cosmological models which yield a redshift and a cosmic background radiation can hardly be refuted. The possibility cannot be excluded that our universe is *not* homogeneous and isotropic, but has those properties only approximately in our neighbourhood. An expanding 'dust star', that is, a section of a Friedmann universe which is surrounded externally by a static Schwarzschild metric (the model of a collapsing star discussed in Section 36.3), may also perhaps be an excellent model of the universe.

Secondly, one also examines solutions of the field equations where it is clear in advance that they do not correctly reproduce the properties of our universe. Every model is of course a great simplification of reality, and only by the study of many solutions can one establish which simplifications are allowed and which assumptions are essential. Exaggerating one can say that there is almost no exact (perfect fluid) solution of the field equations to which one could not attribute the name 'cosmological model'.

A special rôle is played now as before by cosmological models which satisfy the cosmological principle to such an extent that the universe (the three-dimensional position space) is homogeneous, that is, that the points on a section $t = $ const. are physically indistinguishable. Besides the Robertson–Walker metrics, these models include all spaces which possess a simply transitive group of motion G_3 and are accordingly to be associated with one of the Bianchi types I to IX (see Section 33.5) or which permit a transitive group G_4 which possesses no transitive subgroup G_3 (Kantowski–Sachs model). We want to go briefly into two examples of such cosmological models.

43.2 Solutions of Bianchi type I with dust

If the three-dimensional space is the rest space of the matter and possesses three commuting Killing vectors, then we are dealing with homogeneous cosmological models of Bianchi type I. Since one can simultaneously transform the three Killing vectors to the normal forms

$$\xi_1^a = (1,0,0,0), \qquad \xi_2^a = (0,1,0,0), \qquad \xi_3^a = (0,0,1,0) \qquad (43.1)$$

in suitably chosen coordinates, the metric depends only upon the time coordinate $x^4 = ct$. By the transformation $x^{4\prime} = x^{4\prime}(x^4)$, $x^{\alpha\prime} = x^\alpha + f^\alpha(x^4)$ one can, without destroying (43.1), always bring the metric to the normal form

$$ds^2 = -c^2\,dt^2 + g_{\alpha\beta}(ct)\,dx^\alpha\,dx^\beta. \qquad (43.2)$$

As one can see, the subspaces $t = $ const. are flat three-dimensional spaces in which – for a fixed t – Cartesian coordinates can always be introduced.

To calculate the curvature tensor of this metric we use the reduction formulae (30.25). They give

$$R_{\alpha\beta\mu\nu} = -\tfrac{1}{4}(\dot{g}_{\beta\mu}\dot{g}_{\alpha\nu} - \dot{g}_{\beta\nu}\dot{g}_{\alpha\mu}), \qquad R^4{}_{\beta\mu\nu} = 0,$$
$$R^4{}_{\beta 4\nu} = \tfrac{1}{2}\ddot{g}_{\beta\nu} - \tfrac{1}{4}\dot{g}_{\beta\alpha}\dot{g}_{\mu\nu}g^{\alpha\mu}. \qquad (43.3)$$

Using the relations $\dot{g}/g = g^{\alpha\beta}\dot{g}_{\alpha\beta}$ and $\dot{g}^{\beta\nu} = -g^{\beta\alpha}g^{\nu\mu}\dot{g}_{\alpha\mu}$, the field equations

$$R_{ab} - \tfrac{1}{2} R g_{ab} = \kappa \mu u_a u_b \tag{43.4}$$

can be written as

$$R_4^4 - \tfrac{1}{2} R = -\tfrac{1}{8} \dot{g}_{\beta\nu} \dot{g}^{\beta\nu} - \tfrac{1}{8} (\dot{g}/g)^2 = -\kappa\mu c^2, \tag{43.5}$$

$$R_\beta^\alpha - \delta_\beta^\alpha \tfrac{1}{2} R - \tfrac{1}{2} \big(\sqrt{-g} g^{\alpha\rho} \dot{g}_{\rho\beta}\big)^{\cdot} / \sqrt{-g} - \delta_\beta^\alpha \tfrac{1}{2} \kappa\mu c^2 = 0 \tag{43.6}$$

(the equations $R_\alpha^4 = 0$ are satisfied identically). Because of the equation of conservation of rest mass (21.87), which always holds for dust, the system of field equations is only integrable if

$$\kappa\mu c^2 \sqrt{-g} = \hat{M} = \text{const.} \tag{43.7}$$

In order to integrate the field equations we take the trace of (43.6), which gives the differential equation

$$\big(\sqrt{-g}\big)^{\cdot\cdot} = \tfrac{3}{2} \hat{M}, \tag{43.8}$$

which we can solve as

$$\sqrt{-g} = \tfrac{3}{4} ct(\hat{M} ct + A). \tag{43.9}$$

The complete system (43.6) can be integrated once, using (43.7), with the result

$$\sqrt{-g} \dot{g}_{\beta\alpha} = \hat{M} ct g_{\beta\alpha} + a_\alpha^\mu g_{\mu\beta}. \tag{43.10}$$

If for a fixed arbitrary time one introduces a Cartesian coordinate system and arranges its axes so that the constant matrix a_α^μ is diagonal, then because of (43.10) the diagonal form of the metric remains preserved for all time. Hence from (43.9) and (43.10) follows

$$\dot{g}_{11} = \big[\tfrac{4}{3}\hat{M}/(\hat{M} ct + A) + 2p_1 A/ct(\hat{M} ct + A)\big]g_{11}, \quad p_1 A = \tfrac{2}{3} a_1^1 \tag{43.11}$$

with the solution

$$g_{11} = \text{const.}(\hat{M} ct + A)^{4/3} \big[ct/(\hat{M} ct + A)\big]^{2p_1} \tag{43.12}$$

and analogous results for g_{22} and g_{33}. Thus we have finally the solution

$$ds^2 = -c^2 \, dt^2 + g_{11} \, dx^2 + g_{22} \, dy^2 + g_{33} \, dz^2,$$

$$g_{11} = (-g)^{1/3} \big[ct/(\hat{M} ct + A)\big]^{2p_1 - 2/3},$$

$$g_{22} = (-g)^{1/3} \big[ct/(\hat{M} ct + A)\big]^{2p_2 - 2/3}, \tag{43.13}$$

$$g_{33} = (-g)^{1/3} \big[ct/(\hat{M} ct + A)\big]^{2p_3 - 2/3},$$

$$\kappa\mu c^2 \sqrt{-g} = \hat{M},$$

$$\sqrt{-g} = 3ct(\hat{M} ct + A)/4,$$

in which, because of (43.6) and (43.5), the three coefficients p_δ must satisfy the conditions

$$p_1 + p_2 + p_3 = 1, \qquad p_1^2 + p_2^2 + p_3^2 = 1 \qquad (43.14)$$

which is guaranteed by, for example,

$$2p_1 - \tfrac{2}{3} = \tfrac{4}{3}\sin\alpha, \qquad\qquad 2p_2 - \tfrac{2}{3} = \tfrac{4}{3}\sin\left(\alpha + \tfrac{2}{3}\pi\right),$$
$$2p_3 - \tfrac{2}{3} = \tfrac{4}{3}\sin\left(\alpha + \tfrac{2}{3}\pi\right), \qquad -\tfrac{1}{6}\pi < \alpha \le \tfrac{1}{2}\pi. \qquad (43.15)$$

For the four-velocity $u^a = (0,0,0,c)$ of the field-producing matter we have

$$u_{a;b} = \tfrac{1}{2}c g_{ab,4}. \qquad (43.16)$$

Thus we are dealing (compare with the definitions (31.18) of the kinematic quantities) with a geodesic ($\dot{u}_a = 0$), rotation-free ($\omega_{ab} = 0$) flow, whose expansion velocity is

$$\Theta = \frac{2\hat{M}ct + A}{t(\hat{M}ct + A)}, \qquad (43.17)$$

and the components of whose shear velocity are

$$\sigma_{\lambda\lambda} = \tfrac{1}{4}Acg_{\lambda\lambda}(3p_\lambda - 1)/\sqrt{-g} \quad \text{(no summation over } \lambda\text{).} \qquad (43.18)$$

The integration constant A is therefore a measure of the shear, while the p_λ characterize its dependence upon direction. The particular case $A = 0$ leads to an (isotropic) Friedmann universe, see Exercise 43.1.

The metric (43.14) describes an anisotropic, homogeneous universe, which is expanding or contracting. The distances between the dust particles (at rest in these coordinates) change in a direction-dependent fashion, as the isotropic case $p_1 = p_2 = p_3$ stands in contradiction to (43.14). For $A > 0$ (which can always be achieved by choice of the time direction) the metric becomes singular at $t = 0$, if we approach the origin from the positive t side.

In the general case $\alpha \ne \pi/2$ ($p_3 \ne 0$) precisely one of the p_λ, namely, p_3, is negative. Because

$$\dot{g}_{33}/g_{33} = \left[ct(\hat{M}ct + A)\right]^{-1}\left(\tfrac{4}{3}\hat{M}ct + 2p_3 A\right), \quad p_3 < 0, \qquad (43.19)$$

then the relative change in distances in the z-direction is very strongly negative at very small times. This collapse comes to a halt for $ct = -3p_3A/2\hat{M}$ and it is followed by an expansion. In the x-direction and the y-direction, on the other hand, the universe expands continuously. If we follow its history backwards from positive t, then from an initial sphere we find a very long, thin, elongated ellipsoid, and in the limiting case $t \to +0$ a straight line – a 'cigar' singularity occurs.

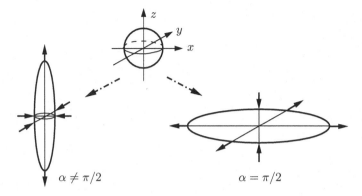

Fig. 43.1. The two types of singularity of a Bianchi type I universe.

It is worth noting that the mass $\hat{M}$ does not affect the behaviour as $t \to 0$; the metric (43.14) can be approximately replaced by the vacuum solution (*Kasner metric*)

$$\mathrm{d}s^2 = (ct)^{2p_1}\,\mathrm{d}x^2 + (ct)^{2p_2}\,\mathrm{d}y^2 + (ct)^{2p_3}\,\mathrm{d}z^2 - c^2\,\mathrm{d}t^2,$$

$$p_1 + p_2 + p_3 = 1, \qquad p_1^2 + p_2^2 + p_3^2 = 1. \tag{43.20}$$

In the exceptional case $\alpha = \pi/2$, that is, $p_1 = 1$, $p_2 = p_3 = 0$, we have

$$\frac{\dot{g}_{11}}{g_{11}} = \frac{\lambda\hat{M}ct/3 + 2A}{ct(\hat{M}ct + A)}, \qquad \frac{\dot{g}_{22}}{g_{22}} = \frac{\dot{g}_{33}}{g_{33}} = \frac{4\hat{M}}{3(\hat{M}ct + A)}. \tag{43.21}$$

A singular behaviour occurs for $t \to +0$ only in the x-direction, and then in such a way that (followed backwards in time) out of a sphere first a strongly flattened, rotating ellipsoid is formed and finally a 'pancake' singularity (see Fig. 43.1).

For large times the metric approaches (independently of α) that of a homogeneous and isotropic Friedmann universe with $\varepsilon = 0$.

43.3 The Gödel universe

The Gödel (1949) universe is a homogeneous, but anisotropic, four-dimensional space whose metric is best be written in either of the two forms

$$\mathrm{d}s^2 = a^2\big[\mathrm{d}x^2 + \tfrac{1}{2}\mathrm{e}^{2x}\,\mathrm{d}y^2 + \mathrm{d}z^2 - (\mathrm{e}^x\,\mathrm{d}y + c\,\mathrm{d}t)^2\big], \quad a = \text{const.}, \tag{43.22}$$

$$\mathrm{d}s^2 = 4a^2\big[\mathrm{d}r^2 + \mathrm{d}z^2 + (\sinh^2 r - \sinh^4 r)\,\mathrm{d}\varphi^2$$
$$- 2\sqrt{2}\sinh^2 r\,\mathrm{d}\varphi\,c\,\mathrm{d}t - c^2\,\mathrm{d}t^2\big]. \tag{43.23}$$

It possesses five Killing vectors, which in the coordinates (43.22) have the form

$$\xi_1^a = (0,1,0,0), \quad \xi_2^a = (0,0,1,0), \quad \xi_3^a = (0,0,0,1),$$
$$\xi_4^a = (1,-y,0,0), \quad \xi_5^a = (y, e^{-2ex} - \tfrac{1}{2}y^2, 0, -2e^{-x}). \tag{43.24}$$

Its gravitational field is produced by the energy-momentum tensor

$$T^{mn} = \frac{1}{2\kappa a^2} g^{mn} + \frac{u^m u^n}{\kappa c^2 a^2}, \quad u^m = (0,0,0,c/a), \tag{43.25}$$

which we can interpret either as the energy-momentum tensor of a perfect fluid with

$$p = \mu c^2 = 1/2\kappa a^2, \tag{43.26}$$

or as an energy-momentum tensor which besides the contribution from the dust also contains the cosmological term Λg_{mn} (see (22.4)):

$$\mu = 1/\kappa c^2 a^2, \quad \Lambda = -1/2a^2. \tag{43.27}$$

Since only the components

$$u_{1;2} = -u_{2;1} = ac\, e^x/2 \tag{43.28}$$

of the derivative $u_{a;b}$ of the four-velocity are non-zero, the matter current is geodesic, shear-free and expansion-free, but rotates with the constant velocity

$$\omega = \sqrt{\omega_{ab}\omega^{ab}/2} = c/a\sqrt{2}. \tag{43.29}$$

The Gödel universe is certainly not a realistic model of the universe, but it does possess a series of interesting properties. It is one of the few cosmological models which contains rotating matter, and it also contains closed timelike lines; that is, an observer can influence his own past, see Exercise 43.1.

43.4 Singularity theorems

Of the cosmological models which we have so far discussed, the physically reasonable ones (Friedmann model, Bianchi type I universes) have a singularity in their evolutionary history, that is, a beginning of the universe or a primeval 'big bang', while the physically less realistic ones (Einstein universe, de Sitter universe, Gödel universe) certainly do not possess such a singularity, but they involve the cosmological constant, or its matter obeys a rather implausible equation of state, show no redshift or else contradict our ideas about causality.

Since a singularity at the beginning of the universe is, however, a rather unwelcome property of cosmological models, one would very much like to know whether this singularity is unavoidable for physically reasonable models. Do singularities perhaps occur only in cosmological models of high symmetry and vanish under the small deviations from symmetry which are always present in reality; or were we unlucky in our selection of the model: are singularities absent in other universes of high symmetry (for other Bianchi types)?

In answer to the last of the questions raised here we shall now show that in gravitational fields which are produced by perfect fluids whose elements move without rotation along geodesics, then under certain plausible assumptions singularities *must* occur. Our starting point is the decomposition (31.18) of the covariant derivative of the velocity field u_m of the fluid, that is, the representation

$$u_{m;i} = \omega_{mi} + \sigma_{mi} + \tfrac{1}{3}\Theta\left(g_{mi} + u_m u_i/c^2\right) - \dot{u}_m u_i/c^2. \qquad (43.30)$$

If we substitute this into the identity

$$(u_{m;i;n} - u_{m;n;i})g^{mi}u^n = -R_{an}u^n u^a \qquad (43.31)$$

which is valid for every vector u_m, and use the field equations, we obtain

$$\mathrm{d}\Theta/\mathrm{d}\tau = -\Theta^2/3 - \sigma_{in}\sigma^{in} - \kappa c^2\left(3p + \mu c^2\right)/2 + \omega_{in}\omega^{in} + \dot{u}^n{}_{;n} \qquad (43.32)$$

(*Raychaudhuri equation*). If we also assume that

$$\mu c^2 + 3p \geq 0 \qquad (43.33)$$

(physically we would of course expect further that $\mu \geq 0$ and $p \geq 0$), then all terms on the right-hand side of (43.32) except $\omega_{in}\omega^{in}$ and possibly $\dot{u}^n{}_{;n}$ are negative. Therefore, if the rotation and the acceleration vanish we have

$$\mathrm{d}\left(\Theta^{-1}\right)/\mathrm{d}\tau \geq 1/3. \qquad (43.34)$$

Accordingly Θ^{-1} was either $(\Theta > 0)$ zero at a finite proper time in the past, or $(\Theta < 0)$ will take the value zero after a finite proper time. Since the expansion Θ is a measure of the relative change in volume, then singularities (with $\Theta = \infty$) are always present in such models. Because $\Theta = 3c\dot{K}/K$ in the Friedmann universe, these singularities just correspond to the zero points of the radius function K.

A similar conclusion can be drawn if the fluid itself is not necessarily non-rotating, but if there does exist a rotation-free, geodesic congruence of timelike world lines (cluster of test particles). Since for two timelike, future-directed, unit vectors u^i/c and V^i/c we always have $u^i V_i \leq -c^2$, and hence the field equations yield

$$R_{an} V^a V^n \geq \kappa c^2 \left(3p + \mu c^2\right)/2, \qquad (43.35)$$

the inequality (43.34) follows also for these geodesics, and the family of test particles shows a singular behaviour. The space-time is therefore singular in the mathematical sense. Whether the physical quantities (pressure, rest-mass density) behave singularly must be investigated separately.

As a generalization of these laws one can show that in every universe which is at some time homogeneous (which possesses a transitive spatial group of motion), for which the associated initial value problem can be solved uniquely on this initial surface and in which the condition $R_{ab} V^a V^b < 0$ is satisfied for all timelike or null vectors V^a, then there exists a singularity. This singularity is characterized by the occurrence of geodesics which although of finite length cannot be extended. Again the type of physical singularity must in every case be clarified separately.

The existence of singularities can be proved under still weaker assumptions; singularities occur, for example, in every spatially closed universe which at some time or other expands or contracts.

Exercises

43.1   Show that for $A = 0$ equation (43.8) leads at once to $\dot{g}_{\beta\alpha} = 4g_{\beta\alpha}/ct - a^\mu_\beta g_{\mu\alpha}/c^2 t^2$, and that the field equations (43.9) can only be satisfied for $a^\mu_\beta \equiv 0$. What kind of universe is this?

43.2   Use the form (43.23) of the Gödel metric to show that – if φ is a angular coordinate – there are curves $r = R$, $v = $ const., $t = -\varepsilon\,\varphi + $ const., which are timelike circles for $\varepsilon = 0$ and which go into the past for small $\varepsilon > 0$.

Further reading for Chapter 43

Hawking and Ellis (1975), Krasiński (1997), Wainwright and Ellis (1997).

Bibliography

Alternative textbooks on relativity and useful review volumes

Anderson, J. L. (1967). *Principles of Relativity Physics* (London, Academic Press).

Bergmann, P. G. (1958). *Introduction to the Theory of Relativity* (Englewood Cliffs, NJ, Prentice-Hall).

Eddington, A. A. (1923). *The Mathematical Theory of Relativity* (Cambridge, Cambridge University Press).

Einstein, A. (1950). *The Meaning of Relativity* (Princeton, NJ, Princeton University Press).

Einstein, A. (1969). *Über spezielle und allgemeine Relativitätstheorie* (Berlin, Akademie-Verlag).

Einstein (1970). *Grundzüge der Relativitätstheorie* (Berlin, Akademie-Verlag).

Hawking, S. W. and Israel, W. (eds.) (1987). *Three Hundred Years of Gravitation* (Cambridge, Cambridge University Press).

Landau, L. D. and Lifshitz, E. M. (1975). *The Classical Theory of Fields* (Oxford, Pergamon Press).

Lichnerowicz, A. (1955). *Théories Relativistes de la Gravitation et de l'Electromagnétisme* (Paris, Masson et Cie).

Misner, C. W., Thorne, K. S. and Wheeler, J. A. (1973). *Gravitation* (San Francisco, Freeman).

Mœller, C. (1972). *The Theory of Relativity* (Oxford, Clarendon Press).

Rindler, W. (1977). *Essential Relativity* (New York Springer).

Straumann, N. (1984). *General Relativity and Relativistic Astrophysics* (Berlin, Springer).

Synge J. L. (1960). *Relativity – the General Theory* (Amsterdam, North-Holland).

Synge, J. L. (1965). *Relativity – the Special Theory* (Amsterdam, North-Holland).

Von Laue, M. (1965). *Die Relativitätstheorie* (Braunschweig, Vieweg).

Wald, R. (1984). *General Relativity* (Chicago, University Press).

Weinberg, S. (1972). *Gravitation and Cosmology* (New York, Wiley).

Weyl, H. (1917). *Space, Time, Matter* (Berlin, Springer).

Monographs and research articles

Ashby, N. (1998). Relativistic effects in the global positioning system. In *Gravitation and Relativity*, proceedings of the GR-15 conference, eds. N. Dalhich and J. Narlikov (Pune, IUCAA).

Ashtekar, A. (1991). *Lectures on Non-perturbative Canonical Gravity* (Singapore, World Scientific).

Bekenstein, J. D. (1973). Black holes and entropy, *Phys. Rev.* **D7**, 2333.

Birrell, N. D. and Davies, P. C. W. (1984). *Quantum Fields in Curved Space* (Cambridge, Cambridge University Press).

Blanchet, L. (2002). Gravitational radiation from post-Newtonian sources and inspiralling compact binaries, *Living Reviews in Relativity*, www. livingreviews.org.

Buchdahl, H. A. (1983). Schwarzschild interior solution and the truncated Maxwell fish eye, *J. Phys.* **A16**, 107–110.

Celotti, A., Miller, J. C. and Sciama, D. W. (1999). Astrophysical evidence for the existence of black holes, *Class. Quant. Grav.* **16**, A3.

Chandrasekhar, S. (1931). The maximum mass of ideal white dwarfs, *Astrophys. J.* **74**, 81.

Chandrasekhar, S. (1998). *The Mathematical Theory of Black Holes* (Oxford, Oxford University Press).

de Sitter, W. (1917). On the curvature of space, *Proc. Kon. Ned. Akad. Wet.* **20**, 229.

Droste, J. (1916). The field of a single centre in Einstein's theory of gravitation, and the motion of a particle in that field, *Proc. Kon. Akad. Wet. Amsterdam* **19**, 197.

Eddington, A. S. (1924). A comparison of Whitehead's and Einstein's formulae, *Nature* **113**, 192.

Ehlers, J. (1961). Beiträge zur relativistischen Mechanik kontinuierlicher Medien, *Abh. Mainzer Akad. Wiss., Math.-Nat. Kl.* Nr.11.

Ehlers, J. (1966). Generalized electromagnetic null fields and geometrical optics. In *Perspectives in Geometry and Relativity*, ed. B. Hoffmann (Bloomington, Indiana University Press).

Ehlers, J. (1971). General relativity and kinetic theory. In *General Relativity and Cosmology*, ed. R. K. Sachs (New York, Academic).

Einstein, A. (1905). Zur Elektrodynamik bewegter Körper, *Ann. Phys. (Germany)* **17**, 891.

Einstein, A. (1915). Zur allgemeinen Relativitätstheorie, *Preuss. Akad. Wiss. Berlin, Sitzber.*, 778–786.

Einstein, A. (1917). Kosmologische Betrachtungen zur allgemeinen Relativitätstheorie, *Preuss. Akad. Wiss. Berlin, Sitzber.*, 421.

Einstein, A. and Strauss, E. G. (1945). The influence of the expansion of space on the gravitation fields sorrounding the individual stars, *Rev. Mod. Phys.* **17**, 120.

Eisenhart, L. P. (1949). *Riemannian Geometry* (Princeton, Princeton University Press).

Eisenhart, L. P. (1961). *Continous Groups of Transformations* (New York Dover Publications).

Ellis, G. F. R. (1999). 83 years of general relativity and cosmology: progress and problems, *Class. Quant. Grav.* **16**, A37.

Finkelstein, D. (1958). Past-future asymmetry of the gravitational field of a point particle, *Phys. Rev.* **110**, 965.

Friedmann, A. (1922). Über die Krümmung des Raumes, *Z. Phys.* **10**, 377.

Frolov, V. P. and Novikov, I. D. (1998). *Black Hole Physics* (Dordrecht, Kluwer Academic Publishers).

Gamow, G. (1948). The evolution of the universe, *Nature* **162**, 680.

Gödel, K. (1949). An example of a new type of cosmological solutions of Einsein's field equations of gravitation, *Rev. Mod. Phys.* **21**, 447.

Green, M. B., Schwarz, J. H. and Witten, E. (1988). *Superstring Theory* (Cambridge, Cambridge University Press).

Hawking, S. (1975). Particle creation by black holes, *Commun. Math. Phys.* **43**, 128.

Hawking, S. W. and Ellis, G. F. R. (1975). *The Large Scale Structure of Space-Time* (Cambridge, Cambridge University Press).

Herlt, E. and Stephani, H. (1976). Wave optics of the spherical gravitational lens, *Int. J. Theor. Phys.* **15**, 45.

Hilbert, D. (1915). Die Grundlagen der Physik, *Königl. Gesellsch. Wiss. Göttingen, Nachr., Math.-Phys. Kl.*, 395.

Kerr, R. (1963). Gravitational field of a spinning mass as an example of algebraically special metrics, *Phys. Rev. Lett.* **11**, 237.

Killing, W. (1892). Über die Grundlagen der Geometrie, *J. Reine und Angew. Math.* **109**, 121–186.

Komar, A. (1959). Covariant conservation laws in general relativity, *Phys. Rev.* **113**, 934.

Krasiński A. (1997). *Inhomogeneous Cosmological Models* (Cambridge, Cambridge University Press).

Kruskal, M. D. (1960). Maximal extension of Schwarzschild metric, *Phys. Rev.* **119**, 1743.

Lemaître, G. (1933). Condensations sphériques dans l'universe en expansion, *Compt. Rend. Acad. Sci. (Paris)* **196**, 903.

Michelson, A. A. (1881). The relative motion of the earth and the luminiferous aether, *Amer. J. Science* **22**, 120.

Neugebauer, G. (1974). Einsteinsche Feldgleichungen und zweiter Hauptsatz der Thermodynamik, *Nova Acta Leopoldina* **39**, No. 212.

Neugebauer, G. (1980). *Relativistische Thermodynamik* (Berlin, Akademie-verlag).

Neugebauer, G. and Meinel, R. (1993). The Einsteinian gravitational field of the rigidly rotating disk of dust, *Astrophys. J. Lett.* **414**, L97.

Nordström, G. (1918). On the energy of the gravitational field in Einstein's theory, *Proc. Kon. Ned. Akad. Wet. Amsterdam* **20**, 1238

Oppenheimer, J. R. and Volkoff, G. (1939). On massive neutron cores, *Phys. Rev.* **55**, 374.

Penrose, R. (1959). The apparent shape of a relativistically moving sphere, *Proc. Cambridge Phil. Soc.* **55**, 137.

Penrose, R. and Rindler, W. (1984). *Spinors and Space-Time, Vol. I* (Cambridge, Cambridge University Press).

Penrose, R. and Rindler, W. (1986). *Spinors and Space-Time, Vol. II* (Cambridge, Cambridge University Press).

Penzias, A. A. and Wilson, R. W. (1965). A measurement of excess antenna temperature at 4800 Mcls, *Astrophys. J.* **142**, 419.

Petrov, A. Z. (1969). *Einstein Spaces* (Oxford, Clarendon Press).

Pound, R. W. and Rebka, G. A. (1960). Apparent weight of photons, *Phys. Rev. Lett.* **4**, 337.

Reissner, H. (1916). Über die Eigengravitation des elektrischen Feldes nach der Einsteinschen Theorie, *Ann. Phys. (Germany)* **50**, 106.

Robertson, H. P. (1936). Kinematics and world-structure. *Astrophys. J.* **83**, 187.

Robinson, I. and Trautman, A. (1962). Some spherical gravitational waves in General Relativity, *Proc. Roy. Soc. Lond.* A **265**, 463.

Rohrlich, F. (1965). *Classical Charged Particles* (Reading, Addison-Wesley).
Rohrlich, F. (2001). The correct equation of motion of a classical point charge, *Phys. Lett.* A **283**, 276.
Rovelli, C. (1998). Loop quantum gravity, *Living Reviews in Relativity*, www.livingreviews.org.
Rowan, Sh. and Hough, J. (2000). Gravitational wave detection by interferometry, *Living Reviews in Relativity*, www.livingreviews.org.
Schäfer, G. (2000). Testing general relativity, *Adv. Space Res.* **25** (200), 115–1124.
Schneider, P., Ehlers, J. Falco, E. E. (1992). *Gravitational Lenses* (Berlin, Springer).
Schöbel, K. and Ansorg, M. (2003). Maximal mass of uniformly rotating homogeneous stars in Einsteinian gravity. *Astron. & Astrophys.*
Schouten, J. A. (1954). *Ricci Calculus* (Berlin, Springer).
Schrödinger, E. (1940). Maxwell's and Dirac's equations in the expanding universe, *Proc. Roy. Irish Acad.* **46A**, 25.
Schwarzschild, K. (1916). Über das Gravitationsfeld eines Massenpunktes nach der Einsteinschen Theorie, *Sitz. Preuss. Akad. Wiss.*, **189**.
Shapiro, I. I. (1964). Fourth test of general relativity, *Phys. Rev. Lett.* **20**, 789.
Shapiro, S. L. and Teukolsky, S. A. (1983). *Black Holes, White Dwarfs and Neutron Stars. The Physics of Compact Objects* (New York, Wiley).
Stephani, H. (1974). Physik in geschlossenen Kosmen, *Nova Acta Leopoldina* **39**, No. 212.
Stephani, H., Kramer, D., MacCallum, M. A. H., Hoenselaers, C. and Herlt, E. (2003). *Exact Solutions of Einstein's Field Equations* (Cambridge, Cambridge University Press).
Sundermeyer, K. (1982). *Constrained Dynamics* (Berlin, Springer).
Taub, A. H. (1965). The motion of multipoles in general relativity. In *Atti del Convengo sulla Relativitá Generale* (Firenze, G. Barbera).
Tolman, R. C. (1934). *Relativity, Thermodynamics and Cosmology* (Oxford, Clarendon Press).
Tolman, R. C. (1939). Static solutions of Einstein's field equations for spheres of fluid, *Phys. Rev.* **55**, 364.
Voigt, W. (1887). Über das Dopplersche Prinzip, *Göttinger Nachrichten,* **41**
Wald, R. M. (1994). *Quantum Field Theory in Curved Spacetime and Black Hole Thermodynamics* (Chicago, The University of Chicago Press).
Walker, A. G. (1936). On Milne's Theory of World-Structure. *Proc. Lond. Math. Soc.*
Wainwright, J. and Ellis, G. F. R. (1997). *Dynamical Systems in Cosmology* (Cambridge, Cambridge University Press).
Weinberg, S. (1996). *The Quantum Theory of Fields* (Cambridge, Cambridge University Press).
Westpfahl, K. (1967). Relativistische Bewegungs probleme, *Ann. Phys.* (Germany) **54**, 117.
Will, C. W. (1993). *Theory and Experiment in Gravitational Physics* (Cambridge, Cambridge University Press).
Yano, K. (1955). *The Theory of Lie Derivatives and its Applications* (Amsterdam, North-Holland).

Index